R·C·CURRAN

COLOR ATLAS OF

HISTOPATHOLOGY

COLOR ATLAS OF HISTOPATHOLOGY

BY R·C·CURRAN

MD·FRCP(Lond.)·FRCPath·FRS(Edin.)·FFPathRCPI

*Leith Professor of Pathology, University of Birmingham,
Honorary Consultant to the Central Birmingham Health District
and the West Midlands Regional Health Authority*

THIRD REVISED EDITION

WITH 804 PHOTOMICROGRAPHS

HARVEY MILLER PUBLISHERS
OXFORD UNIVERSITY PRESS

Originating Publisher HARVEY MILLER LTD
20 Marryat Road · London SW19 5BD · England

Published in conjunction with OXFORD UNIVERSITY PRESS
Walton Street · Oxford OX2 6DP

London · Glasgow · New York · Toronto · Melbourne · Auckland · Kuala Lumpur
Singapore · Hong Kong · Tokyo · Delhi · Bombay · Calcutta · Madras · Karachi · Nairobi
Dar es Salaam · Cape Town · and associates in Beirut · Mexico City · Nicosia

Published in the United States by
OXFORD UNIVERSITY PRESS · NEW YORK

First published 1966
Reprinted with minor revisions 1967, 1968, 1969, 1970
Second edition 1972
Reprinted 1973, 1975 (twice), 1976, 1977, 1978, 1979,
1981, 1982, 1983
Third edition 1985, reprinted 1988, 1990, 1993, 1995

British Library Cataloguing in Publication Data

Curran, R. C. (Robert Crowe)
　　Colour atlas of histopathology.—3rd rev. ed.
　　1. Man. Tissues. Diseases—Illustrations
　　I. Title
　　611'.018'0222

　　ISBN 0–19–921058–6
　　ISBN 0–19–261794–X (Pbk.)

Library of Congress Cataloging in Publication Data

Curran, R. C.
　　Colour atlas of histopathology.

　　Includes index.
　　1. Histology, Pathological—Atlases. I. Title.
[DNLM: 1. Pathology—atlases.　QZ 17 C976c]
RB33.C99　1985　　616.07'583'0222　　85–13634
ISBN 0–19–921058–6

Printed in England by Clifford Press Ltd, Coventry

Contents

Preface

THROUGHOUT its long history, pathology has been concerned with the study of the derangements of tissue structure and function that occur in disease, and the correlation of these changes with clinical signs and symptoms. This clinico-pathological approach, which made pathology the foundation of clinical practice, remains as valid as ever. In recent years the rate of advance of the subject has accelerated greatly with the introduction of a steadily widening range of techniques of great sensitivity, many applicable to paraffin sections, which allow the histopathologist to identify, with a high degree of specificity, cells and their products. At the same time new clinical methods (endoscopy, needle biopsy, etc.) have provided the pathologist with samples of fresh tissues and cells, sometimes repeatedly, from virtually every part of the body. The contribution that the pathologist can make to clinical practice and the understanding of disease has been enormously enhanced by these twin developments. There can be no doubt that for the foreseeable future pathology will retain its place as a basic part of the undergraduate curriculum.

Histopathology has the unique advantage of making visible the body's many complex systems and their interactions and malfunctions in disease; and the first edition of this Atlas was produced with the intention of providing the student, in a vivid and readily assimilable form, a means of acquiring a clear understanding of the basis of the many diseases that he or she will encounter in clinical practice, whether in hospital or in general practice. The Atlas was first published in 1966 and it was well received throughout the world. It was translated into six languages and it has been reprinted each year. On each occasion the question of revision was considered and in 1972 about one-twentieth of the contents were changed. However histopathology is (or was until recently) a slowly-evolving science and it was decided to wait until a thorough revision could be justified. The time for this has arrived and this third edition is virtually a new book in both format and content.

An important feature of previous editions was the large size of the page, which enabled the student to compare 18 illustrations at one time. The format of the new edition, however, allows for larger illustrations which provide increased information; and at the same time the student has the greater convenience in using, carrying and storing a smaller-sized volume. The number of pictures has been increased from 765 to 804 and two-thirds (545) are new. The area of each picture has now almost been doubled and the advantages of this will, it is hoped, be immediately obvious.

The great majority of the illustrations are based on paraffin sections stained with hematoxylin and eosin, the method in routine use. Other techniques, in more or less common use (some using cryostat sections), have been included, where they make a specific contribution; and the potential of the newer immunohistochemical methods will be apparent from the examples included.

The general arrangement of the contents has been retained, with a chapter on each of the main systems or organs of the body. There is also an introductory chapter of a general nature which aims to demonstrate the more important reactions of the tissues in disease and at the same time teach the student the basic language of histopathology, thereby enabling him or her to read and assess the significance of changes in the tissue as revealed by microscopy. This proved to be a popular feature of previous editions and it has been extended. The text has been re-written, and along with a description of the contents of each picture a limited amount of clinical information about the lesion and its pathogenesis is given. Most of the conditions are common or fairly common diseases, but occasional examples of rare lesions have been included when they illustrate a pathological process with particular clarity. It must be emphasised however that the book remains an Atlas, the primary purpose of which is to convey information in a visual form. It is meant to complement existing textbooks, and prior study of the Atlas will, it is hoped, make the better textbooks both more intelligible and more pleasurable to read.

A comprehensive Index has been provided, and a limited number of cross-references have been inserted in the text, mainly in Chapter 1, to augment it and to integrate its contents with the rest of the book.

The book is intended primarily for undergraduate students but experience with its predecessors suggests that it is likely to prove useful to postgraduate students training in pathology or another clinical discipline.

Most of the illustrations in the Atlas are based on cases dealt with in the course of the routine hospital service provided for the Hospitals of the Birmingham Central Health District, and I wish to pay tribute to Mr. K.J. Reid, Senior Chief Medical Laboratory Scientific Officer and his colleagues for the consistently high quality of the preparations which they have produced. Mr. Sidney Whitfield, Chief MLSO, was directly responsible for much of the work and he also supplied the section of schistosomiasis of liver (**5.18**) from his personal collection. Two other individuals merit special mention: first Mrs Mary Guibarra, who prepared many of the sections of tumours; and Mr. John Gregory who performed the immunohistochemical techniques.

Some of the illustrations in Chapter 5 are from sections kindly loaned by Professor R.S. Patrick (Glasgow), and some in Chapter 12 are from slides provided by the late Dr. C.W. Taylor, pathologist to Birmingham Women's Hospital for many years. The section of the Stein-Leventhal ovary was provided by Professor J.R. Tighe (St. Thomas's Hospital, London).

A number of illustrations are based on cases referred to the Department of

Pathology and I am grateful to the following pathologists for access to these cases: Dr. T.G. Ashworth, Walsgrave Hospital, Coventry; Dr. B.W. Codling, Gloucestershire Royal Hospital, Gloucester; Dr. N.D. Gower, Hallam Hospital, Sandwell, Birmingham; Dr. P.S. Haslcton, Wythenshawe Hospital, Manchester; Dr. F. Kurrein, Royal Infirmary, Worcester; Dr. A.M. Light, Good Hope District General Hospital, Sutton Coldfield; Dr. J. Martin, Tawam Hospital, Abu Dhabi, U.A.E.; Dr. J. Rokos, Staffordshire General Infirmary, Stafford; Dr. T.P. Rollason, Maelor General Hospital, Wrexham; Dr. D.I. Rushton, Maternity Hospital, Birmingham; Dr. W. Shortland-Webb, Dudley Road Hospital, Birmingham; and Dr. Carol M. Starkie, Selly Oak Hospital, Birmingham.

The pathologist to whom the cases were referred was usually my colleague, Professor E.L. Jones. Professor Jones also gave very generously of his time and advice in many other ways during the preparation of the Atlas, and without his help it is doubtful whether this new edition would have appeared.

The manuscript was typed by Mrs. Valcrie Adkins and Miss G.L. Parkinson, and I thank them for their hard work and patience.

Finally, I must pay a special tribute to Harvey and Elly Miller. As on previous occasions their advice and professional skills were invaluable at all stages of preparation of the volume and I am pleased to have this opportunity to acknowledge my indebtedness to them.

Methods

Brooke's stain: stains growth hormone-containing (acidophil) cells in the pituitary orange and prolactin-containing cells red.

Dialyzed iron method: stains acid mucins, and particularly those of the connective tissues, blue.

Gough–Wentworth technique: thin (300 μm) unstained slices of whole lung (inflated and fixed).

Grimelius's silver method: stains argyrophil materials (such as 5-hydroxytryptamine) black.

Grocott's methenamine silver method: stains fungi black.

Hematoxylin and eosin (HE): combination of basophil (bluish-purple) nuclear stain and acidophil (pink) cytoplasmic stains in routine use.

Indirect antibody immunoperoxidase method: for detecting antigens in cryostat or paraffin sections by means of polyclonal or monoclonal antibodies. Examples shown are based on the peroxidase-antiperoxidase (PAP) sequence and a brown reaction product indicates the site of the antigen in the tissues.

Loyez (iron hematoxylin) method: stains myelin blue-black.

Periodic acid-Schiff (PAS) sequence: stains carbohydrates, and particularly epithelial mucin and glycogen, purplish-red (magenta).

Periodic acid-silver (PA silver) method: stains basement membrane (particularly in kidney) and fungi, black.

Perls' method (Prussian Blue reaction): stains ferric iron, and particularly haemosiderin, blue.

Thin resin-embedded (½–1 μm) sections: facilitates study of the finer details of cell and tissue structure.

Gordon and Sweet's method: silver deposition method staining fine connective tissue fibres (including basement membranes and newly-formed collagen), black.

Solochrome cyanin: stains myelin blue.

Sudan IV: stains lipid orange-red.

Toluidine blue: reveals periodic striations in muscle cells (e.g. to confirm that a tumour is a rhabdomyosarcoma).

Van Gieson: stains collagenous fibrous tissue purplish-red. Sometimes combined with Weigert's method for elastic fibres (Weigert–van Gieson).

Von Kossa: stains bone salts black (in sections of undecalcified tissue).

Weigert's method (Miller's modification); stains elastic fibres blue-black.

Ziehl-Neelsen (ZN) technique: stains mycobacteria, (e.g. *M. tuberculosis*) purplish-red.

Atlas of Histopathology

1.1 Acute inflammation

This is an acutely inflamed mucous membrane. The patient had acute inflammation of the colon (ulcerative colitis) for some months, with recent involvement of the anal canal. The anal canal (left) is lined with stratified squamous epithelium. There are many polymorph leukocytes within the epithelium, most of them close to the surface. The submucosa (centre and right) is hyperemic, with marked dilatation of the small blood vessels. Many polymorphs are visible within these vessels, adjacent to the endothelium. The connective tissues between the vessels are pale-staining, from the presence of inflammatory exudate (e.g. centre). In addition to many polymorphs, lymphocytes and plasma cells are present, showing that the acute inflammatory reaction is superimposed on a subacute or chronic lesion. HE ×230

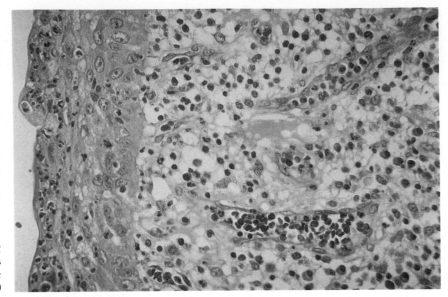

1.2 Acute appendicitis

In acute inflammation the small blood vessels dilate and the blood flow to the inflamed area is greatly increased. The blood flow eventually slows and the polymorphs move towards the periphery of the stream, where they come into contact with the endothelium to which they tend to adhere. The small vessel shown here, a capillary or post-capillary venule in the mesentery of an acutely inflamed appendix, has dilated enormously and an almost continuous layer of polymorphs is adherent to the endothelial surface – 'pavementing' of the endothelium. These polymorphs migrate through the wall of the blood vessel between the endothelial cells, and then through the surrounding tissues towards the cause of the inflammation (chemotaxis). This vessel is so dilated that the blood flow had probably stopped i.e. stasis had developed. HE ×335

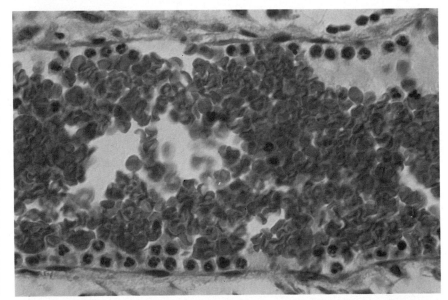

1.3 Pustule: skin

When the agent causing an acute inflammatory reaction is persistent (e.g. pyogenic staphylococci) emigration of polymorphs to the site of the inflammatory stimulus (often bacterial) may continue until they form a mass in the tissues. Some surrounding tissue is invariably killed (necrosis) by the inflammatory agent and enzymes released by polymorphs. An abscess thus forms containing polymorphs, necrotic tissue elements and serous fluid. When the collection forms in the skin it constitutes a pustule (pus-filled vesicle). This is an early stage in the formation of a pustule. Large numbers of polymorph leukocytes have migrated from the small blood vessels in the dermis (bottom) into the stratified squamous epithelium of the skin, and they are starting to form a collection within the epithelium (above centre). HE ×335

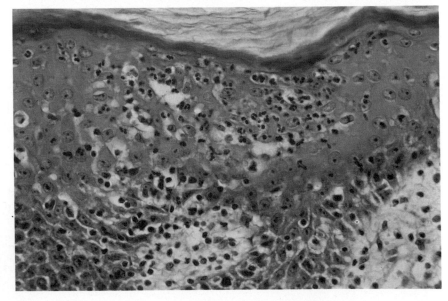

1.4 Pustule: skin

This is a fully-formed pustule. It is an ovoid abscess (centre top) within the upper epidermis and it is causing the surface to bulge upwards. There is a fairly thick layer of keratin on the surface of the skin over the pustule but it is starting to break down over the centre of the pustule. Polymorphs are the main constituent of the pustule but there are also many eosinophilic keratinous 'squames' mingling with the polymorphs. At higher magnification necrotic epithelial cells were also found to be present. The connective tissues of the dermis (below) are also inflamed and infiltrated with inflammatory cells.   HE ×150

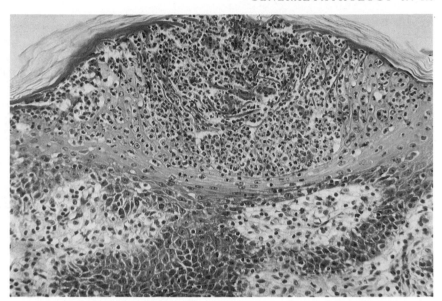

1.5 Pyemic abscess: myocardium

In pyemia, pyogenic organisms are present in significant numbers in the bloodstream, and, in addition to the primary focus, there are pus-filled abscesses in various tissues. An abscess is always accompanied by destruction of tissue. This abscess is in the wall of the left ventricle. It is an ovoid mass (centre) destroying and displacing the fibres of the myocardium. The muscle fibres nearest the abscess are necrotic with no nuclei, having been killed by the toxins from the two deeply basophilic colonies of *Staphylococcus pyogenes* in the centre of the abscess which contains also degenerate polymorphs and macrophages, fibrin (thin arrows) and red cells. There is also an infiltrate of acute inflammatory cells and red cells in the interstitial tissues of the myocardium (top). The small blood vessel (thick arrow) is dilated.
                                                HE ×150

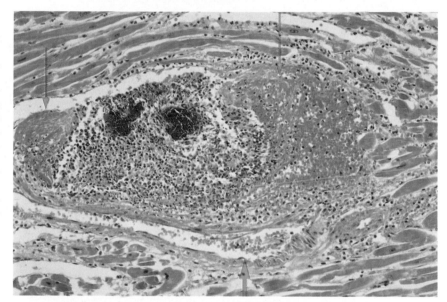

1.6 Pyemic abscess: myocardium

This is another of the many small abscesses that were present in the wall of the ventricle. The pus consists of two deeply-stained colonies of *Staphylococcus pyogenes* (left), polymorph leukocytes (most of which are necrotic and disintegrating) and macrophages. Fragments of necrotic heart muscle (arrows) killed by the toxins from the bacteria and the polymorph enzymes are also present. The serous fluid component of the pus is unstained. The small blood vessels (right) adjacent to the abscess are dilated. Although the myocardial fibres at the edge of the abscess retain their nuclei, they are probably necrotic. Pyemia is usually a very serious illness, the patient surviving for only a few days; and there is no evidence of encapsulation of the lesion by granulation tissue or fibrous tissue.   HE ×235

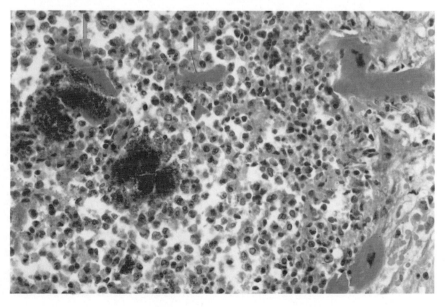

1.7 Acute abscess (surgical wound)

Sometimes pyogenic (pus-forming) micro-organisms get into a wound and cause acute inflammation which may go on to suppuration and abscess formation. This is particularly liable to happen if foreign material (including fibrin and blood clot) is present in the wound. This is a surgical wound, 3 weeks old. The epithelium (left) has healed. In the dermis however there is a collection of pus consisting of polymorph leukocytes, macrophages and dead tissue cells. The eosinophilic material (right of centre) consists of fibrin and keratin squames, and the unstained clefts contained foreign material which dissolved during processing of the tissues. The microorganisms responsible for the formation of pus are unstained. Suppuration causes delay in healing and leads to the formation of increased amounts of fibrous tissue.   HE ×215

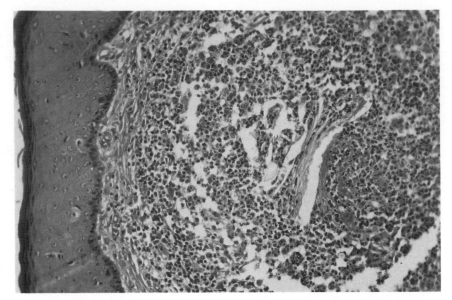

1.8 Fungal abscess: hand

A man of 40 developed a lump in the subcutaneous tissues of his right hand. After it had persisted for several months, it was resected. The excised specimen (5 × 3 × 2cm) was composed of firm, white fibro-fatty tissue in which there were several hard irregular yellow areas. This is one of the yellow areas. It is an abscess containing several colonies (right and left of centre) of the fungus *Madurella grisea*. The colonies lie in pus consisting of darkly-staining polymorphs and an outer zone of larger paler-staining macrophages. Infection of the tissues with Madurella species (Madura disease) usually produces multiple abscesses and considerable destruction of tissue (including bone). The organism is found in the soil or on plants and lesions are accordingly located more often in the foot than in the hand.   HE ×150

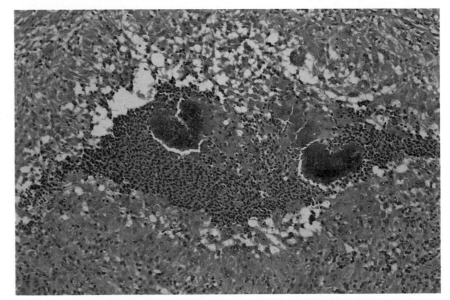

1.9 Infective (mycotic) aneurysm: cerebral artery

This patient had infective endocarditis, in which vegetations consisting of platelets and fibrin and containing microorganisms are present on the valve cusps of the heart. Material tends to break away from the vegetations and impact in small blood vessels, including the vasa vasorum of arteries. The bacteria set up a destructive inflammatory lesion in the wall of the artery, which may weaken and bulge, to form a mycotic aneurysm. The wall of the artery (right) has been destroyed, only a thin layer of degenerate muscle remaining. The thick muscle of the wall has been largely replaced by a pus-like exudate of polymorph leukocytes and macrophages. A similar exudate is present in the lumen along with a mass of red-staining thrombus (left) in which there are colonies (blue) of bacteria.   HE ×135

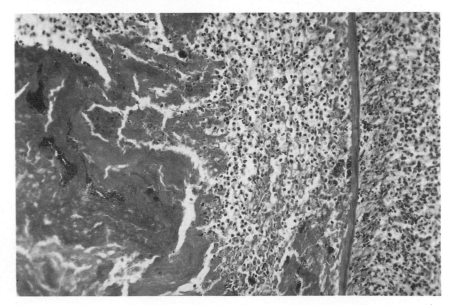

1.10 Eosinophil leukocytes

The role of eosinophil leukocytes is linked to the body's immune response, perhaps through the ingestion of antigen-antibody complexes. Eosinophil leukocytes are characteristically associated with the response of the tissues to parasites; and in this example, the parasite is the filarial worm, *Onchocerca volvulus*. This shows the connective tissues adjacent to a necrotic worm, which is not visible in this section. Around the small blood vessel (centre) there are lymphocytes and plasma cells but they are greatly outnumbered by the eosinophil polymorphs peripheral to them. Unlike neutrophil polymorphs eosinophil leukocytes do not release enzymes which destroy the tissues and so cause the formation of an abscess. *See also 13.58, 14.13.*   HE ×200

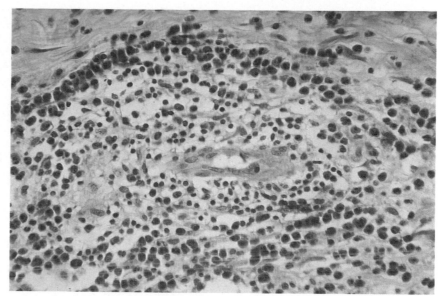

1.11 Macrophages: gall bladder

After an acute inflammatory reaction subsides, removal of necrotic debris is carried out by macrophages, the lysosomal enzymes of which can destroy a very wide range of substances. This is the gall bladder of a woman of 71 who had suffered from cholecystitis for many years. The wall of the gall bladder was much thicker than normal and there were many chronic abscesses in the wall. With the exception of a few degenerate polymorphs (thin arrows) the cells are large macrophages, each with a single vesicular (sac-like) nucleus and abundant granular and vacuolated cytoplasm. Whole cells (polymorphs and red cells) and cell fragments are present in several macrophages (thick arrows).   HE ×575

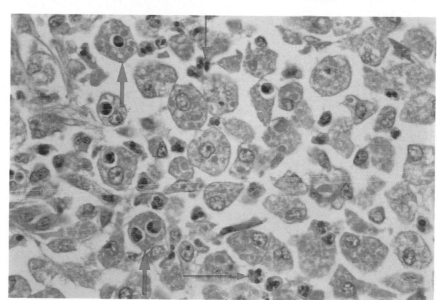

1.12 Cholesterol granuloma: chronic abscess

When an abscess persists for a long time, there is continued destruction of cells. The content of neutrophil polymorphs reduces and the macrophage and lymphocyte population steadily increases. Lipid derived from the degenerated (necrotic) cells and possibly from the blood also increases. It consists of cholesterol crystals and sometimes also globules of neutral fat, and it may be sufficient to colour the lesion yellow. This is material from a chronic abscess in the mastoid process. In addition to many macrophages, lymphocytes and extravasated red cells, there are multinucleated foreign-body giant cells (arrows) which line the elongated spaces occupied by cholesterol crystals.   HE ×235

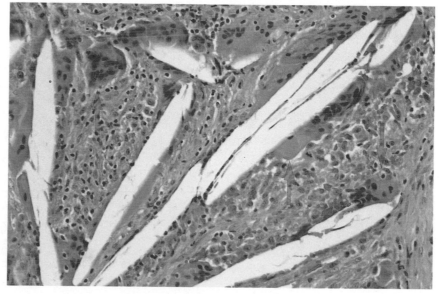

1.13 Hypersensitivity reaction to penicillin: skin

In the skin the pattern of reaction to drugs varies widely, but is generally a non-specific dermatitis. In the more acute lesions, vesicles and bullae may form. In this case a woman of 52 developed a skin rash after treatment with penicillin. The epidermis (left) is edematous, with separation of the epithelial cells (spongiosis) and formation of a vesicle beneath the keratinous surface. There are polymorph leukocytes within the epithelium; and eosinophil leukocytes and detached epithelial cells within the vesicle. The upper dermis is edematous, from the presence of pale-staining exudate. There is an infiltrate of inflammatory cells, mainly around the dilated small blood vessels (right). The inflammatory cells are eosinophil leukocytes, with small numbers of lymphocytes and macrophages.   HE ×235

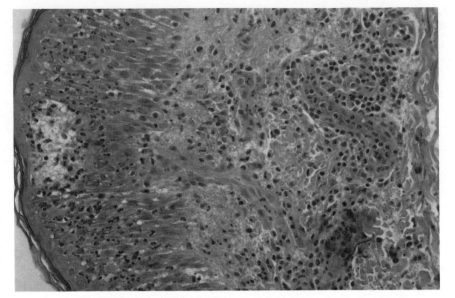

1.14 Hypersensitivity reaction to penicillin: skin

This is a more advanced lesion than that shown in **1.13**. A large vesicle (bulla) has formed (left). It is roughly spherical and contains eosinophilic fluid and small numbers of polymorphs (mostly eosinophil leukocytes). The bulla is located within the epidermis, with a layer of epithelial cells both superficial and deep to it. For a bulla of this size to form within the epidermis, there must be loss of coherence and separation of the keratinocytes (acanthocytes), that is, acantholysis; and detached epithelial cells are present within the bulla. The dermis (right) is fairly heavily infiltrated with inflammatory cells (eosinophils, lymphocytes and macrophages) but does not appear hyperemic.   HE ×150

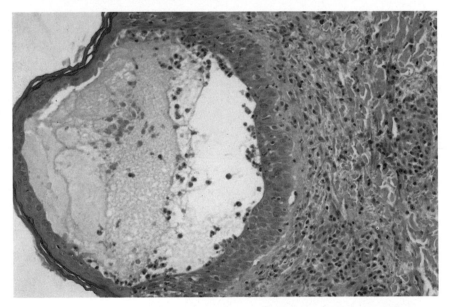

1.15 Herpes zoster: skin

Herpes zoster affects adults, particularly the elderly. It is a recrudescence of a latent Varicella infection as immunity to the virus falls. The virus attacks a posterior root ganglion, generally in the lower cervical or dorsal region. Many ganglion cells are destroyed and the ganglion is infiltrated by lymphocytes. In the skin supplied by the sensory nerve related to the ganglion, there is pain and hyperalgesia, followed by erythema and formation of vesicles filled with a serous or even a hemorrhagic fluid. This is a cutaneous vesicle, typically intra-epidermal and multilocular, with strands of necrotic cells and cell walls traversing the vesicle. Swollen and necrotic cells (balloon cells, arrows) lie free within the vesicle. The dermis beneath the vesicle is infiltrated at one point by lymphocytes (lower right). *See also 14.18.*

                HE ×150

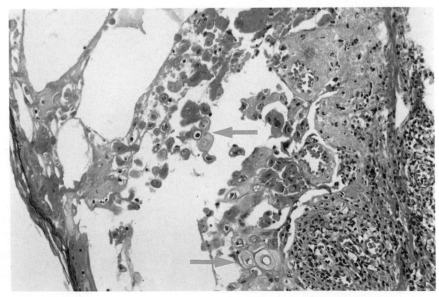

1.16 Acute anterior poliomyelitis: spinal cord

Although there may be an initial polymorph response, viruses characteristically evoke a lymphocytic response in the tissues. The poliomyelitis virus is neurotropic, destroying the large motor neurons in the anterior horns of the spinal cord and brainstem. Paralysis of the muscles supplied results. This shows a small blood vessel (cut in two places) in the lumbar cord, close to an acutely inflamed anterior horn. The vessel is dilated, reflecting the hyperemia of the inflamed tissues, and it is surrounded by a 'cuff' of darkly-staining small lymphocytes. The adjacent white matter (bottom) is degenerate, showing extreme edema and vacuolation. The lymphocytic infiltration of the tissues may persist for some months after the acute phase of the illness. *See also 9.17, 9.18.* HE ×135

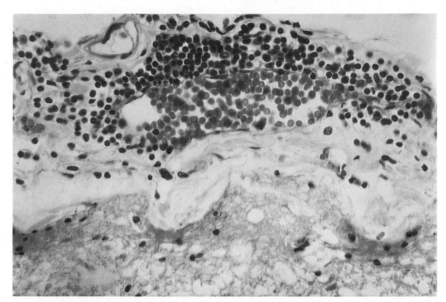

1.17 Rheumatoid nodule: skin

Rheumatoid arthritis is one of the range of overlapping diseases which share some form of disorder of the autoimmune system. A common manifestation is necrosis of collagenous tissue, particularly in rheumatoid arthritis, and nodules frequently form in the subcutaneous tissues. This is part of a small cutaneous nodule. There is marked proliferation of small blood vessels (right), a feature of the early stages of formation of a rheumatoid nodule. On the left, necrosis of the collagenous tissue has occurred. It stains deeply like fibrin, and the change is described as fibrinoid necrosis. The fibrinoid material is formed from disintegrating collagen, but fibrin can be demonstrated in it. The tissues are infiltrated by lymphocytes and macrophages. *See also 6.36, 14.29.*

HE ×235

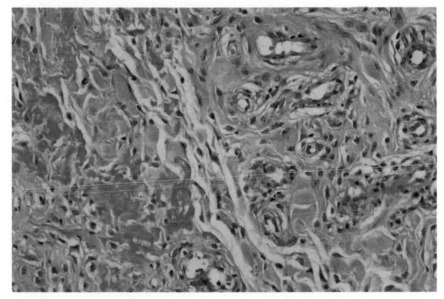

1.18 Gumma: brain

Gummas occur in the tertiary stage of syphilis. Although few spirochetes (*Treponema pallidum*) are present, there is extensive necrosis of tissue, and the cells are mostly lymphocytes and plasma cells, the latter often in considerable numbers. Healing tends to lead to extensive fibrous tissue formation. This shows part of a gumma of brain. There are mature plasma cells (arrows) lying in collagenous fibrous tissue. Small lymphocytes and macrophages are also present. The plasma cells have moderately sized nuclei, with clumps of chromatin at the periphery (a 'clock-face' or 'cartwheel' distribution) and densely staining cytoplasm, in which the nucleus is located at one pole. Plasma cells synthesize immunoglobulin and the bluish-pink colour is caused by rough endoplasmic reticulum in the cytoplasm. *See 5.13, 9.14.* HE ×375

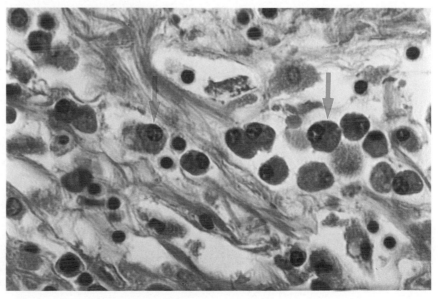

1.19   Fat necrosis: skin

When fat cells die, the fat in their cytoplasm is released extracellularly, where it is ingested by macrophages. Necrotic fat can excite a marked inflammatory reaction, with the eventual production of much fibrous tissue. In this fatty tissue, all the fat cells (adipocytes) have died and released their fat which appears as large clear spaces. These are droplets of 'free' (i.e. extracellular) neutral fat, and each droplet is surrounded by a ring of macrophages which are ingesting the fat. The cytoplasm of the macrophages has a pale granular or 'frosted-glass' appearance, from the presence in it of the many small droplets of fat undergoing digestion by lysosomal enzymes. One macrophage appears to be in mitosis (arrow).

HE ×135

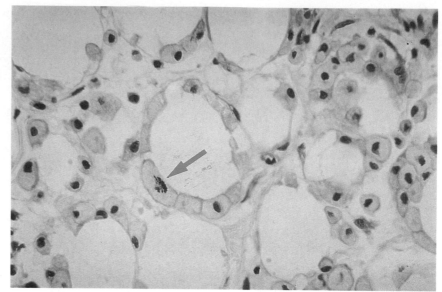

1.20   Fat necrosis: skin

To allow lipids to be demonstrated in histological sections it is necessary to avoid reagents which dissolve fat. 'Frozen' (cryostat) sections are therefore used. This is a cryostat section of the same lesion as that shown in **1.19**, stained with a solution of Sudan IV in alcohol (the dye does not dissolve in water). The small droplets of fat in the cytoplasm of the macrophages have taken up the dye from the solution and stain a bright orange-red colour. However the large droplets of 'free' fat are not stained, since the fat in these, being extracellular, is readily soluble in the alcohol of the staining solution. *See also 12.1, 12.2.*

Sudan IV ×135

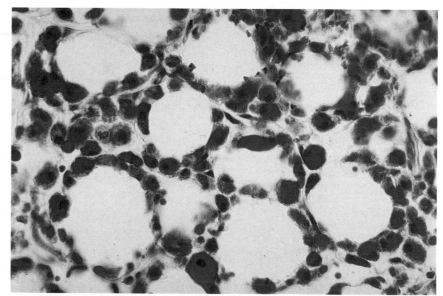

1.21   Melanosis coli

A woman of 63 had the proximal half of the colon resected for carcinoma. The mucosa of the ascending colon was dark brown, almost black, in colour. The lumen (not visible) is on the left and the muscularis mucosae is on the right. The deeper parts of several colonic glands are shown, each gland lined by a single layer of mucus-secreting columnar cells. There is a collection of mononuclear macrophages with very abundant brown cytoplasm lying between the glands. The brown colour of the cytoplasm is produced by the presence of melanin-like pigment which is thought to be synthesized by enzymic action in the wall of the colon from amino acids absorbed from the lumen of the colon. Chronic constipation is believed to predispose to melanosis coli. The condition is symptomless and of no pathological significance.   HE ×255

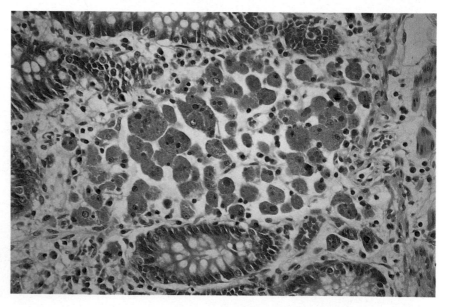

1.22 Macrophages: lung

Particles of carbon-containing dust (smoke) are present in the atmosphere of many cities and are consequently inhaled by the inhabitants. Small particles (1μm or less) readily reach the alveoli where they are ingested by the pulmonary macrophages and carried by the lymphatics to the lymph nodes. Many carbon-laden phagocytes remain within the lung and pleura however, and if the load of dust is heavy, anthracosis results. In this case the amount of dust present was moderate. The pleura (right) is thicker and more fibrous than normal and within it there are collections of macrophages laden with carbon particles. Several dilated pulmonary alveoli are also shown. The normal flat lining cells have been replaced by cuboidal cells and within the lumen of the alveoli there are numerous round macrophages with brown-coloured cytoplasm.   HE ×310

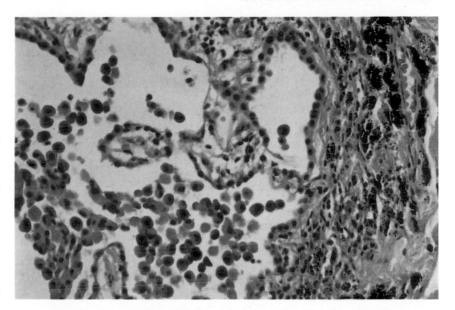

1.23 Sarcoidosis: lymph node

Sarcoidosis is typified by the presence of rounded collections (follicles) of epithelioid cells. Sarcoid follicles do not undergo necrosis. Healing may however lead to considerable fibrosis. The clinical manifestations of the disease reflect the sites of greatest involvement by the lesions. Lymph nodes, particularly in the mediastinum and cervical region, are often affected. The lymphoid tissue of this node has been largely replaced by two follicles composed of large cells with eosinophilic cytoplasm and vesicular nuclei. Several multinucleated giant cells (arrow) of the Langhans type are also present. A large laminated mass, stained blue-black because of its content of calcium, is present in the larger follicle (left). This is a Schaumann body. Fibrous tissue has formed at the periphery of the follicles. *See also 2.32, 7.4, 7.48, 14.10.*   HE ×235

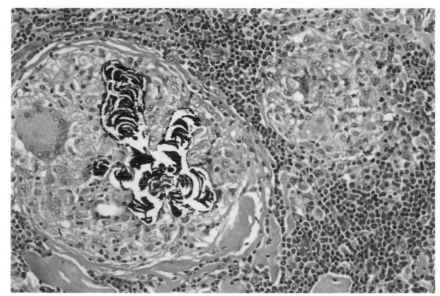

1.24 Phagocytosis of cells: lymph node

Removal of effete or damaged cells of all types is one of the functions of macrophages, and they do so very effectively by means of the wide range of hydrolytic enzymes in their lysosomes. In this case the patient had a deposit of metastatic tumour in the axilla, from an unknown primary. This is a lymphatic sinus of an axillary lymph node and it shows very large phagocytic cells in the sinus. The cells have large pale vesicular nuclei (arrow) and in its very abundant cytoplasm each cell has a considerable number of ingested cells. Some of the ingested cells appear to be necrotic with pyknotic or fragmenting nuclei but others have well-preserved nuclei. Their precise nature is uncertain. Some may be lymphoid cells but some are almost certainly neoplastic.

   HE ×860

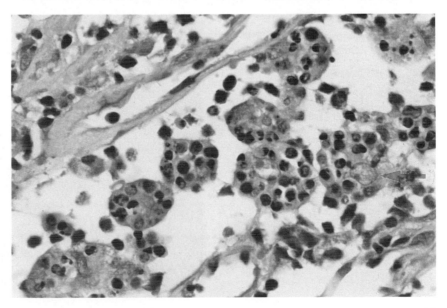

1.25 Healing ulcer: polyp of rectum

A man of 20 suffered from rectal bleeding, and a small ovoid mass (2 × 1cm) attached by a short stalk to the rectal mucosa was resected. It proved to be a juvenile polyp, a benign lesion. It consists of vascular granulation tissue, containing large numbers of dilated thin-walled blood vessels, large mononuclear phagocytes (thin arrow) and small numbers of lymphocytes and plasma cells. A single layer of pleomorphic epithelial cells covers the surface (top) but it is lacking at one point (top left). In the tissue beneath the area of ulceration the number of inflammatory cells is much increased, with polymorph leukocytes predominating. The nuclei of the epithelial cells are slightly pleomorphic and the cells are proliferating and heaped up at the edge of the ulcerated area (thick arrow).   HE ×235

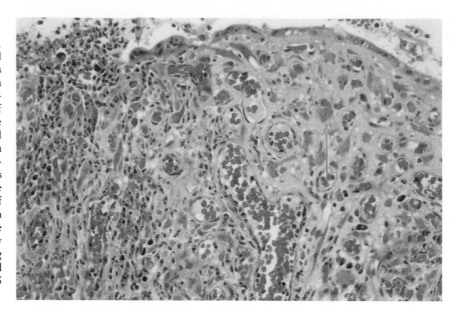

1.26 Foreign-body giant cells: healed wound of skin

A woman of 40 developed small soft irregular elevated nodules in the scar of a healed surgical incision of the skin. This is one of the nodules. It consists of granulation tissue, with many large and greatly dilated thin-walled blood vessels, lymphocytes and plasma cells, and a few elongated fibroblasts. The most notable feature however is the presence of many large phagocytes, most of which are multinucleated (arrows), containing pale-staining fragments of foreign material. The material was birefringent when examined by polarized light and proved to be nylon suture material from the original surgical incision. The nodules therefore were 'stitch granulomas'. *See also 1.37.*   HE ×235

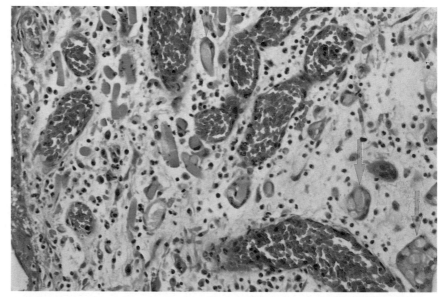

1.27 Pyogenic granuloma: skin

A pyogenic granuloma is a soft red swelling sometimes with an ulcerated surface which grows rapidly and tends to recur on removal. It may sometimes originate in a capillary hemangioma which has been traumatized but in some instances it is probably derived from exuberantly-growing granulation tissue. This lesion is in the skin and consists mainly of thin-walled blood vessels, some very dilated but others small, with little or no lumen. The predominant cells are endothelial cells and the resemblance to capillary hemangioma is close. Few lymphocytes and plasma cells are present in the edematous connective tissue between the vessels. The epidermis (left) is flattened and stretched over the lesion, but the surface, covered with a thick layer of keratin, is intact. The end result of healing is a fibrous scar.

HE ×150

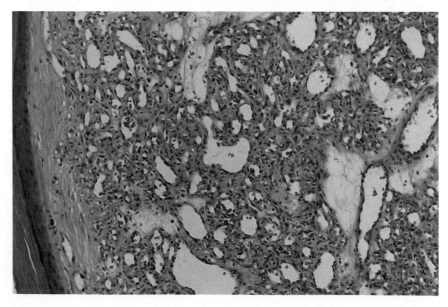

1.28 Organization of thrombus: vein

Thrombus in the lumen of a blood vessel is 'foreign' material, and the body removes it by a process of organization. Small thin-walled blood vessels penetrate the thrombus along with fibroblasts and macrophages. The macrophages ingest the fibrin, red cells and platelets, and the fibroblasts form connective tissue. The organizing tissue is sterile and few polymorphs are present, in contrast to the large numbers in the granulation tissue of a healing wound. This shows a small vein in which organization of the thrombus is well advanced. The wall of the vein is on the right. The lumen is occupied by eosinophilic thrombus which has been widely penetrated by thin-walled blood vessels and thus become a spongy mass. Eventually fibroblasts will convert the thrombus into a fibrous cord. *See also 6.19, 6.20.*

HE ×215

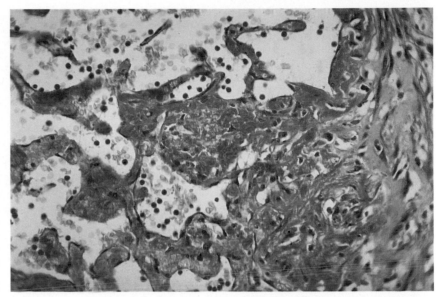

1.29 Healing ulcer: ileum

There was extensive ulceration of the mucosa of the ileum caused by Crohn's disease, and this shows an area where healing of the ulceration is occurring. The lumen of the intestine is at the top and the muscularis mucosae at the bottom. The normal villous structure of the mucosa has been destroyed, and replaced by a thin layer of granulation tissue in which there are large numbers of dilated blood vessels and inflammatory cells. The latter include large macrophages (arrow), lymphocytes and plasma cells. Covering the surface of the granulation tissue there is a layer of closely-packed tall columnar epithelial cells. Some restoration of the normal architecture of the ileum may take place in time but villi are unlikely to form in an area as severely damaged as this.

HE ×360

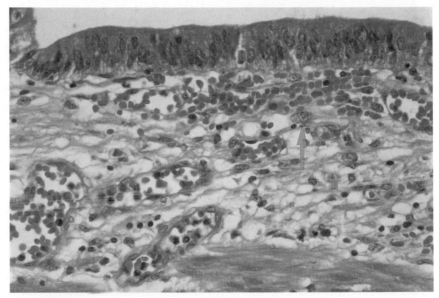

1.30 Healing wound: skin

A woman of 55 developed a melanoma of calf which was removed surgically. Some weeks later a small area of persistent ulceration was resected. This shows the margin of the ulcer. The tissue in the floor of the ulcer (centre and bottom) consists of mature granulation tissue in which there are abundant collagen fibres and thick-walled blood vessels. Small groups of inflammatory cells are present (right). A layer of stratified squamous epithelium growing from top left covers part of the granulation tissue, but part of the ulcerated area is visible (top right). The sheet of epithelium is covered with necrotic cells and keratinous debris (top). The epithelium is stratified but rete ridges are not present. HE ×235

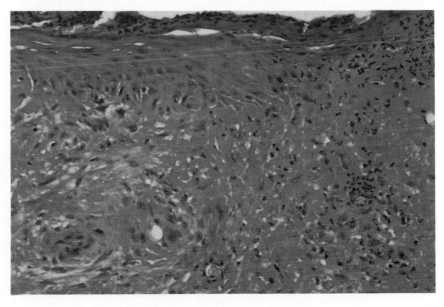

1.31 Healed wound: cornea

The external surface of the cornea (top) is covered by stratified squamous epithelium which rests on a thick hyaline basement membrane, Bowman's membrane (arrow). Beneath Bowman's membrane is the avascular collagenous stroma of the cornea. The cornea had been incised surgically two months prior to removal of the eye, and the healed wound is visible as a 'gap' in the stroma filled with cellular connective tissue (centre). The new tissue also appears avascular but the collagen fibres in it lack the regular orientation of the large fibres of the stroma. There is also a gap (top centre) in Bowman's membrane and the epithelium covering this gap is much thinner than the normal epithelium on each side of the wound.   HE ×134

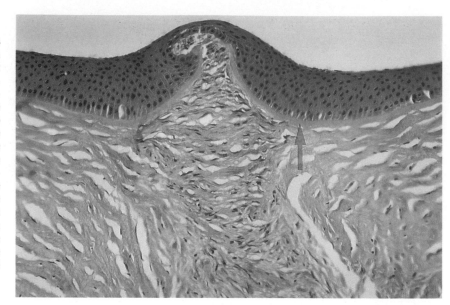

1.32 Healed wound: skin

A wound of skin that heals quickly and completely eventually becomes relatively inconspicuous, and even histologically restoration of the epithelial and collagenous tissues of the dermis may be so complete as to make the site difficult to identify. However elastic fibres do not regenerate as a rule, and stains for elastic tissue generally reveal a gap in the dermis, even in old wounds. This is a surgical wound (top half) which had been healed for some months. The surface epithelium (left) is intact and fairly normal-looking, apart from the fact that it is relatively flat and lacks well-formed rete ridges. The normal dermis (bottom half) adjacent to the wound is rich in dark-staining elastic fibres whereas in the scar of the healed wound (top half) fibres of this type are completely lacking.   Elastic stain ×55

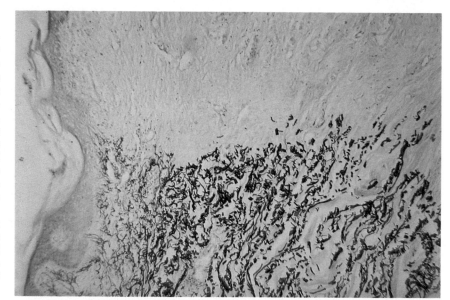

1.33 Keloid in healed wound: skin

The main structural component in a healed wound of skin is collagen. Occasionally however abnormally large amounts of collagen form, to such an extent that the scar bulges above the surface of the adjacent skin. The epidermis (left) and the upper dermis appear normal. In the deeper dermis and subcutaneous tissues, however, there are very broad hyaline bands of collagen, much larger and thicker than e.g. the collagen fibres in the dermis (left). Even at this magnification it is obvious that the fibroblasts associated with the keloid fibres are also very large and active-looking in comparison with normal fibroblasts. Black people are more liable to develop keloid than whites, and the basal layers of the skin in this case contains a high concentration of melanin. *See also 14.5, 14.6.*   HE ×60

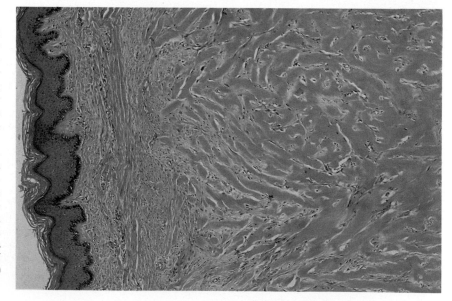

1.34 Healing of fracture: rib

In bone the osteoblast is the counterpart of the fibroblast of collagenous tissue and it plays the major role in restoring the structure of a bone that has been fractured. The osteoblasts are derived from the 'resting' osteocytes of the periosteum and endosteum, and they quickly respond by laying down bone and cartilage in somewhat haphazard fashion to form callus which effects temporary repair. The normal cortical bone is on the right and the other tissue (centre and left) is the 'external' callus, i.e. the callus on the external aspect of the bone. It consists of vascular connective tissue and thick trabeculae of woven bone (centre). The surfaces of the bone trabeculae are covered with large osteoblasts which will lay down dense lamellar bone. There is also some cartilage (left) superficial to the woven bone.

HE ×95

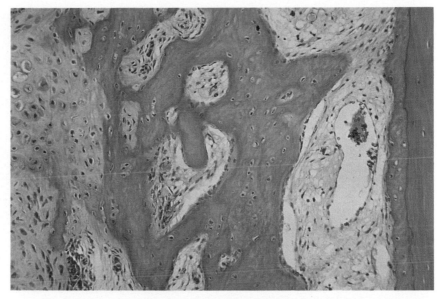

1.35 Healing of fracture: rib

This shows external callus in more detail. It consists of cartilage (left and centre) and trabeculae of new bone (right). The chondrocytes in the cartilage vary considerably in size and their arrangement is somewhat disorderly. The new bone is coarse-fibred ('woven') bone, with prominent osteocytes within the matrix and also osteoblasts on part of the surface of the trabeculae. During the process of remodelling of the callus (internal and external) the cartilage is removed and trabeculae of dense lamellar bone are laid down in orderly fashion to reform the cortex of the bone. In this way, if the ends of the broken bone have been correctly apposed, the normal bone structure can be restored completely. HE ×200

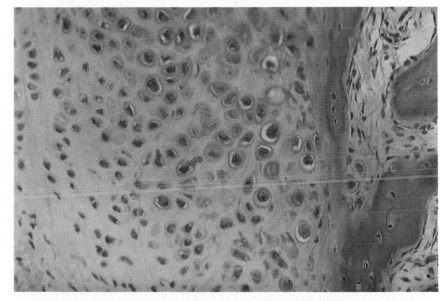

1.36 Hyperostosis (reactive bone formation): Ewing's tumour in femur

The bone-forming cells of the periosteum or endosteum may be stimulated in various ways other than by a fracture, e.g. by the presence of a low-grade inflammatory lesion or a slowly growing neoplasm, most commonly a meningioma (rapidly growing tumours destroy the bone). In this case the patient had a Ewing's tumour of the femur which was invading and disrupting the cortical bone of the shaft. The periosteum (top) has reacted and become very cellular with outgrowth of osteogenic connective tissue in which numerous trabeculae of woven bone have formed, at right angles to the cortex. The osteocytes in the woven bone are typically large and round or ovoid in shape. In places there are many osteoblasts (arrow) on the surface of the lamellae. HE ×150

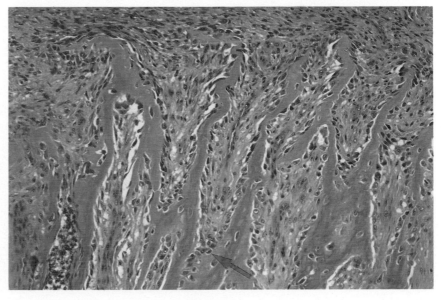

1.37 Foreign-body giant cells: pilomatrixoma of skin

Pilomatrixoma is a benign tumour of hair matrix in which the neoplastic cells mature and die, becoming keratinized 'shadow' or 'ghost' cells without nuclei. The necrotic cells are foreign to the tissues and phagocytes attempt to remove them. This tumour was in the skin of the upper arm of a man of 31. An eosinophilic mass of 'ghost' cells, lacking nuclei (top), has evoked a vigorous response by phagocytes, mostly multinucleated foreign-body giant cells (bottom). The ghost cells frequently calcify and are then even more resistant to phagocytosis. *See also 1.26.*   HE ×440

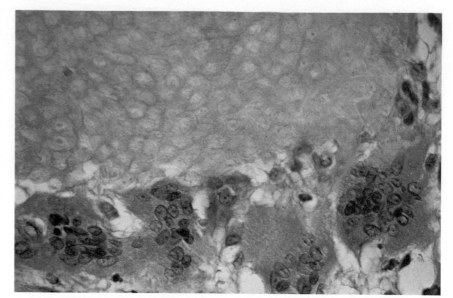

1.38 Plasmacytoma: chest wall

A man of 58 developed a swelling on his chest wall. A mass (15 × 12 × 10cm), which appeared to be encapsulated, was removed surgically, along with part of a rib. Histologically it consists of neoplastic plasma cells, of varying degrees of differentiation, and phagocytes, most of which are multinucleated giant cells. The giant cells contain large quantities of amorphous material which proved to be amyloid, staining with Congo Red and then showing green (anomalous) birefringence when examined with polarized light. Some of the amyloid material is extracellular. The amyloid is derived from immunoglobulin and since neoplastic plasma cells secrete an abnormal immunoglobulin, amyloid is frequently present in plasmacytomas (solitary myelomas) and in multiple myeloma. *See also 2.16, 2.17, 3.19, 3.20.*   HE ×360

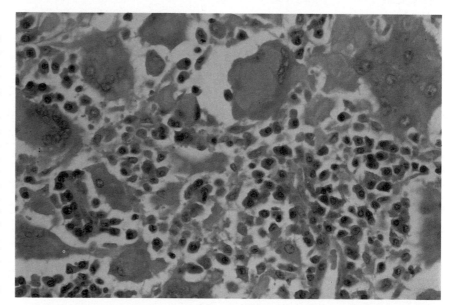

1.39 Schistosomiasis: rectum

Schistosomiasis of the intestine is caused by *Schistosoma mansoni* which lodges in the tributaries of the portal vein. Most of the ova laid by the female worm pass through the rectal mucosa and are excreted, but some remain in the tissues. This shows the granulomatous reaction to the ova in a biopsy specimen of rectal mucosa. Between the bases of three mucosal glands there is a dense infiltrate of inflammatory cells, consisting of collections of eosinophil leukocytes (arrow), lymphocytes and large macrophages with abundant pale-staining cytoplasm (centre and right of centre). No ova are visible. Lymphocytes are also present. The granulomatous reaction tends to produce fibrosis in, and occasionally very marked fibrous thickening of, the wall of the intestine. *See also 5.18, 10.54.*   HE ×150

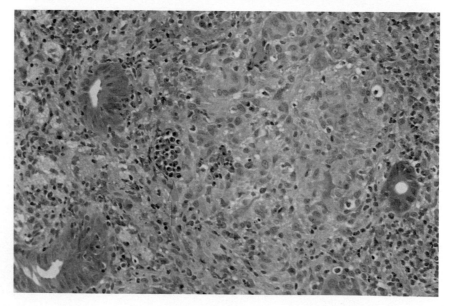

1.40 Phagocytosis of lipid: skin

The patient was a man with a considerably raised level of plasma lipids (hyperlipidemia). Deposits of lipid had formed in his skin (xanthomas). One of these had been traumatized and the lipid dispersed in the dermis. The lipid has been phagocytosed by macrophages, many of them foreign-body giant cells (arrows). The smaller macrophages (e.g. bottom right) have finely granular (frosted glass) cytoplasm. The large clefts in the cytoplasm of the giant cells contained cholesterol crystals. The cytoplasm of the giant cells is also vacuolated in places. Three of the giant cells contain asteroid (star-shaped) bodies. The significance of asteroid bodies is not known.   HE ×360

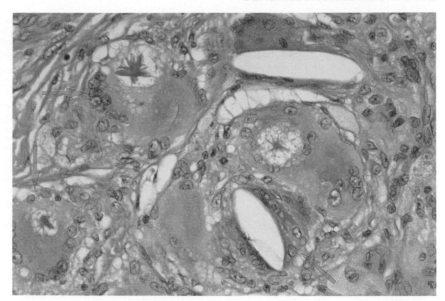

1.41 Malakoplakia: retroperitoneal tissues

Occasionally in chronic cystitis rounded plaques (1–2cm dia) form in the bladder wall. They consist of very cellular granulation tissue, containing large numbers of plasma cells, lymphocytes and large macrophages. Many of the macrophages contain distinctive inclusions. The macrophages (thin arrows) have abundant eosinophilic cytoplasm in which there are a variety of 'inclusions'. One of these is basophilic and spherical (thick arrow), the deep blue-black colour being caused by its content of calcium and iron. Inclusions of this type which were numerous throughout the tissue are termed Michaelis–Gutmann bodies. The smaller cells are mature plasma cells and small lymphocytes in roughly equal numbers. Although malakoplakia usually affects the bladder, this lesion was in perinephric fat.
     HE ×360

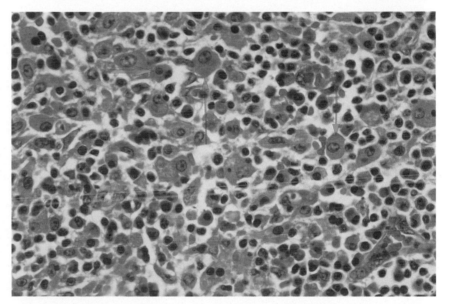

1.42 Pseudotumour: orbit

Pseudotumour of the orbit is a granulomatous lesion. Its etiology is unknown but the process may be of the same type as that which occurs in retroperitoneal fibrosis. An autoimmune basis has been postulated but without clear evidence. This is a fairly early lesion and it is therefore very cellular with minimal fibrosis. The cells are mainly eosinophil polymorphs, leukocytes and lymphocytes but a few macrophages and plasma cells are also present. A lesion of this structure may be mistaken for Hodgkin's disease. The 'onion-skin' thickening of the adventitia of the small artery (right) is a nonspecific inflammatory response but reminiscent of the changes that affect small arteries in the spleen in systemic lupus erythematosus. Older lesions tend to become progressively more fibrous.   HE ×135

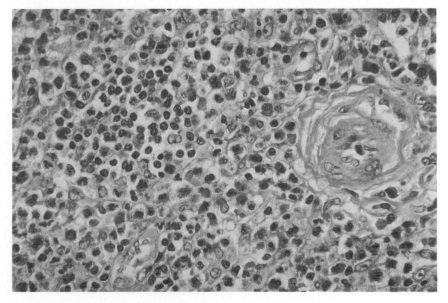

1.43 Hyaline (Mallory) bodies: liver

This is a cirrhotic liver. Almost all of the hepatocytes are swollen with edematous vacuolated cytoplasm (ballooning degeneration). Some hepatocytes, however, have large round 'clear' spaces in their cytoplasm which were occupied by droplets of neutral fat; and several contain a deeply eosinophilic rounded body (arrows). Liver cells are prone to accumulate droplets of fat in their cytoplasm from various causes, including abuse of alcohol, chronic malnutrition and obesity. The eosinophilic bodies are hyaline (Mallory) bodies and are characteristically found in livers damaged by chronic alcoholic abuse. They consist of effete cytoplasmic organelles. Similar bodies are however occasionally seen in non-alcoholic cirrhosis and they can be produced in livers of animals experimentally by toxins.

HE ×135

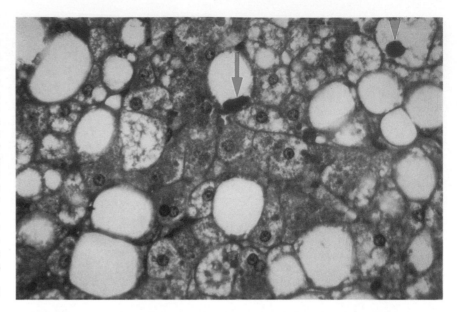

1.44 Elastotic degeneration of collagen: skin

A man of 62 had a basal cell carcinoma removed surgically from his forehead. This shows the dermis some distance from the tumour. Most of the collagen fibres have lost their nuclei and are fragmenting into slender basophilic fibres (arrows). The slender fibres are tortuous and resemble elastic fibres. They also stain with special stains for true elastic fibres, and the change in the collagen is therefore termed 'elastotic' degeneration. This form of degeneration is commonly seen in skin exposed to the sun, particularly the face, and especially in fair-skinned individuals. It is probably caused by the ultraviolet component of sunlight. The same factors are also involved in the development of basal cell carcinoma.

HE ×360

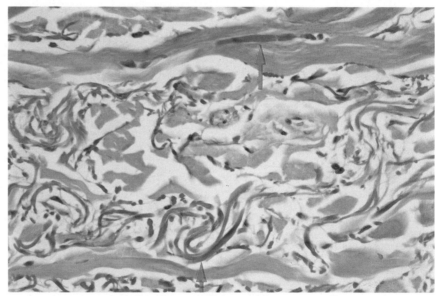

1.45 Radiodermatitis: skin

This is skin which had been exposed to comparatively large (therapeutic) doses of ionizing radiation (X-rays) 3 months prior to excision. The epidermis (left) is thickened and edematous, with wide separation of the epithelial cells. There is a failure of keratinization. Hemorrhage has also occurred into the epidermis from the blood vessels in the dermis, which are greatly dilated and surrounded by a densely eosinophilic cuff of fibrinous exudate. The dermal connective tissues are diffusely and heavily infiltrated by inflammatory cells, and there is no clear separation into upper and deeper dermis. Skin as severely damaged as this, and lacking a protective layer of keratin, is very prone to ulceration.

HE ×55

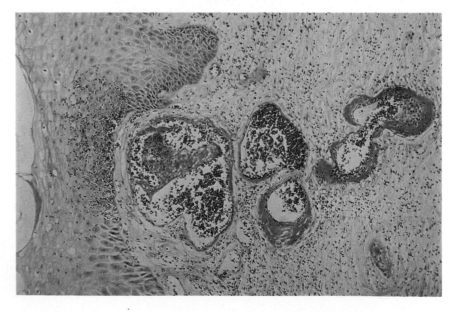

1.46 Hypercellularity (hyperplasia): bone marrow

This is a specimen of bone marrow from the iliac crest of a person with myelosclerosis. The disease is at an early stage, with marked increase in the various types of hemopoietic cells, erythrocytic, granulocytic and mega-karyocytic. The cells are closely packed and the amount of fat is less than normal. Elsewhere in the marrow the fat cells had been almost completely replaced by hemopoietic cells. The increase in the number of megakaryocytes is particularly obvious, and some of these are morphologically abnormal. As the disease progresses, the cellularity of the marrow decreases and fibrous tissue increases; and even at this stage an increase in reticulin fibres was detectable by special stains. *See also 2.61, 2.62.* HE ×360

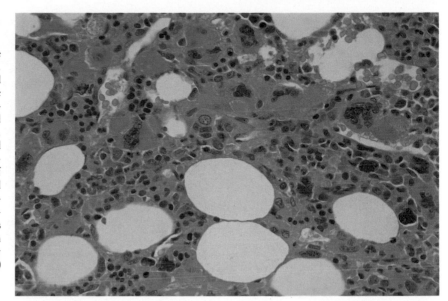

1.47 Squamous metaplasia: bronchus

Metaplasia is the transformation of one type of tissue to another. Thus in a gall bladder containing a calculus the mucus-secreting columnar epithelium tends to undergo metaplasia to the reputedly more resistant stratified squamous epithelium; and when the bronchial mucosa is 'irritated' by e.g. carcinogenic substances (benzpyrene, etc.) it too tends to undergo metaplasia. This shows the bronchial epithelium from a man who was a heavy cigarette smoker. The normal ciliated mucin-secreting columnar epithelium has been replaced by a thick stratified squamous epithelium, the surface cells of which are tending to keratinize. The cells in the basal layers show marked pleomorphism with large hyperchromatic nuclei, and there is increased mitotic activity (arrow)　　　　　　　　HE ×135

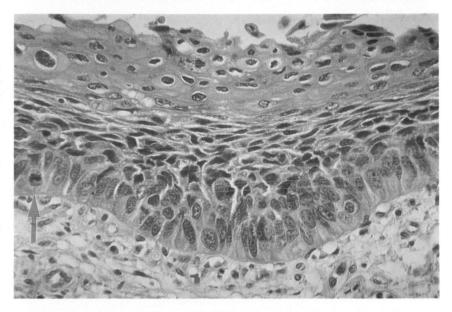

1.48 Metaplasia and dysplasia: bladder

A man of 67 who had worked for many years in the rubber industry developed severe recurrent cystitis, for which he underwent total cystectomy. The epithelial lining (top) is markedly dysplastic. The epithelial cells are large, with eosinophilic cytoplasm and large nuclei with prominent central nucleoli. They show only a slight tendency to stratification. The submucosa (lower half) is very vascular and there is an inflammatory infiltrate of lymphocytes and plasma cells, as well as some eosinophilic exudate. The dysplastic changes in the epithelium were almost certainly produced by carcinogenic substances such as beta-naphthylamine previously used in the rubber industry. Careful search however revealed no evidence of carcinoma in the bladder. *See also 4.48, 11.10.*　　HE ×360

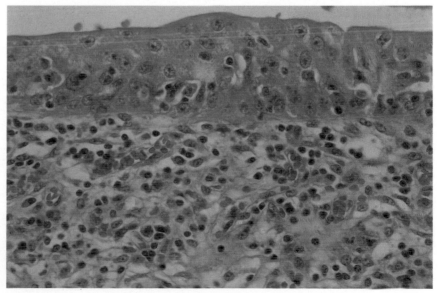

1.49 Hemangioma: skin

This was a small red 'birthmark' in the skin of the eyelid. Histologically it is sponge-like, consisting of large blood-filled spaces, with walls of fibromuscular tissue and lined by flat endothelial cells. The vascular spaces connect with each other. This type of structure is characteristic of the cavernous type of hemangioma. They are not tumours but hamartomas, i.e. malformations in which there is localized over-production of a tissue. Hamartomas are present at birth, grow in step with the other tissues of the individual, and stop growing when body growth is complete. *See also 5.5, 14.78.*

HE ×150

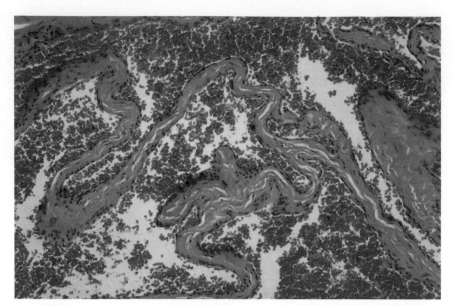

1.50 Epidermal (keratinous) cyst: skin

A cyst is a fluid-filled cavity which arises from dilatation of a pre-existing structure and is generally lined by epithelium. This is a trichilemmal (pilar) type of cyst which is almost always located on the scalp and is never found on hairless skin. The cyst cavity (right) is filled with densely eosinophilic desquamated keratinized cells. The cyst is lined by stratified squamous epithelium (left). The cells adjacent to the contents of the cyst are large cells with abundant pale granular cytoplasm (arrow). Underlying these are much smaller eosinophilic cells with deeply-staining nuclei. The basal layer is a single row of small cuboidal cells (extreme left). There is no granular layer of cells, the pale cells undergoing very rapid keratinization without that stage of maturation.

HE ×135

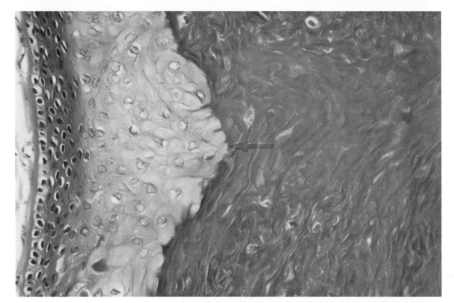

1.51 Squamous papilloma: tonsil

This is the right tonsil of a man of 35. The tonsil is out of sight on the right and this shows part of a warty mass 4cm in height which was growing from its surface. It consists of papilliform processes of vascular connective tissue, seen here in longitudinal- and cross-section (thin arrows) and covered with a thick layer of stratified squamous epithelium. There are deep clefts, some containing cell debris (thick arrow), between the finger-like processes. There is a fairly intense infiltrate of chronic inflammatory cells in the connective tissue of the cores of the papilliform processes and in the underlying tissues (right). The structure shown here is that of papilloma of stratified squamous epithelium. Other types of epithelium also give rise to papillomas. *See also 3.10, 3.11, 7.5, 7.6.*

HE ×60

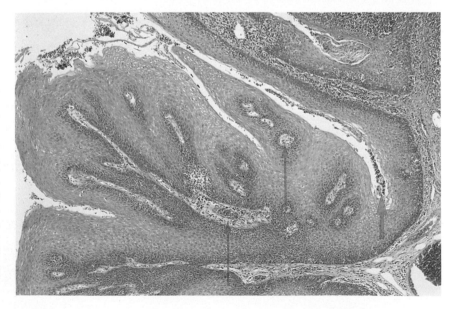

1.52 Papillary adenoma: rectum

Papillary adenomas of the colon are generally polypoid, often with a slender stalk, and show a small but definite tendency to undergo malignant change. This is more marked in villous papillomas. This lesion is from the rectum of a man of 79. Normal mucosal glands are seen in cross-section, lined by mucus-secreting epithelium. The papilliform processes of the adenoma (centre and left) consist of a slender core of vascular connective tissue (arrows) covered with a thick layer of tall columnar epithelial cells which show considerable dedifferentiation and dysplasia. The nuclei are large and basophilic and apparently form several layers; mucus secretion is greatly reduced; and at higher magnification, occasional mitotic figures were evident. No invasion of the connective tissue of the stalks however had occurred. HE ×100

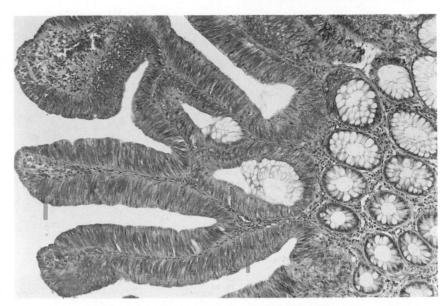

1.53 Leiomyosarcoma: skin

A sarcoma is a malignant neoplasm of connective tissues, and a leiomyosarcoma is a malignant neoplasm of smooth muscle cells. It is a comparatively rare tumour, which tends to spread by the bloodstream to distant sites to form metastases. This tumour was in the scrotum. Histologically it is a highly cellular tissue, consisting of interlacing bundles of very elongated cells, with ovoid vesicular nuclei and abundant cytoplasm, in a somewhat haphazard arrangement that does not occur in normal tissues. The elongated nuclei of the tumour cells are pleomorphic, some being very large. All have prominent nucleoli. There are occasional mitoses. The cells are poorly differentiated smooth muscle cells and with special techniques myofibrils can be demonstrated in their cytoplasm. *See also 4.20, 11.33.*

HE ×135

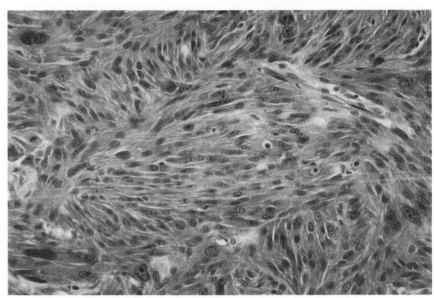

1.54 Carcinoma in lymphatics: caecum

A carcinoma of caecum which was removed surgically along with part of the colon. This shows the macroscopically-normal mucosa of the caecum several cm from the tumour. The mucosa (left) is normal and the muscularis mucosae (centre) is intact. In the submucosa (right) there are collections of basophilic malignant (carcinoma) cells. Some lie close to the muscularis mucosae but the larger collections are within distended lymphatic channels (thin arrows). The malignant cells are forming gland-like spaces (thick arrows). The tumour cells are likely to be part of finger-like growths of neoplasm extending along the lymphatics from the primary growth. Lymphatic permeation is one reason for excising a considerable portion of normal-looking bowel around a carcinoma. *See also 2.26, 7.56.* HE ×80

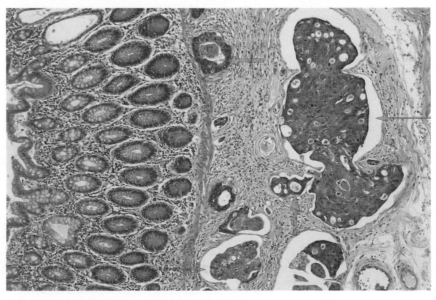

2.1 Gaucher's disease: spleen

Patients with the adult form of Gaucher's disease have a deficiency of beta-glucosidase in the lysosomes of their macrophages. The deficiency leads to the storage of large amounts of glucocerebrosides (from the breakdown of erythrocytes) in macrophages in many tissues. This spleen weighed more than 5kg. The great increase in the size of the organ is caused by the presence within the distended sinusoids of macrophages with abundant cytoplasm (Gaucher cells). Most Gaucher cells have a single nucleus, but a few are binucleate. The cytoplasm stains irregularly, tending to appear vacuolated, but the striated appearance sometimes present is not evident. Small numbers of lymphocytes are visible but the massive infiltrate of Gaucher cells had caused atrophy of the lymphoid elements of the organ, including the Malpighian bodies.   HE ×360

2.2 Portal hypertension: spleen

In portal hypertension the spleen is intensely congested and generally considerably enlarged (weighing up to 1.5kg). These changes are more marked than in systemic venous congestion. In this case the capsule of the organ was thickened and adherent to the neighbouring tissues. The cut surface was firm and 'meaty' in appearance. Histologically the venous sinuses (the red pulp) are dilated and the endothelial cells lining them are prominent. The walls of the sinusoids are thicker and more fibrous than normal, the increase of fibrous tissue being more obvious with stains for reticulin. The Malpighian bodies were atrophic.

HE ×235

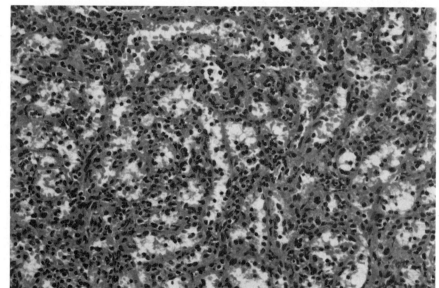

2.3 Portal hypertension: spleen

The increased venous pressure in the portal vein tends to cause repeated hemorrhages from the congested vessels within the spleen. When the extravasated blood is organized it gives rise to fibrous nodules rich in hemosiderin, the so-called siderotic nodules or Gamna-Gandy bodies. These range up to 3 or 4mm dia. and macroscopically appear as brown spots on the cut surface ('tobacco' nodules). This nodule consists of dense collagenous fibrous tissue heavily impregnated with iron and calcium salts and therefore deeply basophilic (bluish-black). Recent hemorrhage is evident around the nodule (left). Deposits of hemosiderin were also present in the fibrous trabeculae throughout the spleen.   HE ×55

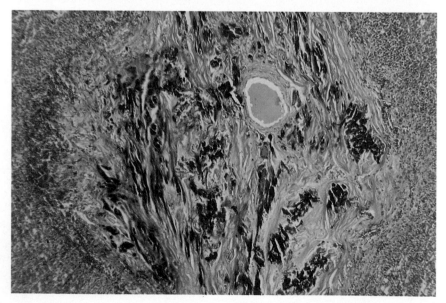

2.4 Systemic lupus erythematosus: spleen

Vascular lesions are prominent in this disease in which there is often widespread and progressive degeneration of the connective tissues. In this case one manifestation of the illness is a great increase in fibrous tissue around two penicillary arteries, with collagen fibres arranged in concentric lamellae: 'onion-skinning'. There is also deposition of eosinophilic material in the wall of the vessels. Sometimes this change is so pronounced as to amount to fibrinoid necrosis.
HE ×135

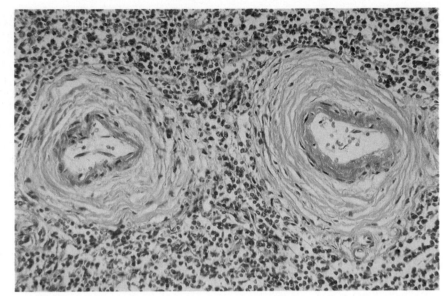

2.5 Myelosclerosis: spleen

In myelosclerosis there is progressive fibrosis of the hemopoietic bone marrow, and extramedullary hemopoiesis frequently occurs, mainly in liver and spleen. The disease was advanced in this case and the spleen, which was dark brownish-red in colour, weighed 820g. Histologically the normal lymphoid elements (the Malpighian bodies) were very small and atrophic, the organ, as shown here, consisting largely of immature cells of the granulocyte and erythrocyte series, some being pleomorphic and abnormal.
HE ×135

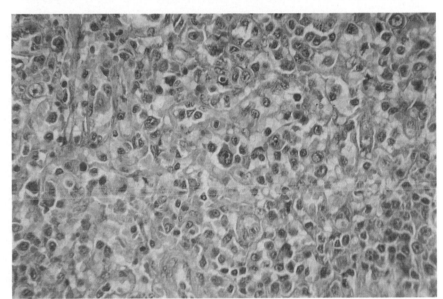

2.6 Secondary carcinoma: spleen

In contrast to the bone marrow and lymph nodes, the spleen, macroscopically at least, is for unknown reasons a rare site for metastatic tumour deposits. In this case the patient had a primary carcinoma of bronchus, and there is extensive permeation of the sinusoids of the spleen by pleomorphic undifferentiated tumour cells with basophilic hyperchromatic nuclei. The Malpighian body (left) is slightly atrophic but otherwise normal. The penicillary artery (arrow) shows marked hyaline change in its wall, with narrowing of the lumen, a not infrequent finding in elderly individuals.
HE ×150

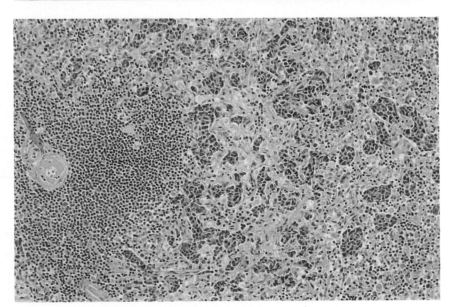

2.7   T cell lymphoma: pleural fluid

The patient who had a lymphoma developed a pleural exudate which was aspirated. This is a smear of the cells in the fluid, reacted for acid phosphatase. The reaction product for the enzyme is orange-yellow, and the cells have been counterstained blue. The nuclei are round and vary only moderately in size. Very little cytoplasm is evident, but in a high proportion of the cells there is a 'dot' of enzyme located on the surface of the cell (arrow). The presence of a 'dot' of acid phosphatase on the cell membrane is a useful marker for T-lymphocytes, and the result confirmed that the patient had a T-lympho-blastic lymphoma with involvement of the pleural cavity. The larger, paler nuclei belong to the mesothelial cells also present in the pleural exudate.

Acid phosphatase-methylene blue ×850

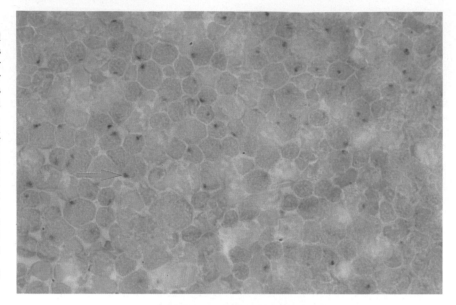

2.8   Atrophy: thymus

The patient, a woman of 57, had a thymoma of mixed lymphocytic and epithelial type resected surgically. This is the adjacent fatty tissue. Within the fatty tissue there are nod-ules of atrophic thymic tissue. Each nodule consists of a central eosinophilic Hassal's corpuscle (arrows) surrounded by a cuff of densely staining small cells. It is just pos-sible to distinguish cortex and medulla, the closely packed outermost cells (cortex) staining more deeply than the more loosely packed and slightly more eosinophilic cells (medulla) nearer the Hassal's corpuscles.

HE ×60

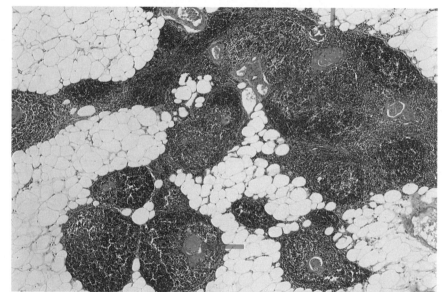

2.9   Atrophy: thymus

Parts of several Hassal's corpuscles are shown. They consist of large squamous epi-thelial cells with abundant eosinophilic cytoplasm, and the centre of the two larger corpuscles contains keratinized cell debris (arrows). The other cells of the adjacent medulla are almost exclusively small lym-phocyte-type cells with deeply basophilic nuclei, very few epithelial cells being evi-dent.   HE ×360

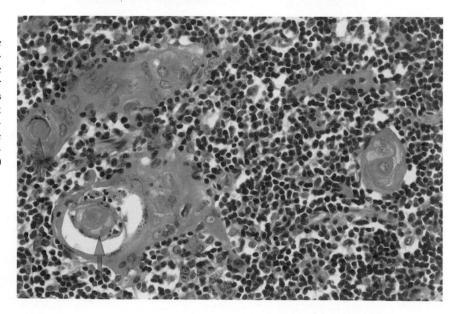

2.10 Myasthenia gravis: thymus

In myasthenia gravis the thymus is almost always abnormal, suggesting that the disease has an autoimmune basis. Germinal centres are generally present but a minority of patients also have a thymoma. Frequently also antibodies to acetylcholine receptors can be demonstrated in the serum. Thymectomy is generally beneficial in the early stages but has little effect later in the disease. In this case, a thymoma is present. It was encapsulated and apparently benign. Part of the fibrous capsule is on the left. The tumour is of mixed type, consisting of a network of very large pale-staining epithelial cells (arrows) and small lymphocytes with deeply basophilic nuclei.　　　HE ×360

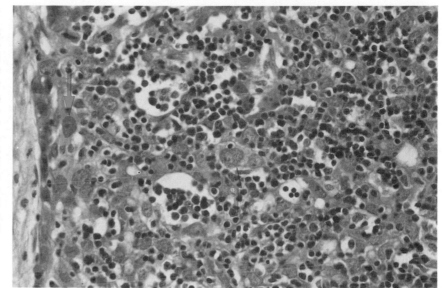

2.11 Thymoma (lymphocytic type): thymus

Most thymomas are benign, do not produce symptoms and are discovered incidentally. They may however cause pressure symptoms or very occasionally myasthenia gravis. Other systemic manifestations such as systemic lupus erythematosus or red cell aplasia may also occur. The majority of thymomas are of mixed composition, containing both epithelial and lymphocytic elements, but may be predominantly epithelial or predominantly lymphocytic. This lesion is predominantly lymphocytic, consisting of nodules of closely packed small basophilic lymphocytic cells, among which there are occasional larger paler cells. Broad dense bands of connective tissue separate the cellular nodules. The presence of fibrous bands is a common feature of thymoma, as is a thick fibrous capsule.　　　HE ×60

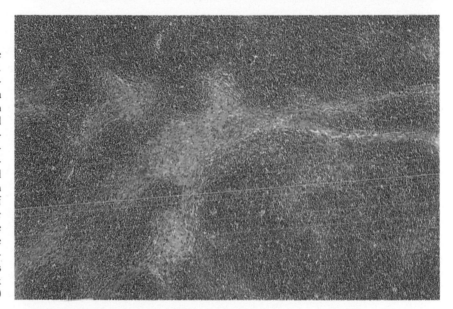

2.12 Thymoma (lymphocytic type): thymus

At higher magnification the tumour is seen to consist very largely of small cells with compact round basophilic nuclei and moderate amounts of eosinophilic cytoplasm. These cells have a marked tendency to form slender cords with one cell behind the other in 'Indian file'. An Indian file pattern of infiltration is a notable feature of T-lymphoblastic lymphomas. No mitoses are evident among these cells. Occasional large histiocytes with pale-staining cytoplasm (arrows) are scattered throughout the tumour.

HE ×360

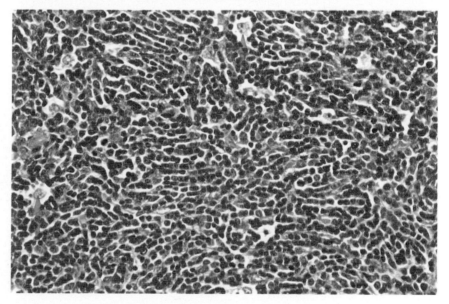

2.13   Thymoma (mixed type): thymus

The patient in this case had a mixed type of thymoma, containing both epithelial and lymphocytic elements. A systemic manifestation was also present in the form of hypercalcemia. The hypercalcemia was relieved by thymectomy. Like the lymphocytic-type lesion shown in **2.11** the tumour consists of deeply basophilic cellular nodules of various sizes, separated by broad bands of eosinophilic collagenous fibrous tissue. The staining of the cellular nodules however is not as uniform as in the lymphocytic type, a large population of paler-staining epithelial cells being present alongside darkly staining lymphocytes.   HE ×60

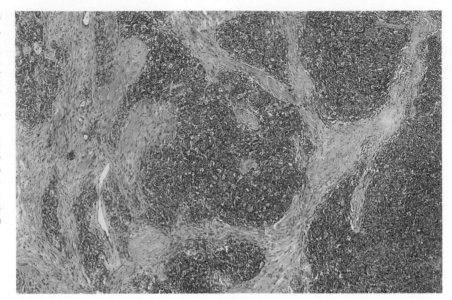

2.14   Thymoma (mixed type): thymus

This shows parts of two nodules of tumour at higher magnification. They are separated by a broad band (right of centre) of dense collagenous tissue (a capsule of even denser fibrous tissue surrounded the tumour). The majority of the tumour cells are fairly large relatively pale-staining epithelial cells and are accompanied by a minority population of smaller deeply staining lymphocytes. The open vesicular structure of the nuclei of the epithelial cells contrasts with the solid pyknotic appearance of the nuclei of the lymphocytic cells.   HE ×360

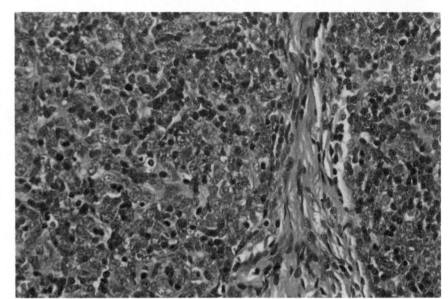

2.15   Secondary deposit of thymoma: marrow

A man of 37 had a thymoma removed surgically. It proved to be a malignant mixed type of thymoma. Nine months later a trephine specimen of bone marrow was taken from his iliac crest. This shows that the normal fat cells and hemopoietic tissue have been largely replaced by a mixture of large pale-staining epithelial cells (thin arrows) and small lymphocytic cells. The epithelial cells have large nuclei with very prominent nucleoli. The structure is that of a mixed type of thymoma. Malignant thymoma is a rare neoplasm and metastasis of a thymoma to bone marrow is extremely rare. Moderate numbers of other mononuclear cells, probably hemopoietic cells, are also present (thick arrow).   HE ×580

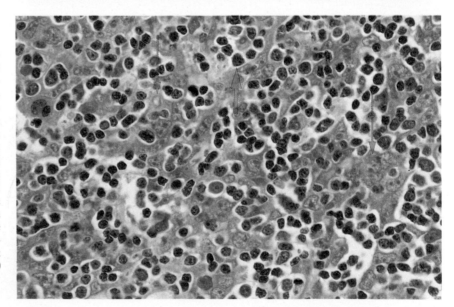

2.16 Multiple myeloma: marrow

Multiple myeloma is a malignant neoplasm of plasma cells, and it is found where there is hemopoietic bone marrow (vertebrae, skull, ribs, ends of long bones). It forms nodules of soft red tissue. It is osteolytic, destroying the bone lamellae, and an X-ray shows characteristic 'punched-out' defects in the bone. The hemopoietic tissue has been replaced by neoplastic plasma cells which vary in their degree of differentiation. Most of the cells appear mature, with a clock-face nuclear chromatin pattern and abundant cytoplasm. The nucleus is typically located at one pole of the cell. The purplish colour of the cytoplasm is produced by its large content of ribosomes, which synthesize the abnormal immunoglobulins (myeloma proteins) which are detectable in the serum in this disease. HE ×335

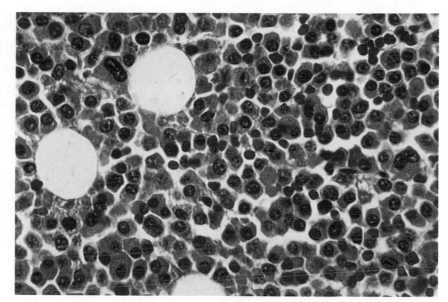

2.17 Extramedullary (solitary) plasmacytoma

Sometimes a mass of plasma cells, which appears to be a solitary deposit, is discovered in bone or in a soft-tissue (extramedullary) site. Excision appears to effect a cure, but multiple myeloma often develops subsequently. In many cells the plasma cell's nuclear chromatin pattern is evident, as are the polar location of the nucleus and pale-staining halo adjacent to the nucleus (thin arrows). The more primitive plasmablasts have a large nucleus with a prominent nucleolus and relatively less cytoplasm (thick arrows). No mitoses are evident, a common finding in myelomatous deposits. To distinguish a plasmacytoma from a granuloma containing larger numbers of plasma cells, it may be necessary to determine whether the cells have a monoclonal origin. *See also 1.38, 3.19, 3.20.* HE ×580

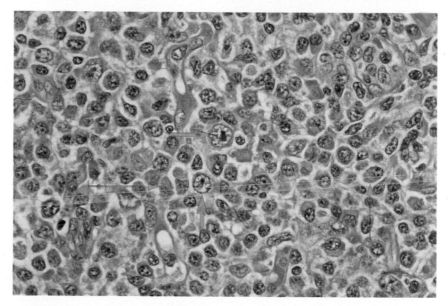

2.18 Secondary carcinoma: marrow

Hemopoietic marrow is frequently the site of metastasis of malignant tumours. This is marrow from the iliac crest of a man of 62. A trabecula of normal lamellar bone is visible on the left. The hemopoietic tissue has been replaced by a vascular cellular connective tissue in which multiple compact nodules of tumour cells are growing (right). Alongside these there are two small trabeculae of new bone, their surfaces covered with osteoblasts (left centre). The formation of new bone (osteosclerosis) has been induced by the malignant cells. The site of the primary tumour was not known clinically but the presence of cytoplasmic granules which were argyrophilic (but not argentaffin) suggested an endocrine-cell origin. Unlike this example, most deposits of metastatic tumour are osteolytic. HE ×235

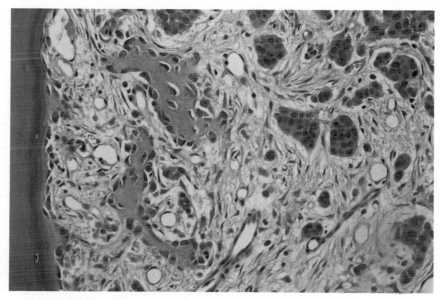

2.19 Crohn's disease: lymph node

This is a lymph node from the mesentery of a woman of 61 who had the terminal ileum and part of the ascending colon removed surgically for Crohn's disease. The capsule of the node is on the left. The node shows marked reactive changes. The sinuses including the subcapsular sinus (thin arrows) are dilated and contain a large population of pale-staining histiocytes. The lymphoid follicles underlying the subcapsular (peripheral) sinus are large, with germinal centres; and the adjacent paracortical zone is well-developed and very vascular, with many prominent high endothelial venules (thick arrows). At higher magnification numerous small granulomas consisting of epithelioid cells and small multinucleated giant cells were evident.   HE ×60

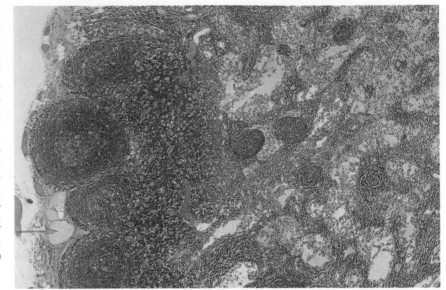

2.20 Whipple's disease: lymph node

In Whipple's disease (intestinal lipodystrophy) large macrophages are present in the villi of the small intestine, and electron microscopy shows bacilliform bodies in their cytoplasm. The nature of the microorganism is not known but treatment with antibiotics is usually beneficial. The bowel lesions disturb fat absorption and neutral fat enters the mesenteric lymph nodes. The patient usually displays the effects of intestinal malabsorption. This shows the fibrous capsule of a node (left margin) and the subcapsular sinus. The lumen of the sinus is occupied by droplets (unstained) of neutral fat and small lymphocytes. The cortical tissue (centre and right) is composed mostly of small lymphocytes, but unstained droplets of fat are also present (e.g. bottom right corner) along with strands of amorphous material (right).   HE ×235

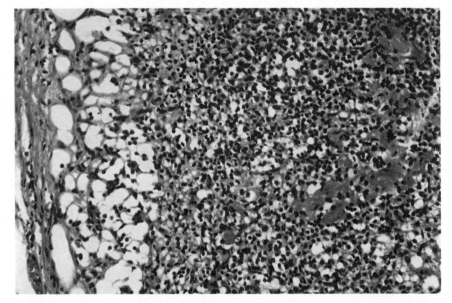

2.21 Whipple's disease: lymph node

This is deeper in the lymph node. The medullary sinus (centre left) is full of macrophages with eosinophilic cytoplasm and droplets (unstained) of neutral fat (thin arrow) and a small channel is full of similar droplets (thick arrow). In the lymphoid tissue adjacent to the sinuses there is a round granuloma (right) consisting of histiocytes with eosinophilic vacuolated cytoplasm and small lymphocytes. No multinucleated giant cells are present. Some of the small round extracellular vacuoles within the granulomatous focus may have contained fat. The changes in the mesenteric lymph nodes usually suggest the diagnosis but they are not pathognomonic as a rule.   HE ×235

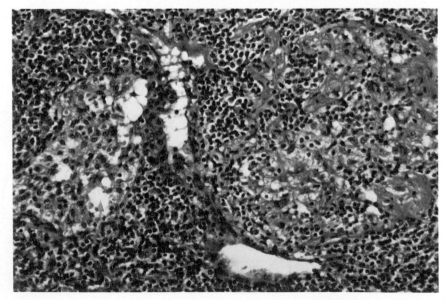

2.22 Reactive sinus hyperplasia (sinus histiocytosis): lymph node

In sinus hyperplasia there is a marked increase in the number of macrophages in the lymph sinuses, including the peripheral sinus. The affected nodes may also show the various changes that occur in an immune response. Two medullary lymphatic sinuses are shown. They are greatly distended, and their lumens filled with large numbers of large eosinophilic histiocytes. Many of these are part of the network of 'fixed' macrophages in the sinuses. The small dark-staining cells accompanying the macrophages are lymphocytes. A fibromuscular septum (arrows) can be seen, which divides the sinus into twin channels. The medullary lymphoid tissue consists of small lymphocytes, plasma cells and histiocytes.

HE ×150

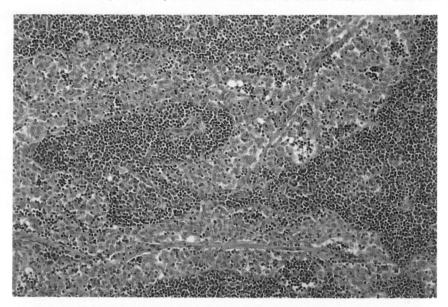

2.23 Reactive sinus hyperplasia: lymph node

This shows a lymphatic sinus in the medulla in more detail, with the fibromuscular septum clearly visible in the centre of the sinus (arrow). The channels on both sides of the septum are filled with macrophages, each with abundant pale-staining cytoplasm and a large vesicular nucleus. Small numbers of lymphocytes are also present. Most of the cells in the medullary lymphoid tissue (bottom) are small lymphocytes, but occasional histiocytes and plasma cells are also evident. Sinus hyperplasia is seen in lymph nodes draining tissues which are chronically inflamed or in which there is continued destruction of cells. It is frequently present in the axillary nodes draining a breast which is the site of a primary adenocarcinoma.

HE ×360

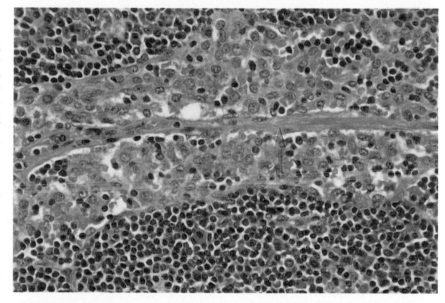

2.24 Sinus hyperplasia with massive lymphadenopathy (Rosai-Dorfman syndrome): lymph node

In this syndrome the patient has massive enlargement of lymph nodes (mainly in the neck), fever, leukocytosis and hypergammaglobulinemia. There is no specific therapy, but complete recovery generally takes place. Characteristically the capsule of the node is thick and fibrous, and the architecture of the node is replaced by dilated lymphatic sinuses. Part of an affected sinus is shown here, full of eosinophilic lymph and a mixture of lymphocytes and macrophages. Some of the latter are relatively enormous (arrows) and contain many phagocytosed small lymphocytes in their cytoplasm. Large cells of this type are a diagnostic feature of this syndrome. The intersinusoidal tissue consists of dark-staining small lymphocytes.

HE ×360

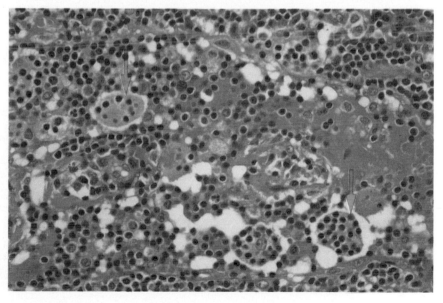

2.25 Lymph node: porta hepatis

Although the liver has no lymphatic system, lipid and bile pigment sometimes drain to the lymph nodes in the porta hepatis and produce a reaction in the lymphatic sinuses. This sinus is dilated, with droplets (unstained) of neutral fat, most of them apparently within vacuoles in the cytoplasm of the macrophages. The macrophages have large vesicular nuclei (thin arrows). Also present are several giant multinucleated phagocytes (thick arrows) containing lipid droplets. These cells have markedly eosinophilic cytoplasm. Bile pigment is not evident here, but is sometimes present along with the lipid within phagocytes. The lymphoid tissue adjacent to the sinus (top left) consists of lymphocytes and histiocytes. HE ×360

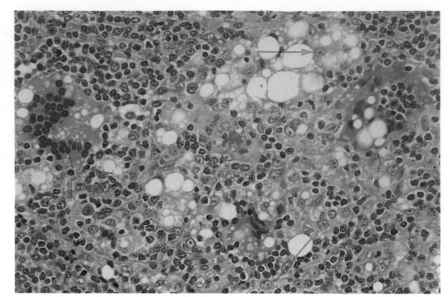

2.26 Secondary (mucoid) carcinoma: lymph node

Lymph enters the peripheral sinus of a lymph node at the capsule, and malignant cells carried to the node in the lymph tend, therefore, to be found within the peripheral sinus, at least initially. A man of 76 had a large mucoid adenocarcinoma of sigmoid colon removed surgically. This is a node in the mesentery. The fibrous capsule is on the left. Within the peripheral sinus and invading the underlying lymphoid tissue (top) are large numbers of round carcinoma cells. Each cell is filled with a droplet of pale-staining mucin which has pushed the nucleus to one side, to produce a signet-ring appearance (arrow). No mitotic activity is evident. Mucoid carcinomas may metastasize less readily than other forms but the evidence is not conclusive. *See also 1.54, 7.56.*

HE ×150

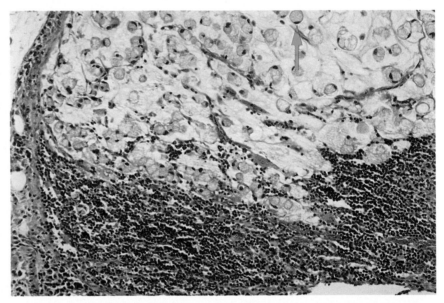

2.27 Lipomelanic reticulosis (dermatopathic lymphadenitis): lymph node

In certain skin diseases, particularly chronic dermatitis, the lymph nodes draining the affected area enlarge. Examination shows reactive hyperplasia. The architecture of the node is preserved and the principal change is in the paracortical zone. The fibrous capsule is on the left. The paracortical zone (top and right), consisting of lymphocytes and large numbers of pale-staining mononuclear cells, is greatly enlarged. Some macrophages are laden with melanin from the inflamed skin. The small blood vessels (arrows) are high endothelial venules. The cortical lymphoid follicles were also enlarged and part of one is shown (lower left). It has a broad corona of small lymphocytes and a large germinal centre in which mitotic figures are visible. HE ×150

2.28 Lipomelanic reticulosis: lymph node

This shows part of the paracortical tissue at higher magnification. The predominant cell is the histiocyte, with its vesicular nucleus (thin arrows) and abundant cytoplasm. Many of them are macrophages containing melanin in amounts which vary from a few granules to dense deposits of coarse granules which fill and distend the cell (e.g. bottom right). A frozen section stained for lipid would demonstrate the presence also of lipid within these macrophages. Some of the histiocytes in this tissue have been shown to be Langerhans cells and interdigitating reticulum cells. The small cells with darkly staining nuclei (right) are lymphocytes (mainly T cells) and the blood vessels (thick arrows) are high endothelial venules.

HE ×360

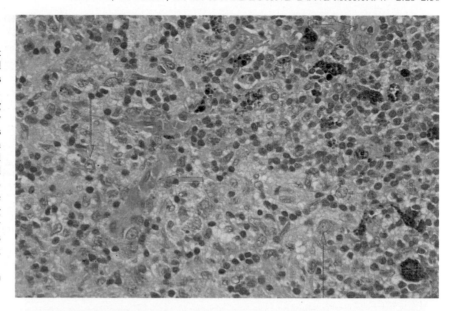

2.29 Cat-scratch disease: lymph node

Some cats carry an infectious agent which, when inoculated by a scratch from the claws of the cat, causes a systemic illness of a benign self-limiting nature. A papule forms in the skin and the regional nodes become inflamed and enlarged. Abscesses often form in the node which tend to be stellate in outline. A woman of 25 developed a tender swelling on the left side of her neck. Part of it (1 × 0.6 × 0.6cm) was removed surgically for diagnostic purposes. Histologically no normal lymph node architecture remained, but several 'abscesses' were present. This shows one abscess. A collection of polymorph leukocytes and necrotic cells (centre and lower right) is surrounded by an eosinophilic band of histiocytes (arrows). Peripheral to this is the lymphoid tissue of the node. HE ×150

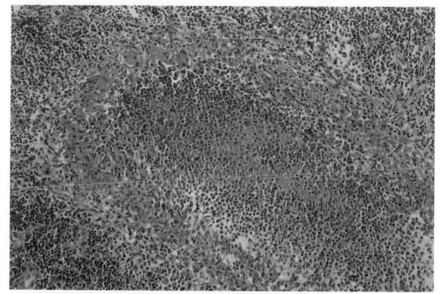

2.30 Cat-scratch disease: lymph node

This shows part of the abscess at higher magnification. The contents of the abscess (right) are pus-like, consisting largely of polymorphs. Enclosing this material are large mononuclear cells with vesicular nuclei and eosinophilic cytoplasm, which show a tendency to palisading. External to this histiocytic 'capsule' are the small lymphocytes of the surrounding lymphoid tissue (left). The agent causing cat-scratch disease is thought to belong to the psittacosis-lymphogranuloma group of microorganisms. HE ×360

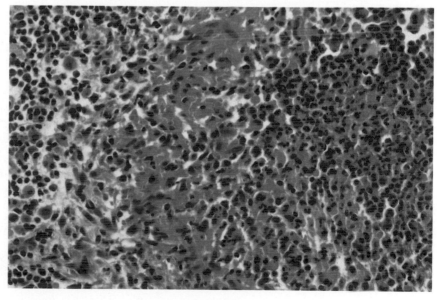

2.31 Toxoplasmosis: lymph node

Toxoplasmosis is a not uncommon systemic infection by the protozoon *Toxoplasma gondii*. Cats are probably the definitive host. Any tissue may be affected, the trophozoites proliferating within cells and destroying them. Lymphoid tissue is often involved, particularly in the cervical region, the nodes becoming enlarged. The architecture of the node is not destroyed, and the most striking feature, shown here, is the presence of clusters of large eosinophilic mononuclear histiocytes, epithelioid cells. The small cells are the remaining lymphocytes of the lymphoid tissue. The blood vessels are high endothelial venules. An occasional Langhans cell may be present. Necrosis is usually absent. The clusters of epithelioid cells may encroach on the lymphoid follicles of the node. Only very rarely can the parasite be demonstrated. HE ×235

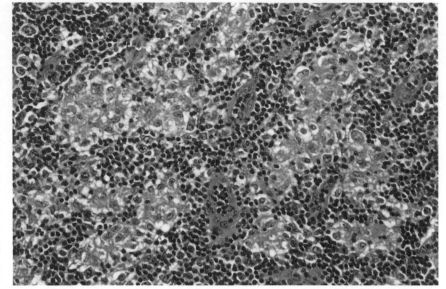

2.32 Sarcoidosis: lymph node

Sarcoidosis is a granulomatous inflammatory disease which affects many tissues, including lymphoid tissue. This is a cervical node, which measured 3.5 × 1.5 × 0.5cm, from a woman of 51 with severe sarcoidosis. The capsule of the node (left) which was thick and fibrous macroscopically, is on the left. The normal architecture of the node has been largely destroyed, with some blue-staining lymphoid tissue surviving beneath the capsule and between the round sarcoid follicles. These follicles vary widely in size, from a few cells to very large collections (right) several mm in diameter. They consist of epithelioid histiocytes, and early fibrosis can just be detected at the periphery of several follicles. There is no necrosis within the follicles, but some contain calcified laminated Schaumann bodies (arrows). *See also 1.23, 7.4, 7.48, 14.10.* HE ×60

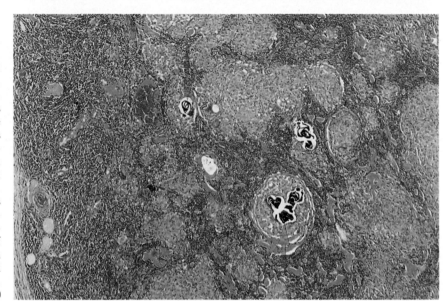

2.33 Infectious mononucleosis (glandular fever): lymph node

Infectious mononucleosis is caused by the Epstein-Barr (EB) virus. It is a systemic illness, but the lymph nodes and particularly the cervical lymph nodes are swollen and tender. The architecture of this node was characteristically obscured by the presence of large numbers of lymphoid blast cells (arrows). These cells have large nuclei in which there are one or more very prominent nucleoli. The cells are considered to be T-immunoblasts. They are accompanied by other cells of the lymphocyte series. The histological features are very suggestive of a malignant lymphoma. Atypical lymphoid cells could also be demonstrated in smears of the peripheral blood. HE ×360

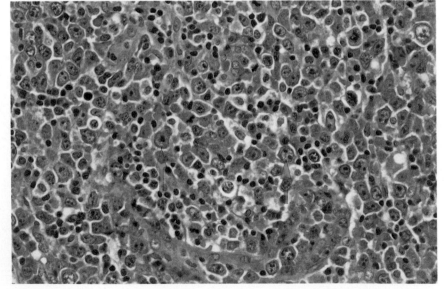

2.34 Hodgkin's disease

Hodgkin's disease differs from other lymphomas in that histologically it is not monomorphic but involves, generally, a variety of different cell types, including histiocytes, lymphocytes, plasma cells and eosinophil leukocytes, as well as a characteristic and diagnostic type of large cell with two or more nuclei, the Reed-Sternberg cell. This is the nodular sclerosis form of Hodgkin's disease characterized by the presence of broad bands of collagenous fibrous tissue enclosing cellular nodules of lymphomatous tissue. The central nodules appear to consist largely of lymphocytes, as does the part of the larger nodule visible on the right; that is, they have a lymphocyte-predominance structure. Numerous lymphocytes are present also throughout the dense collagenous tissue.   HE ×65

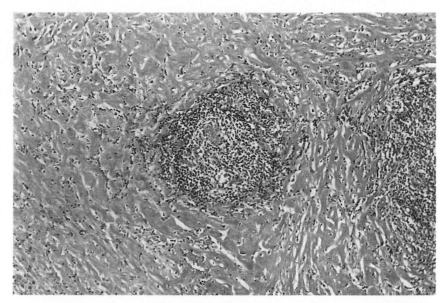

2.35 Hodgkin's disease: lymph node

This is the mixed cellularity form of Hodgkin's disease. Of the several different types of cell present, the most prominent are the many pleomorphic histiocyte-type cells with their large vesicular nuclei (thin arrows). One of these is a Reed-Sternberg cell, a small multinucleated giant cell (thick arrow). The small cells with the deeply staining nuclei are lymphocytes, and eosinophil polymorphs with eosinophilic cytoplasm are also present. On the right is a large area of necrosis, staining deep red. Necrosis is often a striking feature of Hodgkin's disease and was widespread in this node.   HE ×335

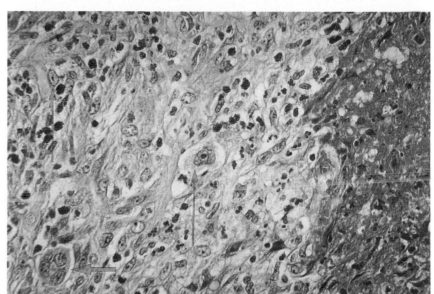

2.36 Hodgkin's disease: lymph node

This is an example of the lymphocyte-predominance type of Hodgkin's disease. Lymphocytes make up the bulk of the tissue and the number of abnormal histiocytes is relatively small. In this field however, along with the small lymphocytes, there are two large cells of classical Reed-Sternberg form: i.e. a binucleate cell, each nucleus forming a 'mirror-image' of the other. The nuclei are vesicular and each has a very prominent eosinophilic central nucleolus. The origin of the Reed-Sternberg cell is uncertain, but its presence is a very important, indeed essential, factor in the histological diagnosis of Hodgkin's disease.   HE ×950

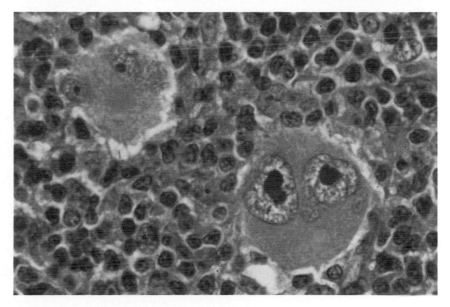

2.37 Angiofollicular lymph node hyperplasia (AFLNH)

The nature of this lesion is obscure. It is located most often in the mediastinum or neck and is usually an incidental finding, e.g. when the thorax is X-rayed. It takes the form, as a rule, of a large encapsulated mass of lymphoid tissue. Young adults are usually affected. A woman of 19 with no systemic manifestations had an encapsulated mass (3.5 × 3 × 2cm) removed surgically from her neck. Characteristically very compact lymphoid follicles with well-defined boundaries (arrows) are present. The two largest follicles have small pale-staining germinal centres and concentric mantles of lymphocytes. The tissue between the follicles consists of lymphocytes and a network of small eosinophilic blood vessels. There are no lymphatic sinuses. HE ×60

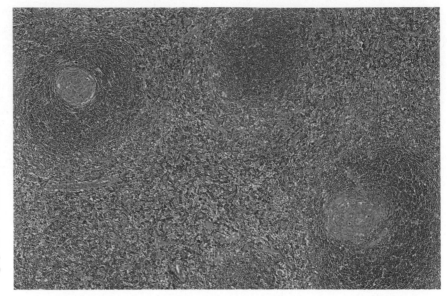

2.38 Angiofollicular lymph node hyperplasia

This shows the interfollicular tissue and a follicle at higher magnification. The follicle (top right) has a mantle of lymphocytes arranged in tight concentric circles, producing an 'onion-skin' appearance, and a small germinal centre of pale-staining cells. Running into the centre and occupying part of it is a hyalinized blood vessel with a thick hyaline wall (arrow). A network of similar small blood vessels with hyalinized walls is present throughout the interfollicular tissue (left). Clusters of small lymphocytes lie between these vessels. HE ×150

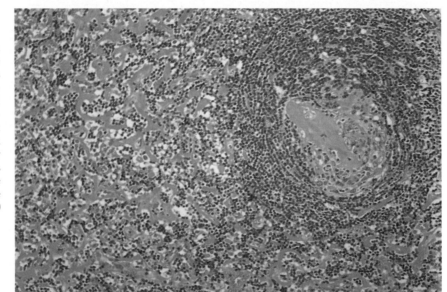

2.39 Angiofollicular lymph node hyperplasia

This is a cryostat section reacted with the indirect immunoperoxidase method, using a monoclonal antibody to an antigen on the surface of T cells (a pan-T cell antibody). The method colours the cell membrane of T cells brown. The capsule of the node is at the top. The follicles are virtually unstained, only a few positively-reacting cells being detectable within them. In contrast the interfollicular tissues are diffusely positive, the unstained areas being small blood vessels. Indirect immunoperoxidase method for T cells ×60

2.40 Angiofollicular lymph node hyperplasia: lymph node

This shows the interfollicular tissue at higher magnification. The small vessels, part of the complex network of vessels in the interfollicular tissue, have the structure of high endothelial venules, being lined by large cuboidal endothelial cells with vesicular nuclei (arrows). In this area the walls of the vessels are only slightly hyalinized. The cell population consists to a considerable extent of lymphocytes which, as shown in **2.39** are T lymphocytes. A few macrophages are also present. In some lesions considerable numbers of plasma cells and eosinophil leukocytes are present along with the lymphocytes.   HE ×360

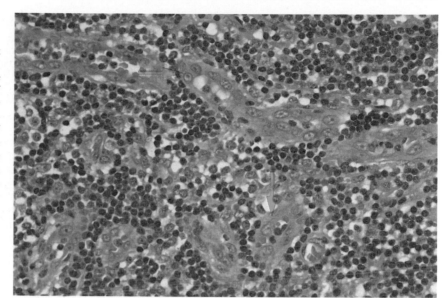

2.41 Angioimmunoblastic lymphadenopathy

Individuals with angioimmunoblastic lymphadenopathy are adult and often elderly. In addition to enlarged lymph nodes the patients generally have fever, hemolytic anemia and a polyclonal hypergammaglobulinemia. The condition may be mistaken for malignant lymphoma but differs in its rapid onset and fluctuant course. Only a few atrophic lymphoid follicles remained in this node. The new tissue consists of small blood vessels resembling high endothelial venules, lined by plump endothelial cells with large ovoid vesicular nuclei (thin arrows). Between the vessels is a cellular infiltrate of small lymphocytes, immunoblasts (thick arrows), plasma cells and eosinophil leukocytes. Multinucleated giant cells are sometimes present and are often mistaken for Hodgkin's cells.   HE ×360

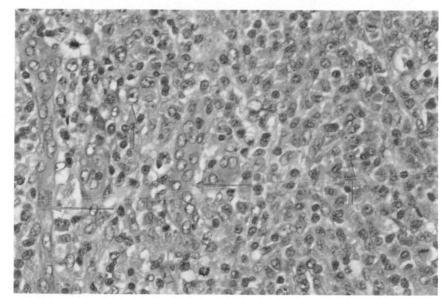

2.42 Angioimmunoblastic lymphadenopathy

The large numbers of arborizing small blood vessels in the lymph nodes form an important diagnostic feature of angioimmunoblastic lymphadenopathy, but they are relatively difficult to detect in ordinary HE sections. The vascular network is made much more obvious by stains for the basement membranes of the vessels (reticulin), as this preparation clearly shows. No lymphoid follicles, which are also clearly demonstrated by this technique, have survived in this part of the node.

Silver method for reticulin ×150

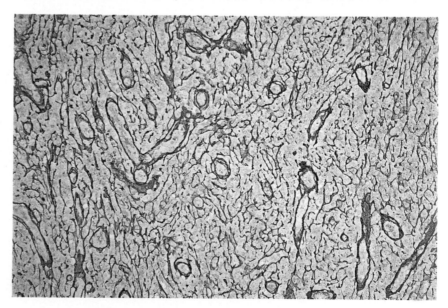

2.43 Non-Hodgkin's lymphoma (NHL): lymph node

Lymphomas other than Hodgkin's lymphomas arise from lymphocytes (B cells or T cells) or from histiocytes (histiocytic lymphomas, malignant histiocytosis). All lymphomas are malignant. This is a follicular lymphoma (centroblastic/centrocytic), one of the commonest forms of low-grade NHL, generally occurring in adults and arising from B cells. It has a better prognosis than diffuse lymphomas which lack a follicular pattern. The tissue consists of a fairly uniform population of cells arranged in round or ovoid clusters. These 'follicles' are bounded by bands of vascular fibrous tissue which stand out because of the presence within them of darker-staining cells. The cells making up the 'follicles' are centroblasts and centrocytes.　　HE ×60

2.44 Follicular lymphoma (centroblastic/centrocytic): lymph node

The lymphoma is composed largely of a mixture of centroblasts and centrocytes, with centrocytes predominating. The centroblasts (thin arrows) are larger than the centrocytes, with large round nuclei which usually contain several prominent nucleoli, located close to the nuclear membrane. They have abundant cytoplasm with a well-defined cell boundary. The more numerous centrocytes are smaller (thick arrows), many comparable in size with small lymphocytes. Their nuclei however are less basophilic than those of small lymphocytes and usually 'deformed' with indentation of the nuclear membrane. Small nucleoli may be seen within some of them. A follicular lymphoma, centroblastic/centrocytic, tends to become diffuse, with centroblasts predominating.　　HE ×360

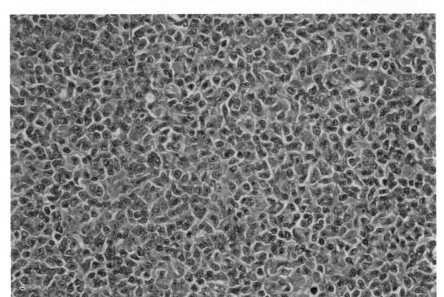

2.45 Lymphocytic lymphoma: marrow

The bone marrow is, rarely, the site of origin of a primary lymphoma. In many instances, however, it is involved secondarily in the form of either a diffuse infiltrate or as focal deposits. Biopsy of the bone marrow is very useful in staging the disease and in determining treatment. Focal or diffuse involvement of the marrow does not, however, reflect the lymph node pattern (follicular or diffuse) of a lymphoma. This is a section of a biopsy of iliac crest, and the normal hemopoietic marrow has been replaced by a population of lymphocyte-type cells, with small basophilic nuclei and very scanty cytoplasm. From the changes in the lymph nodes the lesion was classified as a low-grade lymphocytic lymphoma of diffuse type. The deposit is closely apposed to the bone trabecula (left).　　HE ×360

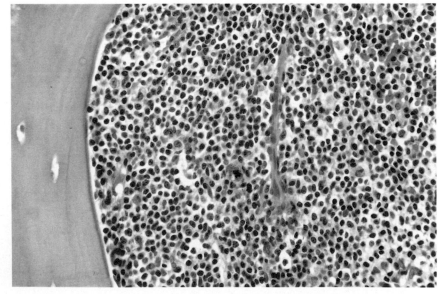

2.46 Lymphocytic lymphoma (diffuse): lymph node

Well-differentiated lymphocytic lymphoma consists of mature lymphocytes and may be regarded as the counterpart in the tissues of chronic lymphocytic leukemia. The normal architecture of the node has been replaced by a diffuse infiltrate of small round cells resembling mature lymphocytes. They are fairly uniform in size and shape, with deeply-staining nuclei and only a sparse rim of cytoplasm. There is no evidence of follicle formation. Scattered amongst them are larger pale-staining histiocytes (thin arrows) with large vesicular nuclei. A few high endothelial venules (thick arrows) are also present.   HE ×150

2.47 Lymphocytic lymphoma (diffuse): lymph node

This is the lymphoma shown in **2.46** at higher magnification. The vast majority of the cells are small lymphocyte-type cells with round basophilic nuclei and a thin rim of cytoplasm. There is no folding or indentation of the nuclei. The histiocytes (arrows) have abundant eosinophilic cytoplasm and their large nucleus has a prominent nucleolus.   HE ×360

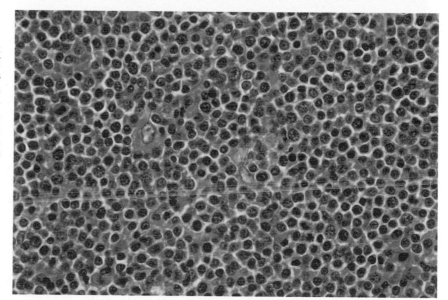

2.48 Lymphoblastic lymphoma: lymph node

Study of the detailed structure of lymphoma cells is easier in thin (0.5–1μm thick) sections than in thicker sections. This is a 1μm, plastic section of a high-grade lymphoma of lymphoblastic type. Lymphoblastic lymphoma occurs at all ages but most often in children and adolescents. Many later develop lymphoblastic leukemia. The normal population of cells of this node has been replaced by a monomorphic infiltrate of cells with ovoid or round nuclei (a small minority are convoluted) in which the chromatin is evenly dispersed. Multiple small nucleoli are visible in some nuclei. Cytoplasm is fairly scanty and not well-defined. There is considerable mitotic activity (arrows). The blood vessel (bottom centre) is a high endothelial venule.   HE ×320

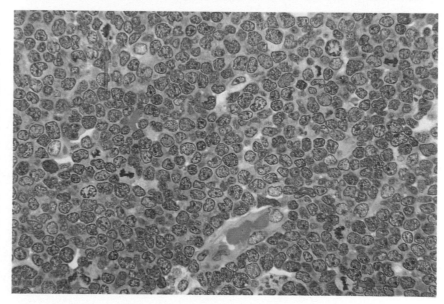

2.49 Centroblastic lymphoma (with immunoblastic transformation): lymph node

Lymphomas of centroblastic type may arise *de novo* or from a follicular centroblastic/centrocytic lymphoma. They have a relatively poor prognosis. Immunoblasts are often present in small numbers in centroblastic lymphomas but in parts of this node they predominated. They have round or ovoid pale vesicular nuclei with the chromatin tending to concentrate at the nuclear membrane; and in most of the nuclei there is a single large prominent central nucleolus (arrows). A few contain several nucleoli. The cytoplasm is abundant, with a well-defined boundary. It tends to stain blue from the presence of many ribosomes, and immunoglobulin can usually be demonstrated in the cytoplasm. The cells on the right are mostly small lymphocytes. HE ×360

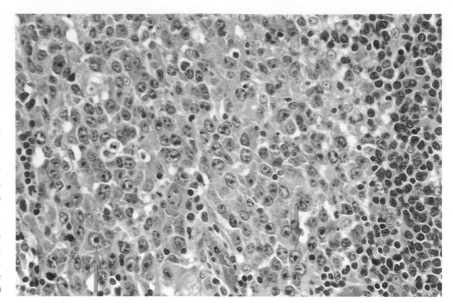

2.50 Lymphoblastic (Burkitt-type) lymphoma: omentum

Most cases of Burkitt's lymphoma arise in children and, at least in Africa, many organs and tissues, including the jaws are involved. Lymph node involvement is often comparatively insignificant. The tumour consists of closely packed cells with round or oval nuclei each containing several basophilic nucleoli which tend to be located near the nuclear membrane (thin arrows). The cytoplasm is scanty. There are numerous pyknotic fragments of necrotic nuclei. Several large macrophages are present (thick arrows). Remnants of ingested cells, including nuclear fragments, are present within their abundant cytoplasm. These macrophages are a feature of lymphoblastic lymphoma, being present throughout the lesion, and are responsible for the 'starry-sky' appearance. HE ×580

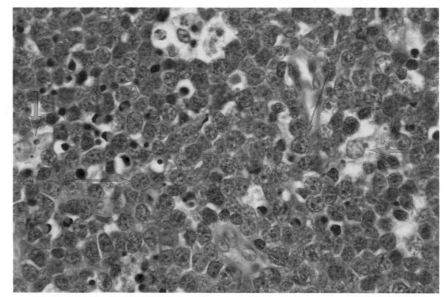

2.51 Malignant lymphoma (lymphoplasmacytic): lymph node

Lymphoplasmacytic lymphoma is a tumour of B lymphocytes which shows differentiation towards plasma cells. It occurs most often in older people and a paraprotein secreted by the tumour cells is usually present in the blood and urine. Histologically it consists of a mixture of small lymphocytes and plasma cells of varying grades of maturity. This section has been treated with specific antibody against IgM, and many neoplastic cells with round nuclei and a 'clock-face' pattern of chromatin resembling that of mature plasma cells contain IgM. Other cells with the same structure and less-well-differentiated cells with large pale nuclei have not reacted for IgM.

Indirect immunoperoxidase method for IgM ×850

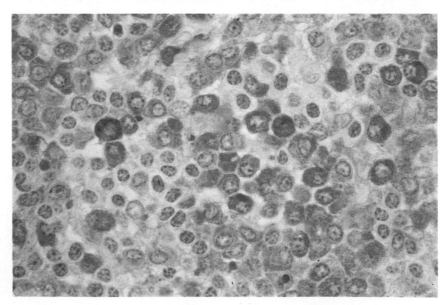

2.52 Malignant lymphoma (lymphoblastic): marrow

Lymphoblastic lymphoma is a high-grade lymphoma which arises from the precursors of B cells or T cells. It is highly infiltrative, the cells tending to infiltrate in Indian-file, displacing the cells of a tissue but leaving the underlying structure intact. Many of the nuclei are round or oval with fine and evenly dispersed chromatin, but a considerable proportion are convoluted, i.e. they have a deep cleft or fold, an appearance suggestive but not conclusive evidence of a T cell origin. Cell marker studies did confirm the T cell nature of this lesion. The cytoplasm is fairly scanty and ill-defined. Typically also, there are numerous mitotic figures (arrows). Scattered throughout the lesion are many reactive macrophages with abundant pale-staining cytoplasm, producing a 'starry-sky' appearance.   HE ×360

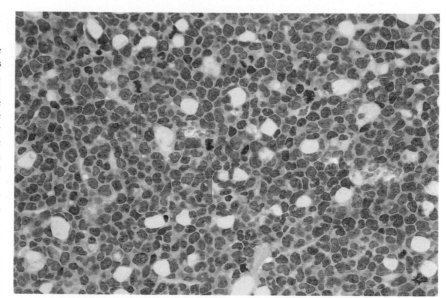

2.53 Malignant lymphoma (Sézary's syndrome): lymph node

Sézary's syndrome produces skin lesions similar to those in mycosis fungoides with involvement of the lymph nodes, but there is in Sézary's syndrome also a leukemic blood picture. The cell of origin is the mature peripheral T cell, and Sézary cells have a distinctive morphology particularly in smears of the peripheral blood. The paracortex of this node is over-run by Sézary cells with their highly-convoluted (cerebriform) nuclei. The nuclear structure is reminiscent of that of centrocytes but the indentations of the nuclei in Sézary cells are much more marked than in centrocytes. Several mitotic figures are present (thin arrows) and typically the Sézary cells are accompanied by large interdigitating reticulum cells with relatively abundant cytoplasm and pale-staining nucleus (thick arrows).   HE ×580

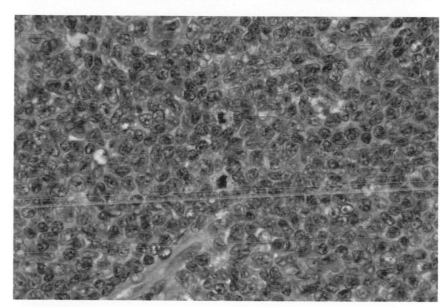

2.54 Chronic lymphocytic leukemia: marrow

In chronic lymphocytic leukemia the number of lymphocytes in the peripheral blood is increased, often very markedly. The leukemic cells are usually of B cell origin but in about 1% of cases they have T cell surface markers. There is often some enlargement of lymph nodes and spleen, and the bone marrow is not infrequently involved. In this trephine specimen of marrow from the iliac crest of a woman of 57, the hemopoietic cells, apart from several megakaryocytes (e.g. centre) have been replaced by a uniform population of small mature-looking lymphocytes. Involvement of the bone marrow may be focal or diffuse. In this specimen there was diffuse invasion with almost complete replacement of the hemopoietic elements.   HE ×360

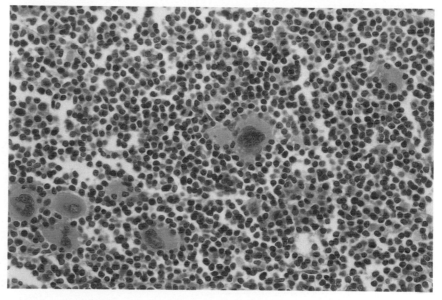

2.55 Histiocytic lymphoma: skin

The monocyte/macrophage system encompasses a wide range of cells of differing structure and function and the term may cover a heterogeneous group of conditions. Moreover, non-neoplastic histiocytes are often present in large numbers in malignant lymphomas of lymphocytic origin. Histiocytic lymphomas have a greater tendency than other types of lymphomas to be located in tissues other than lymph nodes, such as the skin, alimentary system and skeleton. This lesion was in the skin. In this thin (1μm) resin-embedded section, it consists of large cells with pleomorphic vesicular nuclei, many of them elongated and some indented. Each nucleus possesses one or more prominent nucleolus. The cells have abundant cytoplasm but their boundaries are ill-defined. HE ×360

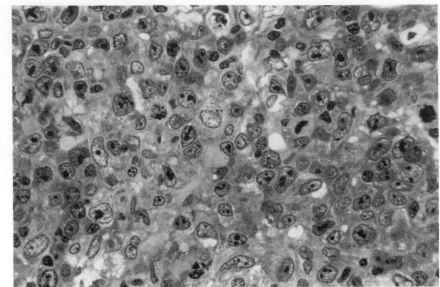

2.56 Histiocytic lymphoma: skin

One useful marker for histiocytes is the lysosomal enzyme cathepsin B which is present in abundance in normal macrophages. This section of a deposit of histiocytic lymphoma which was located in the abdominal wall has been reacted with the indirect immunoperoxidase method using an antibody specific for cathepsin B. The brown reaction product is present in a high proportion of the cells of the lymphoma, thereby demonstrating the presence of the enzyme in their cytoplasm. Morphologically, the lymphomatous cells show very considerable pleomorphism of both nucleus and cytoplasm.

Indirect immunoperoxidase method for cathepsin B ×360

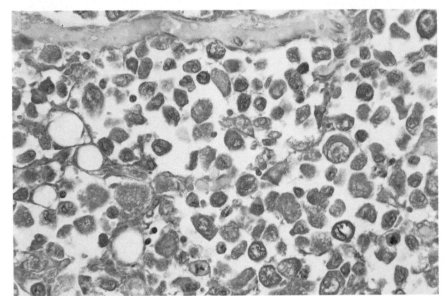

2.57 Malignant fibrous histiocytoma: marrow

Malignant fibrous histiocytoma in bone marrow has the same structure as pleomorphic malignant fibrous histiocytoma of soft tissue. Areas with a similar histological structure may be found in osteosarcomas and fibrosarcomas and care should be taken to exclude these before making the diagnosis. The lesion tends to be located in long bones, and this one was in the proximal femur. The space between the two bone trabeculae (right and left) is occupied by a population of cells which is characteristically extremely pleomorphic, ranging from large multinucleated giant cell forms (arrows) to elongated fibroblastic types. A reticulin stain showed the presence of an extensive network of connective tissue fibres. There is some erosion of the bone trabeculae by the tumour cells. HE ×235

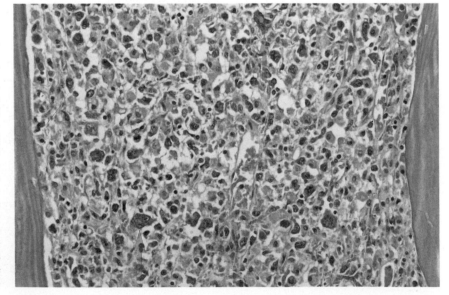

2.58 Malignant histiocytosis (histiocytic medullary reticulosis): lymph node

This is a malignant and rapidly fatal lesion of the histiocytes of the bone marrow which usually spreads to lymph nodes, spleen and liver. In lymph nodes the malignant histiocytes are usually located, at least initially, within the lymphatic sinusoids. They frequently, but not invariably, phagocytose erythrocytes, and this shows a number of atypical histiocytes (arrows). They have vesicular nuclei which vary in size and shape. Some of the cells have many erythrocytes in their abundant cytoplasm. The erythrocytes are being broken down and some are pale and 'ghost-like'. At high magnification remnants of erythrocytes could be detected in a high proportion of other histiocytes. Nuclear pleomorphism is very marked. HE ×580

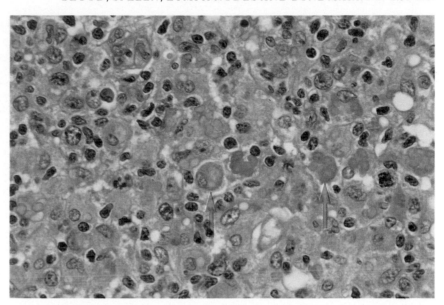

2.59 Monocytic leukemia

Monocytic leukemia is a malignant lesion of the monocyte/macrophage system which originates in the marrow. It is thus related to malignant histiocytosis but the morphology of the malignant cells is more uniform and infiltration of the bone marrow is usually more diffuse than in malignant histiocytosis. The leukemia is usually acute or subacute. This is a trephine specimen of marrow from a woman of 55 who had acute leukemia of monocytic type. A bone trabecula is visible on the left. Closely applied to it and filling the medullary cavity is a solid sheet of cells with ovoid or indented nuclei and fairly abundant cytoplasm. Several mitotic figures are present (arrows). Almost all the cells were shown to have lysozyme and cathepsin B in their cytoplasm.

HE ×470

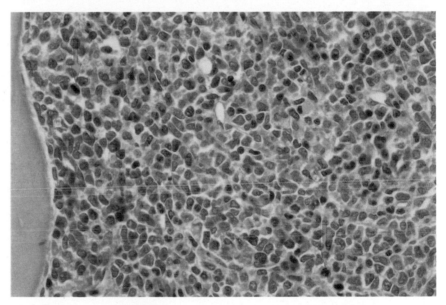

2.60 Monocytic leukemia: peripheral blood

The total white cell count is not very high as a rule in monocytic leukemia and most of the cells are monoblasts or monocytes. This is a smear of the peripheral blood, showing six of the characteristic cells, with their large 'folded' nucleus and abundant very finely granular cytoplasm. The anemia in this case is the Schilling type, the purely monocytic form, as opposed to the Naegeli type in which the cells are a mixture of myeloid cells and monoblasts. Leishman's stain ×1150

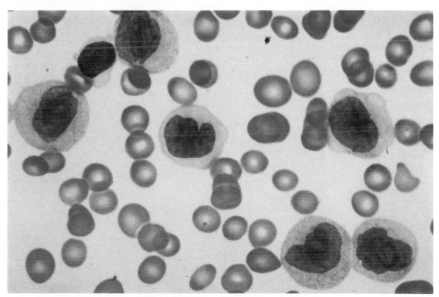

2.61 Myeloproliferative syndrome: marrow

This term covers several inter-related disturbances of hemopoietic tissue. These include polycythemia rubra vera, myelofibrosis and hemorrhagic thrombocythemia. This is the marrow from a woman of 81. Histologically it was hypocellular with hypercellular foci. This is one of the cellular areas. The cells are precursors of the granulocytic and erythrocytic series, and megakaryocytes (right) are also present. Megakaryocytes were numerous elsewhere and many were structurally abnormal. There is however evidence of fibrosis, a stain for reticulin showing a very marked increase in reticulin throughout the marrow (myelofibrosis). The bone trabecula has been eroded at one point (lower right) and osteoblasts (arrows) are laying down osteoid and new bone.   HE ×360

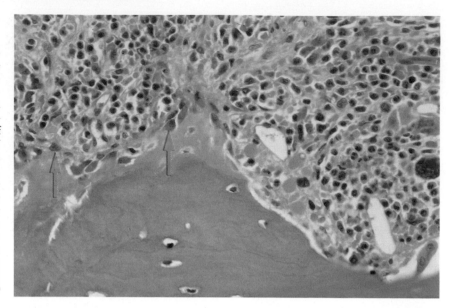

2.62 Myelofibrosis: marrow

The myeloproliferative lesion has now progressed to a late stage. The normal cellular hemopoietic tissue has been replaced by fibrous connective tissue in which there is a scattered population of cells, which at higher magnification were seen to be fibroblasts, lymphocytes and plasma cells. The bone trabecula on the right is part of the normal spongy bone of the medullary cavity and there was no increase in bone in this case. Formation of new bone is sometimes a marked feature and the term myelosclerosis is then applied. There was very active extramedullary hemopoiesis in the liver and spleen, the spleen weighing 3kg.
See also 1.46, 2.5.   HE ×120

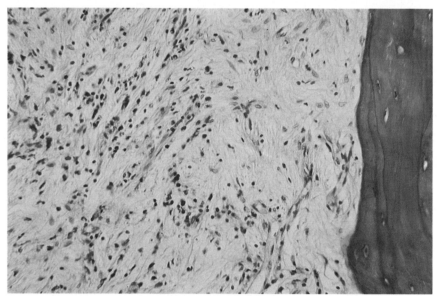

2.63 Chronic myeloid leukemia: marrow

Chronic myeloid leukemia affects mainly the middle-aged and elderly and the number of leukocytes in the peripheral blood is often very large (300×10^9/l or more), most of them mature polymorphs. A more acute phase often develops eventually, with increase in the less mature forms. The liver and spleen are usually greatly enlarged. The normal red hemopoietic marrow is replaced by soft pale pink tissue which may appear almost pus-like. This tissue fills the medullary cavity and the bone trabeculae may be resorbed. The tissue is composed of cells of the granulocyte series, including promyelocytes (with nucleoli), myelocytes (one full of basophil granules), metamyelocytes and neutrophil polymorphs.
          Leishman's stain ×1200

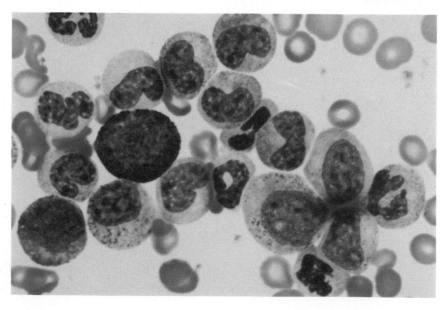

2.64 Chronic myeloid leukemia: spleen

The large size (3kg or more) of the spleen in chronic myeloid leukemia is frequently a source of considerable discomfort to the patient. The organ is firm and the cut surface pale and mottled. Pale infarcts are often present. The increase in size is largely caused by packing of the red pulp with leukemic myeloid cells, but extramedullary hemopoiesis is usually evident, in the form of non-leukemic erythroid cells and megakaryocytes. In this example, most of the cells are immature leukocytes, many with eosinophilic cytoplasm. The other cells include normal erythroid precursors.

HE ×360

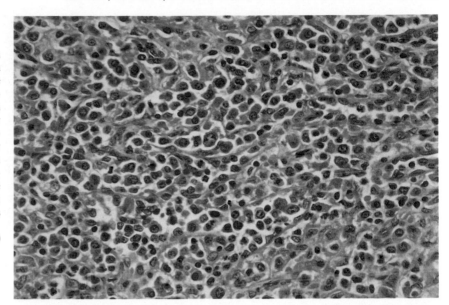

2.65 Eosinophilic myeloid leukemia: lymph node

In most cases of chronic myeloid leukemia most of the cells in the peripheral blood are neutrophil leukocytes. Occasionally however eosinophil leukocytes are the dominant cell. The presence of abundant eosinophilic granules in their cytoplasm is evidence of maturation of the cells and the leukemia is chronic. This is a lymph node from a man of 54 with eosinophilic leukemia. The node contains a dense infiltrate of eosinophil leukocytes, round cells with abundant red-stained cytoplasm and ovoid fairly inconspicuous nucleus. Less mature forms, with larger nuclei and paler staining cytoplasm, are also present. A trephine specimen of marrow showed that much of the hemopoietic tissue had been replaced by the same type of cellular infiltrate.

HE ×360

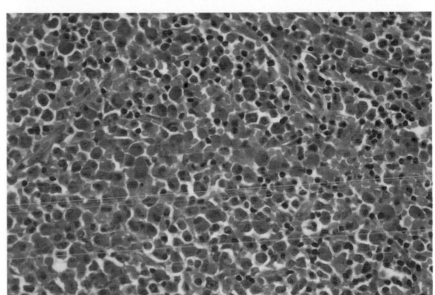

2.66 Acute leukemia: marrow

In acute leukemia the marrow is generally hypercellular from the presence of leukemic cells. The normal hemopoietic elements are reduced and the main clinical features are caused by failure of hemopoiesis with anemia, hemorrhage (from thrombocytopenia) and infection. The marrow infiltrate consists of 'blast' cells, in this case myeloblasts. They are large cells with a round or oval nucleus and a moderate amount of cytoplasm. They form a compact solid sheet which has completely displaced the normal hemopoietic cells, including megakaryocytes. There is however no evidence of resorption of the bone trabecula (right).

HE ×120

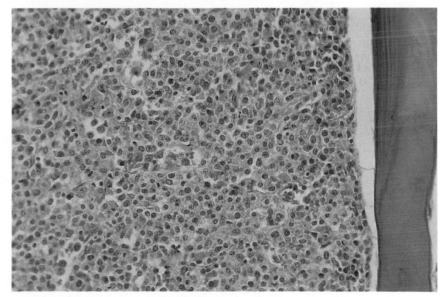

3.1 Branchial cleft cyst

Branchial cysts are located in the antero-lateral part of the neck, most commonly at the angle of the jaw anterior to the sternomastoid muscle. They are painless swellings which may originate in epithelial remnants of the cervical sinus, into which the second, third and fourth branchial clefts open in the embryo. An origin from epithelial inclusions within cervical lymph nodes has also been suggested. The cyst is lined by stratified squamous epithelium (left) and contains keratinous material. The wall consists of lymphoid tissue in which there is a small pale-staining germinal centre and a layer of connective tissue (right). The epithelium lining the cyst is freely permeable to lymphocytes (like the epithelium of the tonsillar crypts) and the small dark-staining cells in the lumen mingling with the keratin are lymphocytes.   HE ×120

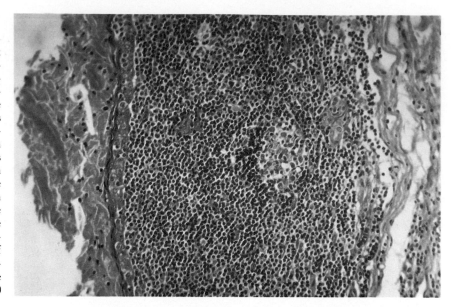

3.2 Branchial cleft cyst

Like the lesion in **3.1**, lymphoid tissue forms part of the wall of this cyst. The cyst is lined however by tall ciliated columnar cells. The epithelium resembles the pseudostratified columnar epithelium of the respiratory tract. Lymphocytes are present at the base of the epithelial cells but they do not penetrate to the lumen (left). The contents of a branchial cyst are usually fluid or semi-fluid and cholesterol is generally present. They are sterile but infection may supervene with pus formation.   HE ×200

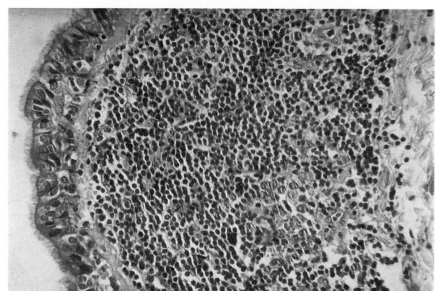

3.3 Internal resorption: tooth

Internal resorption starts centrally within the tooth, with an inflammatory lesion of the pulp, which may be caused by infection following carious exposure. This shows the inner (pulpal) surface of the dentine, the characteristic structure of which is evident (e.g. lower right). A layer of vascular connective tissue containing large numbers of chronic inflammatory cells (upper right) covers the extensively eroded surface of the dentine. Most of the inflammatory cells are plasma cells and they include Russell bodies, which are plasma cells with intensely eosinophilic cytoplasm (arrows). The connective tissue appears to be extending into the dentine and resorption may continue until both the dentine and enamel of the crown of the tooth are absorbed.

HE ×150

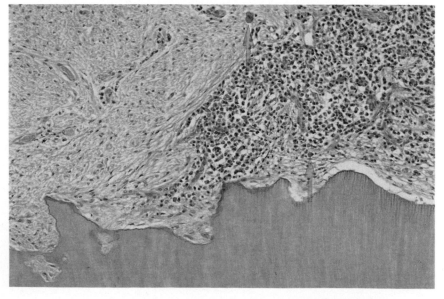

3.4 Dental (radicular) cyst

The radicular cyst is the commonest type of cyst of the jaws. It is benign and usually symptom-free, although it may expand slowly to distort the jaw. It is a consequence of inflammatory disease of the teeth, arising in a granuloma at the apex of a tooth, from epithelial 'rests' derived from structures involved in the formation of the tooth. Any tooth may give rise to the cyst. The lumen is on the left. The cyst is lined by a thick layer of non-keratinized stratified squamous epithelium and the wall consists of vascular connective tissue (right).   HE ×150

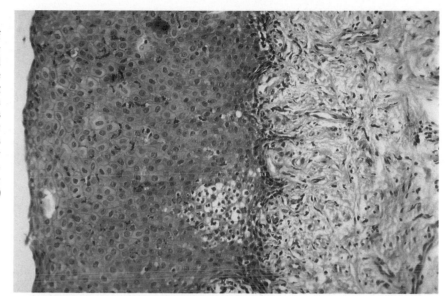

3.5 Giant-cell epulis: mandible

Giant-cell epulis is found at all ages. It is not a neoplasm but an inflammatory reaction to injury and hemorrhage, i.e. a reparative granuloma. It is usually a single lesion and may be situated within bone (either maxilla or mandible) or in the soft tissues of the gingiva. This particular lesion was within the mandible. It consists of a cellular stroma of connective tissue in which there are many multinucleated giant cells of osteoclast type. The stromal connective tissue is relatively mature-looking. The nuclei of the fibroblast-type cells show only slight pleomorphism and no mitotic activity. Giant-cell epulis destroys bone and may be difficult to distinguish from the true giant-cell tumour which also may occur in the same sites.   HE ×135

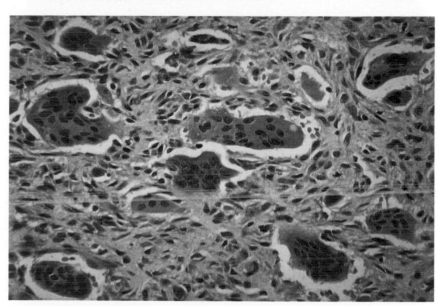

3.6 Calcifying epithelial odontogenic tumour

Calcifying epithelial odontogenic tumour is a rare neoplasm which tends to be invasive and to recur locally after removal. It does not however metastasize. It consists of closely-packed polyhedral cells with large round fairly pleomorphic nuclei and abundant eosinophilic cytoplasm. There are no mitotic figures. There are large numbers of round vacuoles filled with pale-staining amorphous material. A basophilic (densely-calcified) laminated body is present on the left. Others were present elsewhere in the tumour which were eosinophilic and it is only after they become calcified that they stain blue.   HE ×135

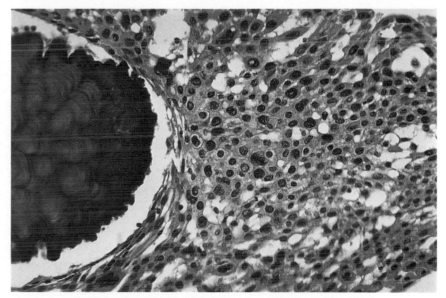

3.7 Inflammatory ('allergic') polyp: nose

Nasal polyps are not true neoplasms but masses of edematous stroma connective tissue covered with respiratory epithelium and usually associated with infection and allergy. They may be large. This specimen, from a man of 43, was a soft piece of greyish-white tissue 1cm dia. It is covered with transitional epithelium (left) and consists of delicate vascular connective tissue, with many dilated small blood vessels in its more superficial parts (left). More deeply (right) the stroma is loose and edematous. There are many plasma cells, eosinophil leukocytes and lymphocytes, particularly towards the surface. A small duct lined with cuboidal epithelium is present (left of centre).

HE ×150

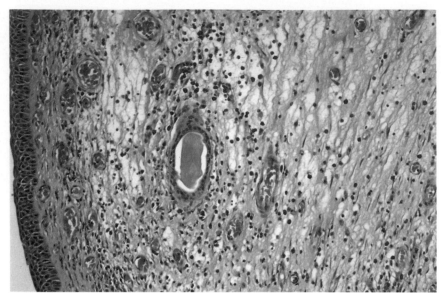

3.8 Inflammatory ('allergic') polyp: nose

This polyp from a woman of 53 measured 2.5cm in its long axis. Its surface (left) is covered with pseudostratified columnar epithelium which resembles respiratory-type epithelium but contains many large mucin-secreting goblet cells. The epithelium rests on a thick hyaline basement membrane, a common feature of nasal polyps. The stroma consists of very edematous connective tissue. The stroma is vascular and contains a considerable number of inflammatory cells (eosinophil leukocytes and plasma cells mostly). The weakly eosinophilic and unstained parts of the connective tissue contain edema fluid.

HE ×360

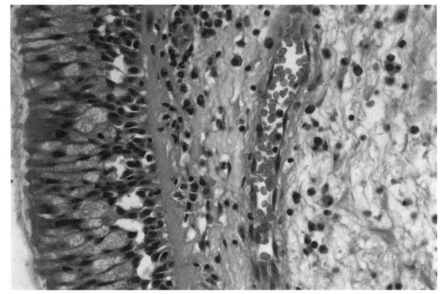

3.9 Inflammatory ('allergic') polyp: nose

The epithelium covering the surface of this polyp has undergone metaplasia to a squamous or very thick transitional form of epithelium. There is no keratinization. The epithelial cells have large hyperchromatic nuclei which show considerable pleomorphism and dysplasia, and nucleoli were prominent at higher magnification. Several mitotic figures are present (arrows). These appearances suggest malignancy. Sometimes also the stromal cells show similar changes. However, malignant change is very rare in a nasal polyp and such a diagnosis is usually erroneous. The underlying stroma is infiltrated with inflammatory cells (polymorphs and plasma cells). HE ×110

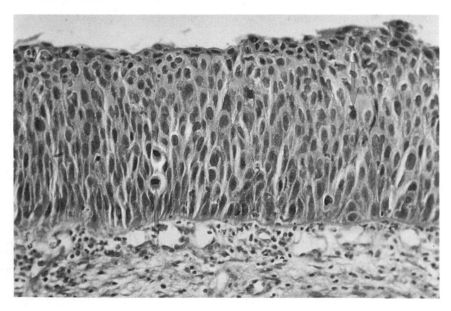

3.10 Squamous (transitional or 'inverted') papilloma: nose

This is a true papilloma of the mucosa of the nasal cavity. It is a benign neoplasm but tends to recur. This specimen, from a man of 63, has an edematous, sparsely cellular connective tissue stroma covered by thick squamous epithelium (left). The epithelium has grown inwards into the stroma, to form large clefts or pits, part of one of which is seen here (top and right). The invaginated epithelial layer is very thick. Inversion of the epithelial covering can be readily mistaken for invasion and malignancy. The stroma contains moderate numbers of chronic inflammatory cells. HE ×60

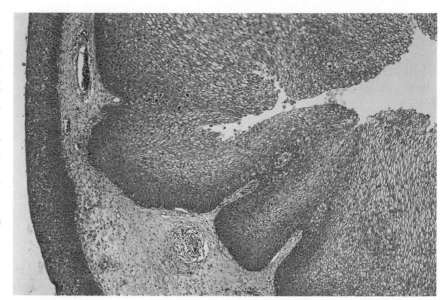

3.11 Squamous (transitional or 'inverted') papilloma: nose

This shows the structure of the papilloma in more detail. The surface epithelium (left) is thick and cellular, and squamous in type. The cells of the basal layers are basophilic and show some dysplasia. Most of the epithelial cells lining the cleft are swollen and vacuolated, but there is some pleomorphism. There is an area of hemorrhage in the stroma but the number of chronic inflammatory cells is small. Unlike inflammatory (allergic) nasal polyps which are very often bilateral, the squamous papilloma is almost always a solitary lesion. *See also 1.51, 7.5, 7.6.* HE ×150

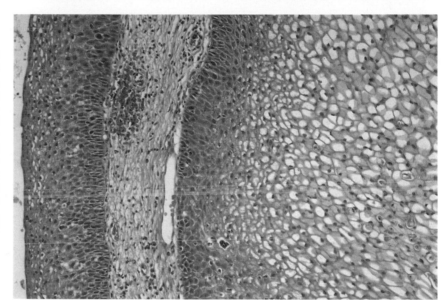

3.12 Lichen planus: mouth

Lichen planus affects mucous membranes as well as skin. The mouth is not infrequently involved, sometimes in the absence of cutaneous lesions. In contrast to the changes seen in skin (**14.20**), there is only a thin layer of keratin on the surface (left) and a granular layer is just detectable. There is moderate hyperplasia of the keratinocytes (acanthosis) but there is a marked change in the rete ridges which are short and pointed, like the teeth of a saw. The basement membrane is intact. An infiltrate of chronic inflammatory cells is present in the submucosa and it extends up to the epithelial basement membrane. A few lymphocytes are also present within the basal layers of the epithelium. Lichen planus may be mistakenly diagnosed as leukoplakia but there is no epithelial atypia in lichen planus.

HE ×190

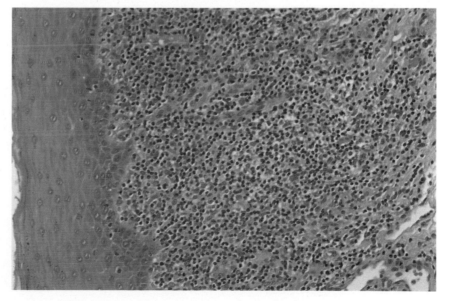

3.13 Leukoplakia: tongue

Leukoplakia is a clinical term which means 'white plaque'. White plaques are not uncommonly present on the oral mucosa of older people. They generally consist of thickened keratotic epithelium which obscures the normal red colour of the underlying vascular tissues of the submucosa. The significance of leukoplakia lies in the fact that it predisposes to malignancy particularly in the floor of the mouth and lateral margins of the tongue. This shows a rete ridge of the mucosa of the tongue of an elderly man. There is hyperplasia and dysplasia of the keratinocytes, with numerous mitotic figures (arrows), and there is some loss of cell polarity. The submucosa (right) is heavily infiltrated with plasma cells and lymphocytes. HE ×360

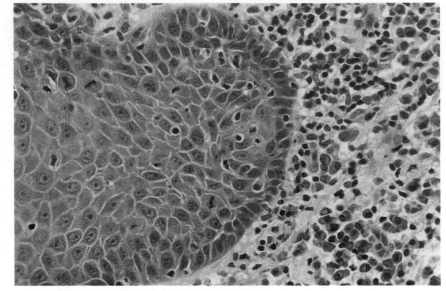

3.14 Squamous carcinoma: tongue

Squamous carcinoma of the tongue constitutes over 90% of the malignant tumours of the oral cavity. The majority are on the lateral margins of the tongue or on the lower lip. Carcinoma of the tongue has a poor prognosis, tending to invade the muscles of the tongue and spread to the regional lymph nodes. This shows several clusters of tumour cells invading the striated muscle fibres of the tongue. Each group of tumour cells consists of a peripheral layer of small basophilic cells and much larger eosinophilic cells centrally. These are forming masses of keratin (arrows) which lie within vacuoles. Chronic inflammatory cells, mostly small lymphocytes, are present in the vicinity of the advancing edge of the tumour (centre right). HE ×135

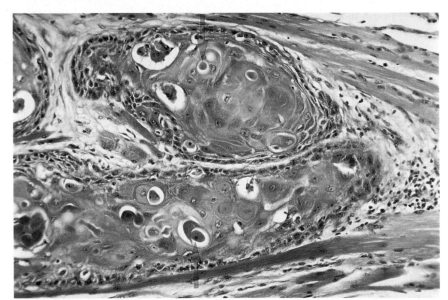

3.15 Granular cell tumour: tongue

Granular cell tumour (myoblastoma) occurs in a variety of sites, including tongue, breast and larynx. This tumour arose in the tongue. The submucosa (centre and right) is occupied by the tumour cells which extend up to the epidermis. They are large with abundant eosinophilic granular cytoplasm and round or ovoid nuclei. There is very striking hyperplasia of the squamous epithelium, with downgrowth of greatly elongated rete ridges (arrows) into the submucosa. Some of the rete ridges cut in cross-section resemble the cell-nests of squamous cell carcinoma. The change is termed pseudoepitheliomatous hyperplasia and it is a more or less constant feature of granular cell tumours. It is readily mistaken for squamous carcinoma, the underlying tumour being overlooked, particularly in the tongue. HE ×150

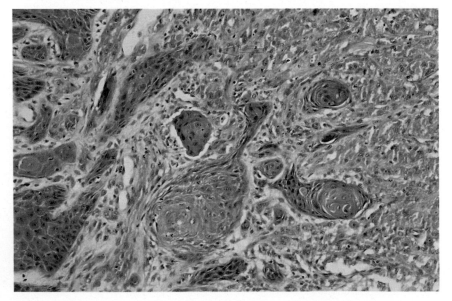

3.16 Granular cell tumour: tongue

The tumour cells are very distinctive in appearance, rather like degenerate or embryonic muscle ('myoblastoma'), with densely eosinophilic granules filling their cytoplasm. Some of the granules are large (up to 5μm dia). The cell borders are indistinct, however and the structure is almost syncytial in form. The nuclei are large and round or ovoid and very uniform in structure. There is no mitotic activity. The blood vessels (arrows) are thin-walled and sinusoidal, with hyaline thickening of the basement membrane in places. Granular cell tumour is a benign lesion which grows very slowly and seldom recurs. Its nature and histogenesis are uncertain but an origin from Schwann cells is probable.   HE ×360

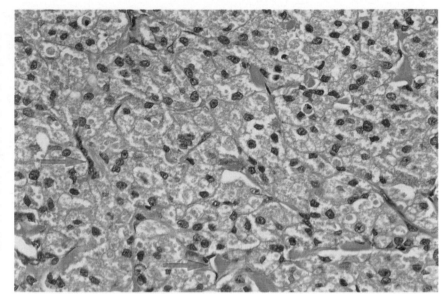

3.17 Verrucous (buccal) carcinoma

This is a well-differentiated squamous carcinoma of the mucosa of the buccal mucosa and lower gingiva. Leukoplakia is usually present and the lesion mainly affects men who chew tobacco. It is a large soft papillary growth which grows slowly, invading the adjacent tissues. It has a distinctive structure and clinical behaviour and may cause extensive destruction. It rarely metastasizes to lymph nodes and never to distant sites.

The buccal mucosa (thin arrow) is thick and hyperplastic. The tumour forms a large round, apparently well-circumscribed mass (top left) which is invading the submucosa on a broad front (thick arrows), pushing aside rather than penetrating the muscles of the cheek. A small mucus-secreting salivary gland is present (bottom right).   HE ×20

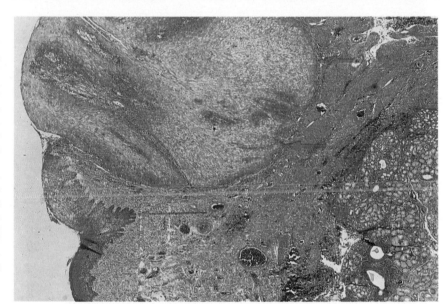

3.18 Verrucous (buccal) carcinoma

This shows the invading edge of the tumour at higher magnification. The tumour cells (centre and left) are squamous epithelial cells, with abundant pale cytoplasm. Nuclei, which are vesicular with a prominent nucleolus, are visible in the more compact cells near the invading edge (right) but are absent from most of the large pale cells. There are no mitoses. Polymorphs, many degenerate, are present within the tumour and there is an infiltrate of chronic inflammatory cells in the tissues adjacent to the invading front of the tumour. The tissue on the right is part of a cheek muscle.

HE ×215

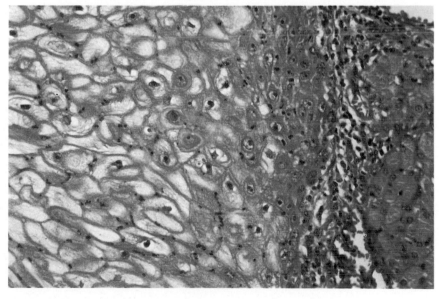

3.19 Plasmacytoma: palate

Solitary plasma cell tumours occur occasionally in the soft tissues. Many of them are in the nasopharynx and upper respiratory tract. It is important to distinguish them from inflammatory lesions rich in plasma cells: plasma cell granulomas. The onset is as a rule insidious and the behaviour of the lesion is unpredictable. Some behave in benign fashion but the majority, cytologically identical, recur locally and eventually become widely disseminated. The lesion may consist of one or more round nodules. The stratified squamous epithelium (left) is stretched over a projecting round, highly vascular mass. It is attenuated but intact. The mass is composed of a homogeneous population of moderate-sized cells, those nearer the surfaces staining more intensely than those deeper in the lesion (right). HE ×150

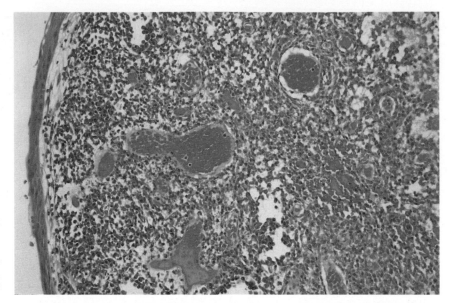

3.20 Plasmacytoma: palate

At higher magnification the plasma cell nature of the cells of the tumour is clearly evident. The nuclei are round and located at one pole of the abundant cytoplasm; the chromatin is distributed in a clock-face or cart-wheel pattern; and there is clear halo adjacent to the nucleus in many cells representing the Golgi apparatus (arrow). There are many small blood vessels. Inflammatory cells such as polymorph leukocytes are not present, a feature which helps to distinguish solitary plasmacytoma from inflammatory lesions. The monoclonal origin of the cells can also be established readily by immunohistochemical methods. *See also 1.38, 2.17.* HE ×360

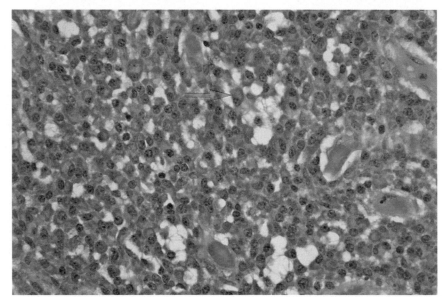

3.21 Carcinoma: nasopharynx

Carcinoma of the pharynx is uncommon in some countries and common in others such as China. There is an association with Epstein–Barr (EB) virus. The tumour is often poorly-differentiated or undifferentiated and contains large numbers of inflammatory cells. The age range varies and this tumour was in a girl of 11. It consists of large anaplastic epithelial cells with vesicular nuclei which contain one or more prominent nucleoli (arrows). The large size of the tumour cells is obvious when they are compared with the many eosinophil leukocytes (top right) and lymphocytes. Some of the tumour cells are binucleate. There was considerable mitotic activity throughout the tumour. Sometimes lymphocytes are the predominant inflammatory cell and tumours with this structure are termed lymphoepitheliomas. HE ×360

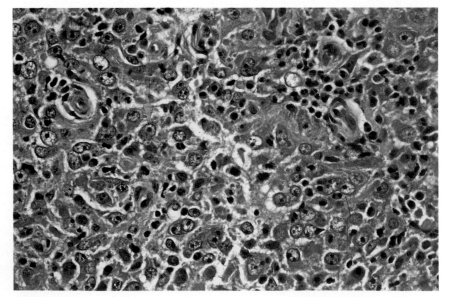

3.22   Chronic sialadenitis: salivary gland

Chronic inflammation of a salivary gland is often associated with obstruction of the duct, usually by a calculus (sialolithiasis). A calculus may form in a previously non-inflamed gland but bacterial infection predisposes to its development. The submaxillary gland is more liable to be affected by calculus formation than the parotid, because of the calcium-rich nature of its secretion. This is a submandibular gland from a woman of 67. It was divided into multiple lobes by broad bands of dense fibrous tissue. Histologically there is extensive loss of acini and only small ducts with their characteristic double layer of lining cells remain. They are surrounded by cellular connective tissue which is infiltrated by chronic inflammatory cells, and there is one follicular collection of small lymphocytes (centre right). HE ×150

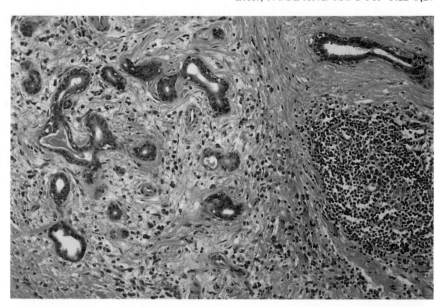

3.23   Chronic sialadenitis: salivary gland

In this part of the submaxillary gland shown in **3.22** glandular acini (thin arrows) survive but are degenerate and show no secretory activity. The small ducts (thick arrows) are dilated and lined by a single layer of attenuated epithelial cells. Their lumen is filled with pus-like material consisting mostly of polymorph leukocytes, evidence of previous acute inflammation. The connective tissue surrounding the ducts and acini is densely fibrous in places. It is infiltrated by lymphocytes and plasma cells and also a few polymorphs.   HE ×235

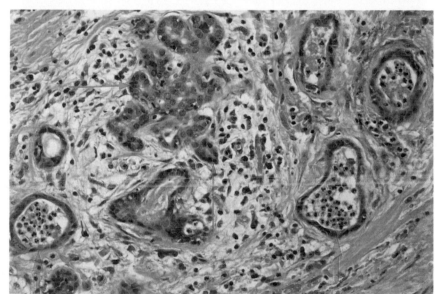

3.24   Sjögren's syndrome: salivary gland

Sjögren's syndrome is a systemic disorder, and as part of this the secretory cells of the lacrimal, conjunctival and salivary glands degenerate. Patients tend to get conjunctivitis (kerato-conjunctivitis sicca) and a dry mouth from lack of saliva. The syndrome affects mainly middle-aged women and it is often accompanied by chronic polyarthritis of the rheumatoid type. The affected glands are swollen. Histologically the glandular acini have disappeared and the gland now consists of fat and fibrous tissue in which only small ducts survive. The connective tissue around the ducts is heavily infiltrated with lymphocytes and plasma cells, and similar cells are present in smaller numbers in the fatty tissue.   HE ×80

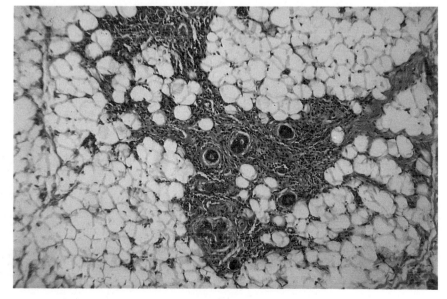

3.25 Mumps: parotid

Mumps is an acute inflammatory lesion of the salivary glands which is caused by the mumps virus. The parotid glands are most commonly attacked. They become swollen, painful and acutely tender. This is an affected parotid. The cellular infiltrate consists of lymphocytes, plasma cells and macrophages, and no polymorph leukocytes are visible. There is extensive destruction of the acinar structure and many acini (thin arrows) are small and disorganized. Many of the small dark nuclei are pyknotic nuclei of necrotic epithelial cells. The epithelium of the small ducts (thick arrow) appears to be less affected than the acinar epithelium. Despite the intensity of the inflammatory reaction, repair appears to be effective and significant permanent damage to the glands is uncommon.   HE ×235

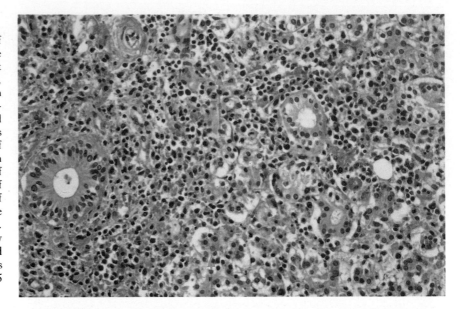

3.26 Benign lymphoepithelial lesion: parotid

Benign lymphoepithelial lesion is an asymptomatic swelling of the salivary glands which is bilateral and symmetrical and may be very pronounced. The parotid seems to be affected more severely than the other salivary glands. This specimen was an irregular lobulated greyish-white mass of tissue (8 × 5 × 3cm) from the right parotid of a 46-year-old woman. It lacked a capsule. Histologically the acinar structure is obscured by a dense infiltrate of lymphocytes and histiocytes. Scattered groups of epithelial cells are also present. They are proliferated myoepithelial cells within the ducts of the glands, and the presence of these so-called myoepithelial 'islands' is an important diagnostic feature. No germinal centres are present in the lymphoid tissue in this field but they were present elsewhere.   HE ×150

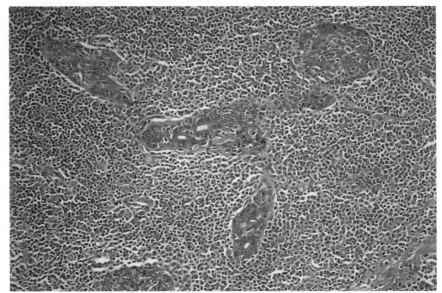

3.27 Benign lymphoepithelial lesion: parotid

At higher magnification the infiltrate consists of small dark-staining lymphocytes and larger pale-staining histiocytes. Two of the myoepithelial islands (thin arrows) are infiltrated by lymphocytes and appear to have no lumen. One (thick arrow) has a central lumen which contains epithelial debris. The precise nature of this condition is uncertain. The systemic features of Sjögren's syndrome, including the polyarthritis, are lacking. Occasionally however non-Hodgkin's lymphoma develops in an affected gland and the condition should be regarded as pre-lymphomatous.   HE ×200

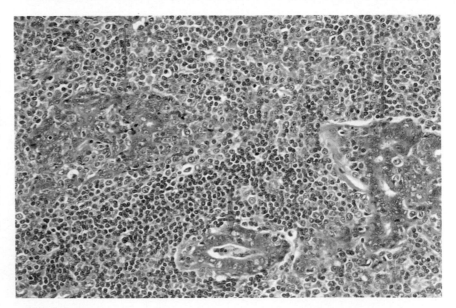

3.28 Adenolymphoma: parotid

Adenolymphoma is often multicentric and sometimes bilateral, affecting men much more often than women. It rarely involves glands other than the parotid. It is benign, usually being encapsulated and easily excised. It consists of numerous irregular cystic spaces into which papillary structures project. It contains a considerable amount of lymphoid tissue and the core of the papilla (left) is filled with small mature lymphocytes. Similar cells are numerous elsewhere in the stroma (arrow). The epithelium covering the papilla and lining the cystic spaces is thick, multilayered and very eosinophilic. The origin of adenolymphoma may be from excretory salivary gland ducts within an intraparotid lymph node (see **3.30**). The cystic space contains cell debris, and mucus-secreting acini of the normal parotid are visible on the right.   HE ×200

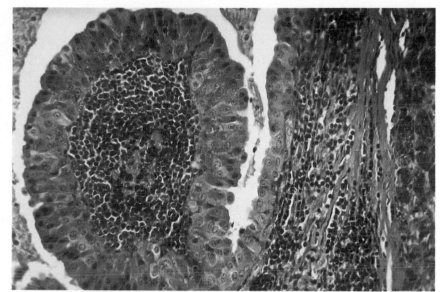

3.29 Pleomorphic adenoma: parotid

Pleomorphic adenomas are the commonest tumours of salivary glands. They are benign but encapsulation is not always complete and they show a marked tendency to recur. The capsule of this tumour is on the left. The tumour consists of irregular groups of cells with deeply basophilic nuclei. The cells, whose cytoplasmic boundaries cannot be distinguished, lie in pale-staining tissue rich in connective tissue mucin (centre and right). The individual cells in this tissue tend to appear stellate. This myxoid tissue is often accompanied by cartilage-like tissue and epithelial mucin and acinar structures are generally also present. Pleomorphic adenoma has been regarded as a 'mixed' tumour, and there is good evidence that both epithelial and myoepithelial cells contribute to its formation.   HE ×200

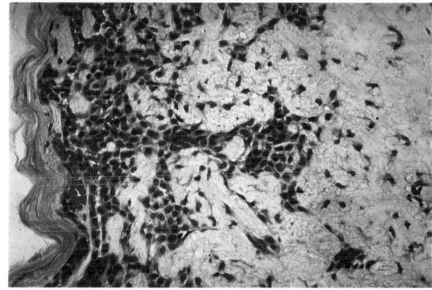

3.30 Oxyphil adenoma: parotid

Oxyphil adenoma (oncocytoma) is a benign slow-growing neoplasm which arises from the epithelium of the excretory ducts. It is generally found in the parotid but may occur in other salivary glands. This lesion was removed from the parotid of a man of 59. It measured $4.5 \times 2.7 \times 2.3$cm and weighed 30g, and consisted of well-formed acini lined by tall columnar epithelial cells. The nuclei of these cells have prominent nucleoli but they show no pleomorphism and mitoses are not present. A notable feature is the large amount of granular eosinophilic cytoplasm that each cell possesses. The appearance of the cytoplasm is produced by the presence in it of large numbers of mitochondria. The stroma between the acini is scanty and contains occasional lymphocytes.   HE ×270

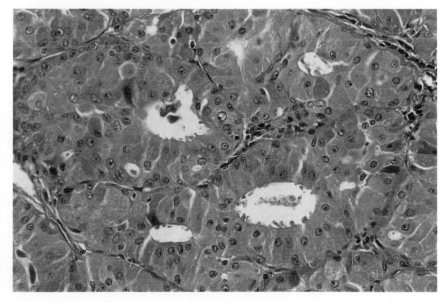

3.31 Acinic cell tumour: parotid

Acinic cell tumour is a rare slow-growing tumour of low-grade malignancy which tends to recur unless adequately resected. It probably represents a variant of the pleomorphic adenoma and generally presents as a round apparently encapsulated mass. Most occur in the parotid gland and this lesion was in the right parotid of a man of 50. It consists of anastomosing cords of mucus-secreting epithelial cells with small densely staining round nuclei located at the bottom of the faintly basophilic vacuolated cytoplasm. The cells resemble the serous cells of the normal gland and form cleft-like spaces, but sometimes the cytoplasm is highly vacuolated and 'clear' (clear-cell carcinoma). There are no mitotic figures. The stroma consists almost entirely of thin-walled blood vessels. HE ×270

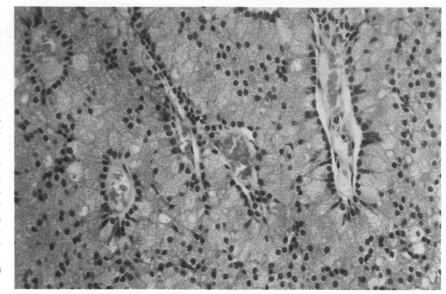

3.32 Adenocarcinoma: parotid

Adenocarcinoma may arise from a pleomorphic adenoma but this happens infrequently. Others arise *de novo*, as in this case, which did not appear to be associated with a pleomorphic adenoma. The malignant epithelial cells form small clumps and a variety of irregular gland-like structures. The cells show considerable pleomorphism, some being small and flattened, others large and columnar. Most of the glandular structures contain cell debris, some of it densely eosinophilic and keratin-like. Mitotic figures are not visible in the epithelial cells but were present elsewhere in the tumour. The stroma is abundant and fibrous. HE ×200

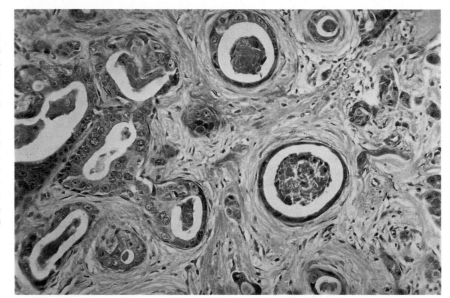

3.33 Adenocarcinoma: parotid

A small minority of salivary gland tumours are highly malignant, poorly differentiated lesions. This is an anaplastic tumour consisting of closely-packed polyhedral cells with large nuclei and very little cytoplasm. Several mitoses are visible and were numerous throughout the tumour. The malignant cells do not form glands or tubules but they have formed elongated spaces (top) containing a little debris. The stroma is scanty and it consists mainly of small blood vessels. The histogenesis of this type of tumour is not known but an origin from the epithelium of the ducts is possible. HE ×335

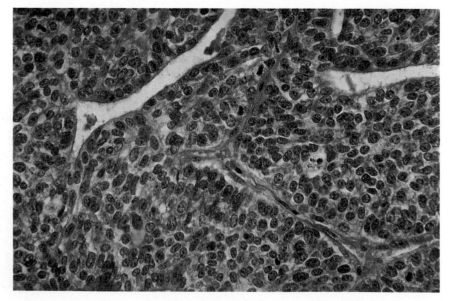

3.34 Adenocystic carcinoma (cylindroma): parotid

Adenoid cystic carcinoma is a highly malignant tumour which has a marked tendency to recur and to invade perineural lymphatics and to metastasize to the lungs. This lesion is from the right parotid of a man of 41. It had infiltrated the parotid extensively. This shows several nodules of tumour, surrounded by fibrous tissue. Each nodule consists of clusters of cells with ovoid basophilic nuclei separated by thick strands of homogeneous eosinophilic material resembling very thick basement membrane. The tumour cells surround vacuoles or small gland-like spaces.   HE ×150

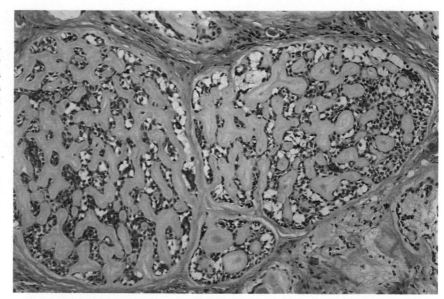

3.35 Adenocystic carcinoma (cylindroma): parotid

Typically the tumour shown in **3.34** recurred and this shows one nodule of the recurrent tumour. The cells have scanty cytoplasm and round or ovoid basophilic nuclei of regular form. There is very little nuclear pleomorphism and no evidence of mitotic activity. The cells are arranged in a round cribriform mass in which the circular areas contain weakly eosinophilic amorphous material resembling basement membrane. The nodule of tumour cells is surrounded by collagenous fibrous tissue. Despite its usually somewhat bland histological structure adenoid cystic carcinoma is a highly invasive tumour. Several normal mucin-secreting parotid acini (stained deep blue) and fat cells are visible on the left.
HE ×360

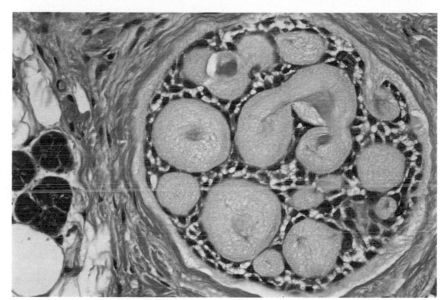

3.36 Mucoepidermoid carcinoma: parotid

This type of tumour is believed to arise from the mucous and basal cells of the ducts of the salivary glands. Some are of low-grade malignancy but others are highly malignant. A low-grade lesion presents as a well-circumscribed mass which is not encapsulated, tending to infiltrate locally and eventually to metastasize. This local recurrence in the submandibular gland in a man of 88 was a well-defined nodule (4 × 3 × 3cm). This group of tumour cells consists of atypical squamous cells with eosinophilic cytoplasm and very large nuclei with prominent nucleoli (arrows). Within the group there are also several small cysts the contents of which stained for epithelial mucin. Despite the macroscopic appearances, the tumour was invading the surrounding tissues and perineural lymphatics.   HE ×235

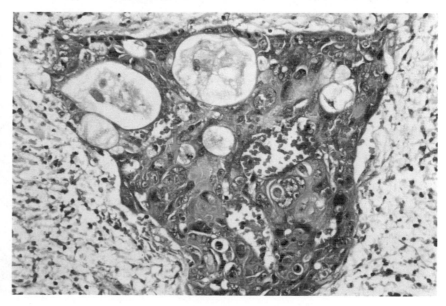

4.1 Candidiasis: esophagus

Candidiasis is caused by the yeast-like fungus *Candida (Monilia) albicans*. It is an organism of low virulence and lesions are confined as a rule to the surface of mucous membranes, invading deeper tissues only when the body's resistance is lowered. This is a section of a plaque on the surface of the esophageal mucosa. The lumen of the esophagus is on the left. The epithelial lining is mildly inflamed and inflammatory edema separates the epithelial cells, forming clefts between them. Inflammatory cells (mostly polymorph leukocytes) are present within all layers of the epithelium and the surface is covered with inflammatory cells and desquamating squamous cells (left). *Candida albicans* hyphae do not stain well in HE preparations, but faintly basophilic hyphae are just detectable.　　HE ×360

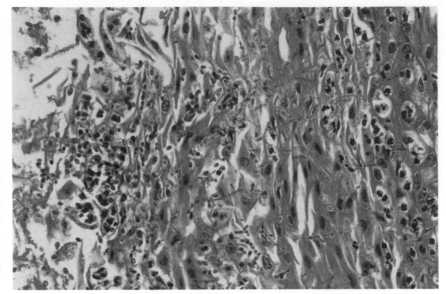

4.2 Candidiasis: esophagus

This section of the lesion shown in **4.1** has been treated with the periodic acid-Schiff (PAS) method which stains the hyphae of *Candida albicans* purplish-red. Dense clumps of hyphae (approx. 1μm thick), present on the surface of the stratified squamous epithelium, are invading the epithelium almost to the basal layer (right). Desquamating cells (left) mingle with the hyphae. The epithelium is tending to separate from the basal layers (right). Inflammatory cells (polymorph leukocytes and lymphocytes) are present throughout the epithelium and in the submucosa (right).

PAS ×360

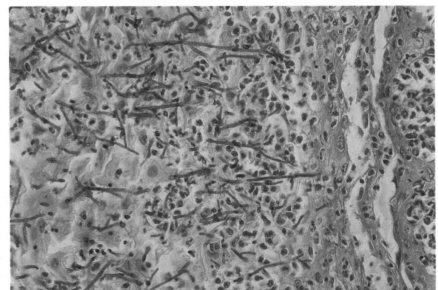

4.3 Mucosal erosion

An erosion of the gastric mucosa does not extend beyond the muscularis mucosae, whereas an acute ulcer does. The acute ulcer does not usually breach the muscle coats. Erosions are often multiple and may be the source of considerable hemorrhage. The lumen of the stomach is at the top and the muscularis mucosae is visible at the bottom. There is a saucer-shaped area of necrosis and ulceration (top centre) affecting the more superficial part of the mucosa with destruction of the glandular tissue. The deeper part of the mucosa survives. The glands on each side of the ulcer are distorted and at the edges (arrows) only the most superficial parts of the glands are necrotic. The ulcer is very shallow and its floor consists of necrotic tissue and cells. The mucosa beneath the ulcer and at both sides is infiltrated with inflammatory cells.　　HE ×60

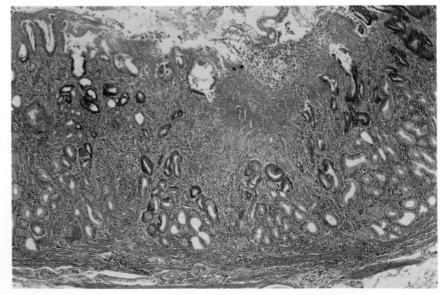

4.4 Chronic peptic ulcer: stomach

Complete disruption of the muscle coats is generally accepted as the criterion for classifying a peptic ulcer as chronic. In this example the ulcer (centre) has penetrated through the gastric mucosa (top) and the muscle coats, parts of which are visible on the left and right of the crater. The surviving mucosa at the edges of the ulcer overhangs the crater but is not heaped up as in a malignant lesion. The floor of the ulcer consists of a densely staining band of necrotic tissue and beneath that a thick layer of collagenous fibrous tissue (bottom).   HE ×7

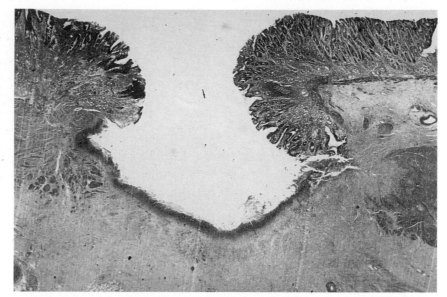

4.5 Chronic peptic ulcer: stomach

This shows the floor of the ulcer at higher magnification. The crater of the ulcer is at the top. The most superficial layer consists of eosinophilic fibrinous exudate and necrotic cells, beneath which there is a thick zone of well-formed granulation tissue. The granulation tissue is very vascular, with many dilated thin-walled blood vessels as well as a large population of macrophages and other inflammatory cells. When the ulcer heals, epithelial cells grow across the surface of the granulation tissue and when continuity of the epithelium has been restored the granulation tissue becomes mature connective tissue.   HE ×135

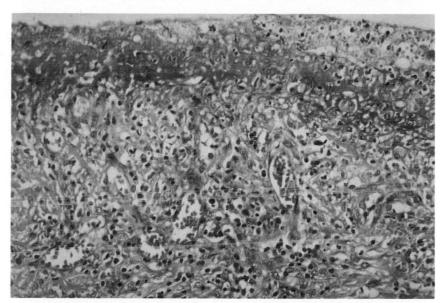

4.6 Chronic atrophic gastritis: stomach

A man of 61 had a partial gastrectomy for chronic peptic ulcer. This is the mucosa several cm from the ulcer crater. There is complete loss of the parietal and chief cells, and the mucosa consists of irregular glands containing large numbers of mucus-secreting goblet cells. The surface of the epithelium had a villous structure, similar to that of the small intestine: 'intestinal metaplasia' of the mucosa. There are numerous lymphocytes in the submucosa, and a large lymphoid follicle. The lamina propria (lower right) is thickened. When the submucosal lymphoid tissue is markedly increased and lymphoid follicles with germinal centres are present, the term chronic follicular gastritis is applied. Follicular gastritis is commonly seen in stomachs in which a chronic peptic ulcer is present.   HE ×60

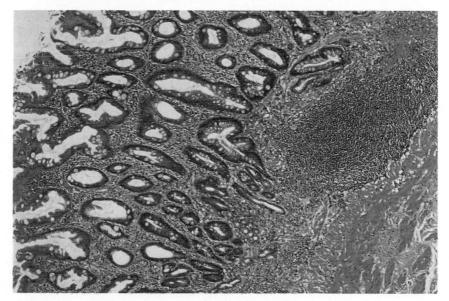

4.7 Chronic atrophic gastritis: stomach

Chronic gastritis is present as a rule in cases of gastric carcinoma, and the larger the tumour the more extensive the lesion. This is the mucosa of the pyloric canal of the stomach of a man of 78 with gastric carcinoma. The lumen is at the top. The mucosa was atrophic and it is composed entirely of small irregular and rather tortuous glands lined by mucin-secreting goblet cells of 'intestinal' type. No oxyntic or chief (peptic) cells remain. There are no villi of small-intestine type and the mucosa resembles more closely that of the colon. The number of inflammatory cells in the stroma is fairly small. HE ×150

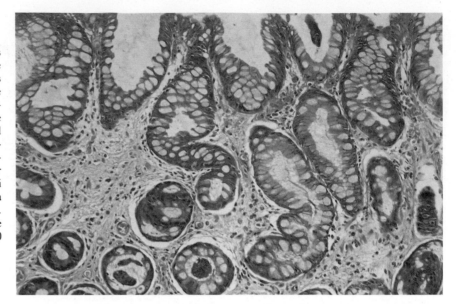

4.8 Chronic superficial gastritis: stomach

A man of 37 had a partial gastrectomy for recurrent duodenal ulcer. The mucosa of the antral region appeared macroscopically flatter than normal. But histologically, there is little evidence of atrophy and this shows only the superficial mucosa, with the lumen of the stomach on the left. The glands are tortuous and most are lined by cells which secrete little mucin and show considerable mitotic activity. These changes are reactive and commonly seen in glands associated with an inflammatory lesion. Between the glands there is a dense infiltrate of inflammatory cells (mainly eosinophils, plasma cells and lymphocytes). Some pyloric glands are also visible (arrows). In peptic ulceration of the duodenum, chronic gastritis when present is usually confined to the antral region of the stomach. HE ×150

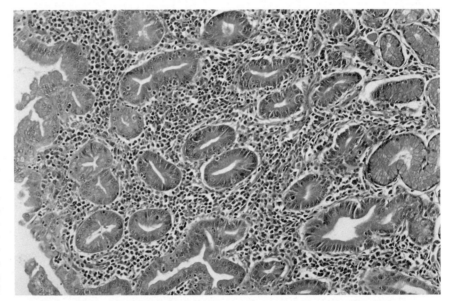

4.9 Chronic hypertrophic gastritis: stomach

In the common forms of chronic gastritis the mucosa is usually atrophic and thinner than normal. In this form of gastritis, however, the mucosa is usually thickened and the rugae are enlarged, the appearance resembling the convolutions of the brain. The lesion can be confused with carcinoma and malignant lymphoma. In this case the patient was a 67-year-old woman suspected of having gastric carcinoma. This is the mucosa from the fundus of the stomach, showing parts of several rugae. The cores of the rugae consist of edematous connective tissue and the greatly elongated glands are lined by tall columnar cells with eosinophilic mucin-filled cytoplasm. Small numbers of chronic inflammatory cells are present in the connective tissue of the rugae. HE ×60

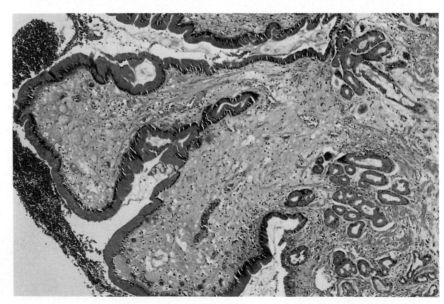

4.10 Epithelial dysplasia: stomach

Dysplasia of the glandular epithelium is seen occasionally in chronic gastritis with intestinal metaplasia, and the difficulty is to distinguish it from intramucosal carcinoma. This is the antral mucosa in the stomach of a woman of 62. The glands are lined by closely-packed columnar epithelial cells with large hyperchromatic nuclei which show considerable pleomorphism. Mitotic activity is also increased. The polarity of the cells is disturbed. The mucin-secreting activity of the cells is reduced, with only a droplet of mucin at the apex of each cell. Many inflammatory cells (polymorph leukocytes, lymphocytes and plasma cells) are present between the glands. HE ×460

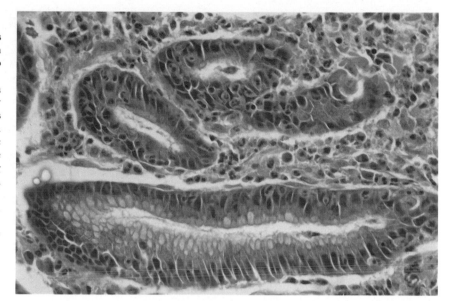

4.11 Adenocarcinoma: stomach

Carcinoma of stomach arises from the mucus-secreting cells of the gastric crypts. The tumour can assume a wide range of forms. This is the mucosa from the lesser curvature of the stomach of a man of 70 who had a partial gastrectomy for carcinoma of stomach. The lumen of the stomach is on the left and the muscularis mucosae on the right. The mucosa is thickened, with papillary projections on the surface. The normal glandular pattern of the gastric mucosa has been replaced by an irregular network of glands lined by basophilic cells. There is probably early invasion of the muscularis mucosae (right). HE ×30

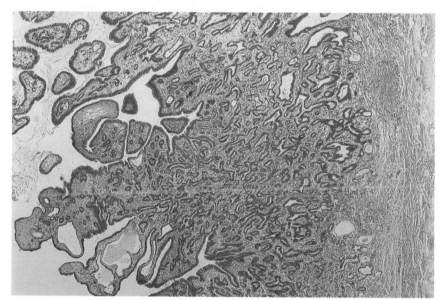

4.12 Adenocarcinoma: stomach

This shows the epithelium lining the glands and covering the surfaces of the papilliform projections. The cells are tall columnar cells, with large basophilic nuclei which form several layers. The nuclei are pleomorphic and show considerable mitotic activity (arrows). Small quantities of mucin could be demonstrated in the cytoplasm of the cells. The connective tissue of the mucosa is infiltrated with chronic inflammatory cells. Although the carcinomatous change is confined to the mucosa in this part of the stomach, and the appearances are those of a superficial polypoid lesion, elsewhere the tumour had invaded the stomach wall deeply. It had not, however, metastasized to the regional lymph nodes.
HE ×360

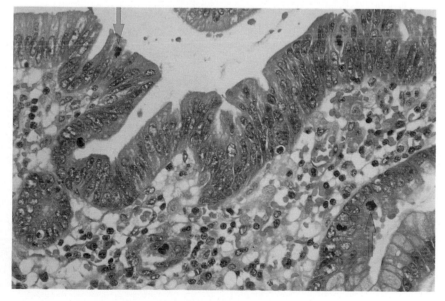

4.13 Adenocarcinoma: stomach

A man of 75 had a partial gastrectomy for carcinoma of stomach. Macroscopically the mucosa was flat but irregularly thickened and superficially ulcerated in places. The lumen (not visible) is on the left and the muscularis mucosae on the right. In the superficial mucosa the glands are slender, lined by atrophic epithelium and reduced in number. The interglandular tissue is filled with eosinophilic carcinoma cells, i.e. this is a superficial spreading form of carcinoma, in which the malignant cells are still confined to the mucosa. The pyloric-type glands in the basal layers of the mucosa (centre) appear relatively normal but there is a dense infiltrate of chronic inflammatory cells around them and in the thickened muscularis mucosae (right). HE ×60

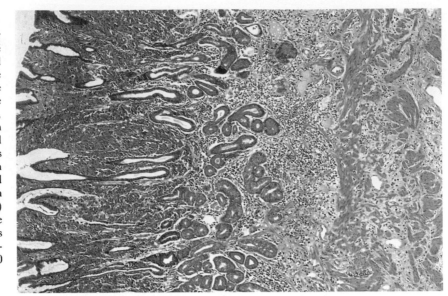

4.14 Adenocarcinoma: stomach

This shows the surface of the mucosa at higher magnification. The lumen of the stomach is at the top. The surviving glands are widely separated and lined by flat atrophic epithelial cells (thin arrows). The tissue between the glands is packed with cells with weakly eosinophilic cytoplasm. Special stains showed that the cytoplasm was full of mucin. The cell nuclei (where visible) have been pushed to one side by the mucin droplet to produce a signet-ring appearance (thick arrows). These are malignant epithelial cells which have spread widely through the upper gastric mucosa but have not yet penetrated even to the muscularis mucosae. Examination of the lymph nodes revealed no metastases. HE ×235

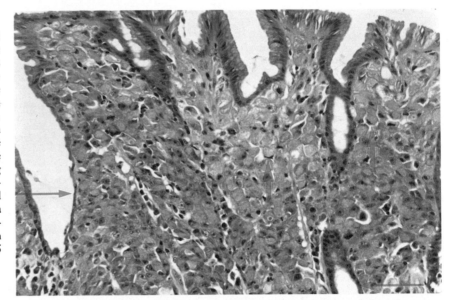

4.15 Adenocarcinoma: stomach

A malignant (carcinomatous) ulcer was present on the lesser curvature of the stomach and this shows the underlying muscle coat. Well-formed malignant glands, lined by a single layer of cuboidal cells with round or ovoid basophilic nuclei, have invaded the muscle. The tumour cells have relatively scanty cytoplasm with no obvious mucin secretion. They show considerable mitotic activity however. HE ×235

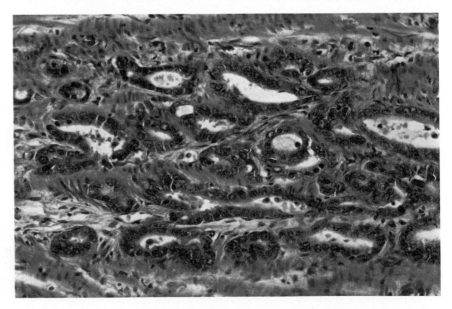

4.16 Adenocarcinoma: stomach

In this case the malignant cells have penetrated to the peritoneal surface of the stomach (top). The serosa (top) consists of a single layer of cuboidal cells (top) and a cluster of malignant cells is lying in the subserosa. They are signet-ring cells with a large droplet of pale-staining granular mucin in the cytoplasm which has pushed the dark-staining nucleus against the cell membrane on one side of the cell. The serosa appears intact but the tumour cells readily penetrate it to reach the peritoneal cavity where they may grow (peritoneal carcinomatosis) or spread to other organs such as the ovaries (transcelomic spread). Gastric carcinoma can also spread to the liver in the portal bloodstream. HE ×400

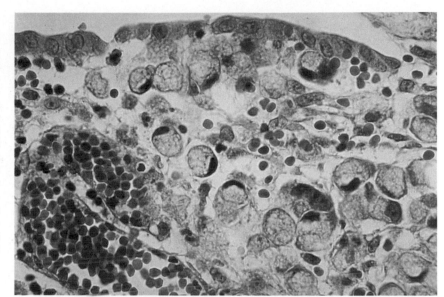

4.17 Adenocarcinoma: ampulla of Vater

Carcinoma can arise in the region of the ampulla of Vater from the terminal part of the common bile duct, from the ampulla itself, or from the duodenal mucosa. Adenocarcinoma of the ampulla tends to form papillary outgrowths into the lumen of the duodenum. Like carcinoma of the bile duct, it may produce clinical signs by obstructing the flow of biliary and pancreatic secretions. The tumour consists of glandular acini which are irregular in size and shape. The epithelial cells form small papillary projections into the lumen of the glands. The stroma consists of fibrous tissue, but smooth muscle fibres which may be part of the wall of the duodenum are also present, and there are many chronic inflammatory cells. The epithelial cells lining the acini are pleomorphic but tend to be cuboidal. HE ×150

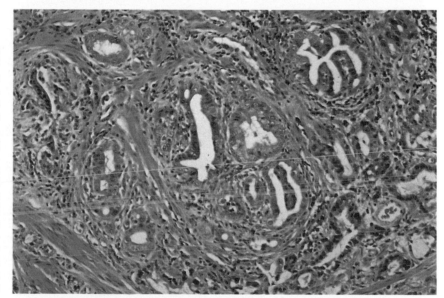

4.18 Adenocarcinoma: ampulla of Vater

At higher magnification the malignant acini are very irregular but the epithelial cells, which tend to be cuboidal or low columnar, have round, fairly uniform nuclei containing a prominent usually central nucleolus. There is no evidence of mitotic activity. Smooth muscle fibres (centre) are present and there are occasional lymphocytes in the fibrous stroma. HE ×360

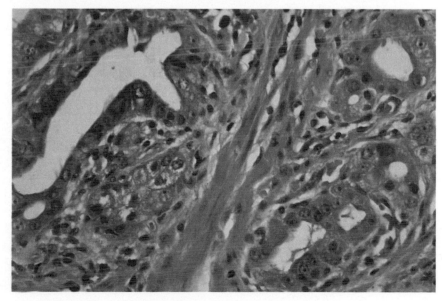

4.19 Leiomyoma: stomach

Leiomyoma of stomach is usually located beneath the mucosa, where it forms a round or ovoid mass. Ulceration may give rise to serious hemorrhage. This lesion was an ovoid mass (5 × 4 × 4cm) in the anterior wall of the stomach of a man of 75. It was apparently encapsulated but there was an area (2 × 1cm) of ulceration of the gastric mucosa stretched over the tumour, and there had been clinical evidence of recurring hemorrhage over a long period. The tumour consists of interlacing bundles of cells with elongated vesicular nuclei and eosinophilic cytoplasm. They resemble smooth muscle cells, and special stains confirmed that little collagen was present. No mitotic figures are present but small numbers were found in other parts of the tumour. Despite their presence and the cellularity of the tumour it was regarded as benign.   HE ×360

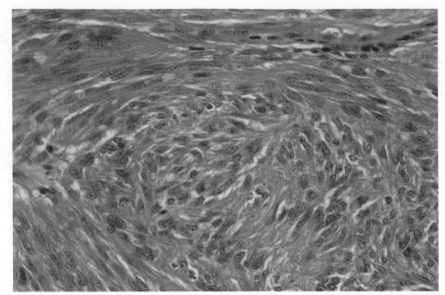

4.20 Leiomyosarcoma: stomach

Most smooth-muscle tumours of the stomach are well-differentiated but some lesions are very cellular, with cells containing bizarre hyperchromatic nuclei present. The most important criterion for malignancy is the mitotic activity of the tumour cells. The cells of this tumour have elongated vesicular nuclei containing several small nucleoli. One mitosis is shown (arrow) and there were between 3 and 6 per high power field throughout the tumour. The cells have eosinophilic cytoplasm which appears fibrillary and special stains confirmed the presence of myofibrils. Multiple secondary deposits were present throughout the peritoneal cavity. Leiomyosarcoma of stomach does not metastasize to lymph nodes but secondary deposits may occur in the liver and lungs. *See also 1.53, 11.33.*   HE ×360

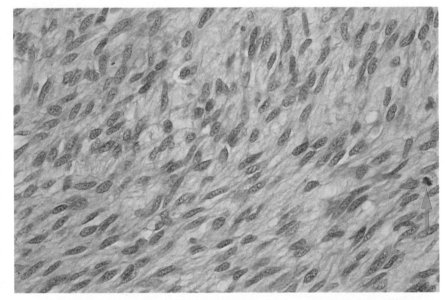

4.21 Infarct: ileum

Infarction of the small intestine can result from either venous or arterial obstruction. The infarction, however is always hemorrhagic, and extensive hemorrhage has occurred into the mucosa and the deeper tissues (right). The muscularis mucosae appears viable, with well-stained nuclei. The villi (left), however, have lost their epithelial covering and have collapsed. The small blood vessels are greatly dilated. The crypts of the glands (left of centre) retain their epithelial cells, although they appear somewhat degenerate.   HE ×235

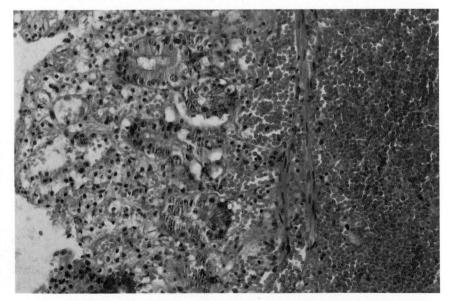

4.22 Gluten enteropathy (celiac disease): jejunum

The patient was hypersensitive to wheat gluten and suffered from the effects of intestinal malabsorption. Examination with the dissecting microscope showed great reduction in the height of the villi (partial villus atrophy). This shows part of one of the atrophic villi. The normal covering of tall columnar epithelial cells has been replaced by a layer of small cells of irregular size and shape, with lymphocytes present between them. The basement membrane is greatly thickened and hyaline, and capillaries lie within it and beneath it. Numerous plasma cells and lymphocytes are also present beneath the basement membrane.

HE ×200

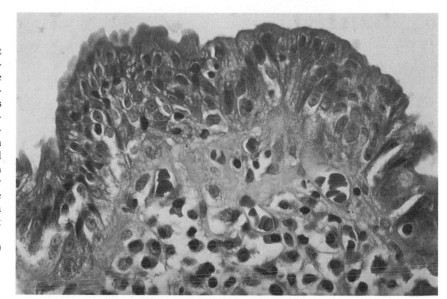

4.23 Intestinal lymphangiectasis: ileum

In this condition there is obstruction of the lymphatic channels draining the small intestine. The lumen of the intestine is to the left. The villus is swollen, from the presence of a greatly dilated lymphatic vessel in its core. The small blood vessels around it are also dilated. The villus is covered with tall columnar epithelial cells which appear normal. All the other villi in the biopsy specimen showed similar features. The patient suffered from malabsorption and protein-losing enteropathy, in which protein-rich fluid is lost from the lymphatics into the lamina propria and then into the lumen of the intestine. HE ×360

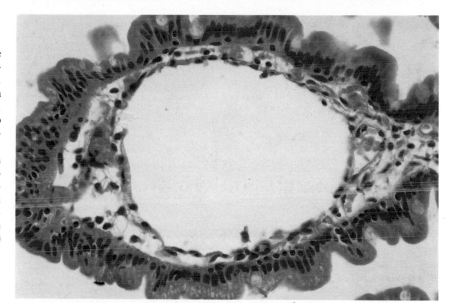

4.24 Typhoid (enteric) fever: ileum

Typhoid fever is a systemic infection caused by *Salmonella typhi*. The portal of entry of the bacillus is the intestine. Peyer's patches and the lymphoid follicles swell and many become necrotic, with ulceration of the overlying mucosa. Hemorrhage and perforation of the wall of the intestine are liable to occur. This is a Peyer's patch in a patient who had been ill for about 10 days. The lymphoid tissue has been replaced by large macrophages, among which necrosis is starting to occur (centre left). Lymphocytes and plasma are also present but there are no polymorph leukocytes. The blood vessel (right) is dilated. A Gram's stain would reveal large numbers of typhoid bacilli. During recovery the necrotic cells are phagocytosed and epithelium rapidly grows over the ulcerated area. There is therefore little scarring. HE ×135

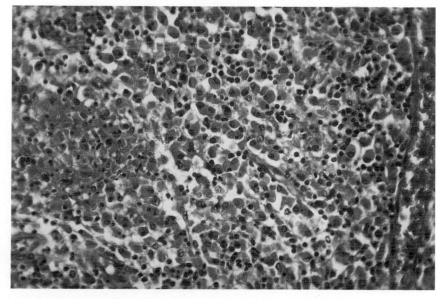

4.25 Crohn's disease

A woman of 51 had 42cm of terminal ileum and 10cm of caecum removed for Crohn's disease. The mucosa of the ileum was inflamed, with extensive patchy ulceration and loss of the normal mucosal pattern. The mucosa of the caecum showed an irregular 'mosaic' pattern. The mesentery of the small intestine was greatly thickened. This is the caecum, with the lumen of the bowel on the left. There is intense edema of the mucosa and submucosa, with diffuse infiltration with eosinophilic fluid. The lymphatics in the submucosa (thin arrows) and the mucosa (thick arrows) are dilated. Chronic inflammatory cells are present in moderate numbers mostly beneath the surface epithelium (left). Red-staining Paneth cells can be detected in the crypts of the colonic glands. HE ×60

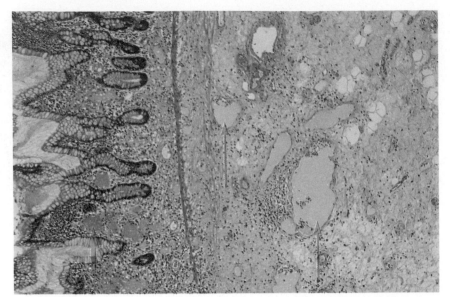

4.26 Crohn's disease: ileum

This is a case of Crohn's disease which was confined to the ileum. It shows part of the mucosa and submucosa. The tissues are infiltrated with inflammatory cells and the crypt of one intestinal gland, partly lined by epithelial cells, is distended with cells, mainly polymorph leukocytes, forming a crypt abscess, a feature associated more with ulcerative colitis than with Crohn's disease. The inflammatory cells on both sides of the intact muscularis mucosae are largely lymphocytes and plasma cells. The inflammatory process in Crohn's disease usually involves the whole bowel wall (transmural inflammation) whereas in ulcerative colitis the lesions are essentially mucosal. Healing produces fibrous tissue which, with the inflammatory edema, is liable to cause intestinal obstruction, a common complication of Crohn's disease. HE ×135

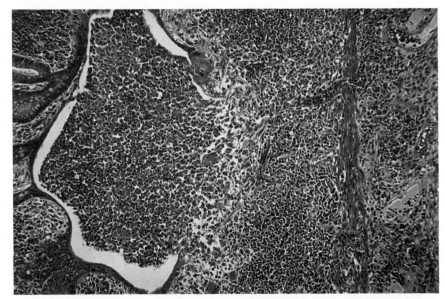

4.27 Crohn's disease: anal canal

The lesions in Crohn's disease are often widespread and the anal canal is sometimes involved, as here. The stratified squamous epithelium lining the canal (left) is infiltrated with polymorph leukocytes and in the submucosa there is a follicular collection of epithelioid cells (centre) surrounded by small numbers of lymphocytes and other chronic inflammatory cells. Multinucleated giant cells are often present along with the epithelioid cells within the follicle. Epithelioid granulomas of this type resembling a sarcoid granuloma are typical of Crohn's disease and may be present in the associated lymph nodes as well as in the bowel wall. Their presence helps to distinguish the lesion from ulcerative colitis but they may be difficult to detect. HE ×200

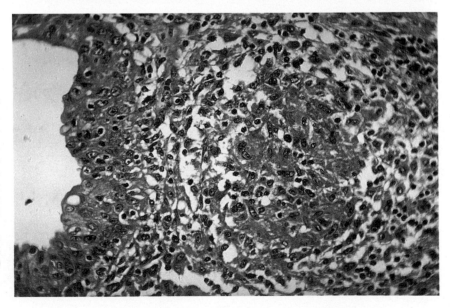

4.28 Ulcerative colitis

In ulcerative colitis the inflammatory process is relatively superficial compared with Crohn's disease. The mucosa may be extensively ulcerated. Hemorrhage may be severe. The mucosa between the ulcerated areas tends to project as pseudopolyps. This shows the mucosa of the colon. The lumen is at the top. There is ulceration and the mucosa is acutely inflamed and intensely hyperemic, with an infiltrate of inflammatory cells which is particularly intense in the region of the muscularis mucosae (bottom). Some of the glands are irregular, appearing compressed by the swollen interglandular tissue. Other glands are distended with mucus and inflammatory cells (arrows). One is filled with pus-like material (crypt abscess). The epithelium lining these glands is atrophic and absent in places.

HE ×60

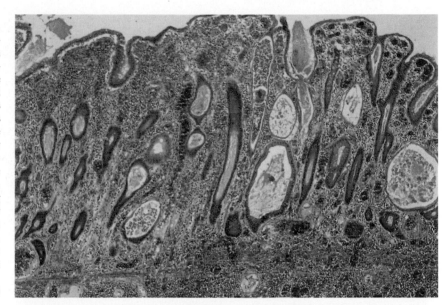

4.29 Ulcerative colitis: colon

This shows part of the mucosa from **4.28** at higher magnification. It is intensely hyperemic with considerable dilatation of the small blood vessels. The surface (left) is covered with low columnar epithelial cells which appear to secrete little mucin. The cells lining the glands are flat and attenuated. Several glands (top) are filled with dense eosinophilic secretion and cell fragments. The lumen of one gland (bottom) is filled with pus-like material which consists largely of polymorph leukocytes, although macrophages and a multinucleated giant cell are also present (crypt abscess). HE ×235

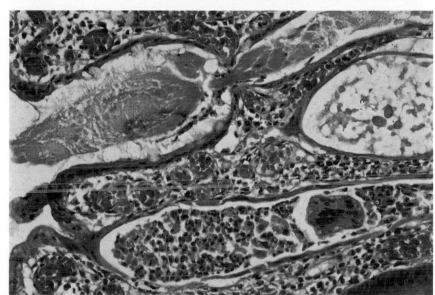

4.30 Ulcerative colitis: colon

This is a section of the colon from a longstanding case of ulcerative colitis. The lumen is just visible on the left. The wall of the colon is thinner than normal. The mucosa and submucosa have been replaced by vascular granulation tissue (left half of picture) containing many plasma cells and lymphocytes; and similar cells are present in small numbers throughout the atrophic muscle coats (right) and in the subserosal connective tissue (extreme right). Although ulcerative colitis is primarily a disease of the colon, in a significant proportion of cases the terminal ileum is also affected ('back-wash' ileitis). In long-standing cases there is an increased risk of carcinoma which may develop at several sites. HE ×60

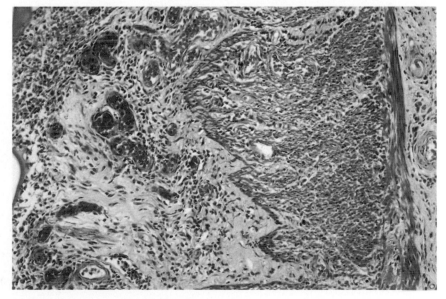

4.31 Carcinoid tumour

Carcinoid tumours arise from cells of the diffuse endocrine (enterochromaffin) system. Those of the alimentary tract, usually located in the ileum, secrete 5-hydroxy-tryptamine (serotonin). Carcinoid tumours of the ileum may metastasize to the lymph nodes and liver where the secondary deposits may give rise to the carcinoid syndrome. This lesion was a small nodule in the wall of the caecum. The crypts of the intestinal glands are visible on the left, and lying in the submucosa between the glands and the muscle coats (right) are compact groups of polyhedral tumour cells of uniform structure (centre and left). Smaller groups of cells are invading the muscle coats. There is no mitotic activity among the tumour cells. In the tissues around the clumps of tumour cells there are many eosinophil leukocytes.

HE ×135

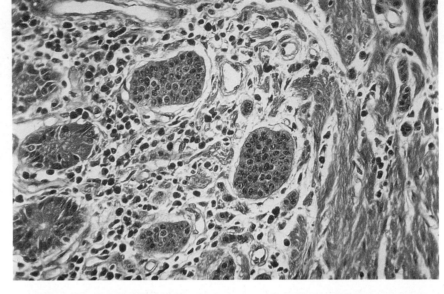

4.32 Carcinoid tumour: lymph node

Carcinoid tumours are slow-growing lesions which invade locally but do not metastasize rapidly. This lesion, however, metastasized to a lymph node in the neck. The site of the primary tumour was unknown. It consists of groups of closely-packed cells of uniform type. They have round nuclei in which the chromatin forms small clumps, a characteristic feature of cells of this type, and in some cells, generally those situated towards the centre of the cords of cells, the cytoplasmic eosinophilia is intense. No mitoses are evident. The cytoplasm was argentaffin with special stains. Delicate bands of stromal connective tissue run between the groups of tumour cells. HE ×360

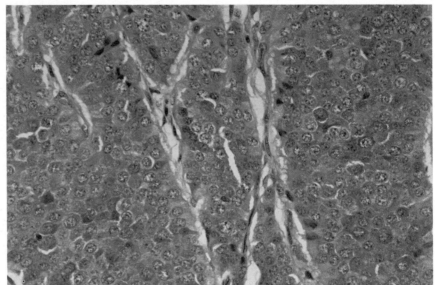

4.33 Carcinoid tumour: appendix

This tumour was found in the wall of an acutely inflamed appendix from a woman of 62. The tumour, which became yellow after fixation in formalin, had penetrated the muscle coats to the serosa. This section has been stained by the Grimelius method which is an argyrophil method, using a silver-containing reagent and another solution to 'develop' the sites of deposition of the silver. The cytoplasm of the carcinoid cells is black, and the nuclei orange-red (from the counterstain). Carcinoid tumours of the proximal parts of the alimentary tract (duodenum and stomach) and of the lung tend to be argentaffin-negative, i.e. they do not react directly with silver nitrate solutions but are usually positive with argyrophil methods such as the Grimelius technique. Grimelius method ×575

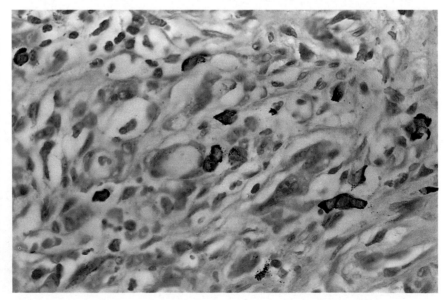

4.34 Adenocarcinoma: small intestine

Primary adenocarcinoma of the small intestine is uncommon. Most arise in the duodenum, in the region of the ampulla of Vater (**4.17**) and form papillary growths but elsewhere carcinoma of the small intestine, like colonic carcinoma, tends to encircle the bowel wall and produce stenosis. This shows neoplastic glands invading the muscle coat of the intestine. They form elongated parallel columns in which there are acini of widely varying size, lined by tall columnar cells with a 'brush' border at their apex and closely resembling the normal columnar epithelium of the glands of the small intestine. The ovoid nuclei are closely packed with some tendency to layering but they are fairly regular in size and shape. Mitotic activity was low among the tumour cells. HE ×235

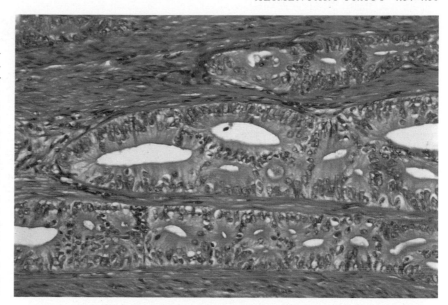

4.35 Lymphoid hyperplasia: ileum

Lymphoid hyperplasia in the small intestine (pseudolymphoma) may be focal, and take the form of a single mass, or more diffuse, as small nodules of lymphoid tissue throughout the intestine. The latter is frequently associated with infection by *Giardia lamblia*. A single large mass of hyperplastic lymphoid tissue, as in this case, may lead to intussusception. The lumen of the ileum is on the left. The mucosa (left) is stretched over lymphoid tissue in which there are large very active germinal centres. These have a 'starry-sky' appearance from the presence of many macrophages. The villi of the mucosa are reduced in height. The epithelial cells are mostly mucin-secreting goblet cells. The cores of the villi are infiltrated with lymphocytes. In one part (top left) the villi have disappeared. Paneth cells are prominent in the crypts of Lieberkühn (arrow). HE ×60

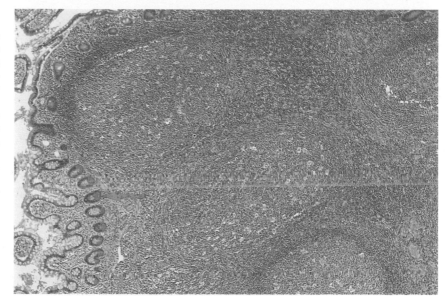

4.36 Immunoblastic lymphoma: small intestine

Lymphomas arising outside the alimentary tract may involve it secondarily but lymphomas also arise in the gut itself, often in association with celiac disease or related malabsorption syndromes. This lesion appeared to be a primary lesion of the small intestine. It is a high-grade immunoblastic lymphoma. Immunoblastic lymphoma consists of immunoblasts of B or T type, and the cells may show a tendency to differentiate to plasmablasts or plasma cells. The tumour cells in this case are shown invading the muscle coats of the intestine. They are very pleomorphic. The nuclei are oval or round and vary widely in size and shape. There is a prominent nucleolus located centrally. The cells have a moderate amount of cytoplasm. They are accompanied by large numbers of eosinophil leukocytes. HE ×360

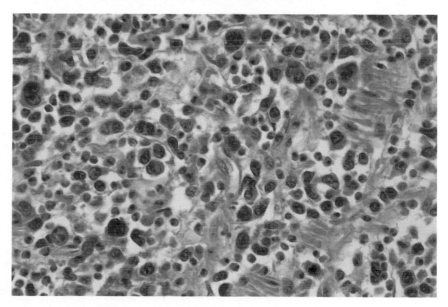

4.37 Oxyuris vermicularis: appendix

Infestation with *Oxyuris* (*Enterobius*) *vermicularis* (threadworm or pinworm) is common in some countries and particularly affects children. It is a small worm (5–12mm long) and may be present in numbers large enough to fill the lumen of the appendix. It is occasionally found in appendices resected surgically but it is not a cause of appendicitis. The mucosal lining of the appendix is visible (top and bottom right). Its lumen contains a number of worms cut in various planes. The outer coat of the worm consists of eosinophilic amorphous chitin within which the viscera are clearly visible. Pus-like material is also present in the lumen of the appendix.   HE ×75

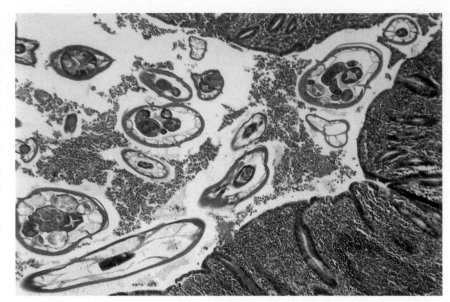

4.38 Oxyuris vermicularis: appendix

Oxyuris vermicularis is an active parasite which may migrate considerable distances, e.g. it may enter the vagina and reach the peritoneal cavity via the uterus and Fallopian tubes. An appendix 9cm long was removed at operation from a woman of 21. It was not macroscopically inflamed but the lumen contained several threadworms. In this section of submucosa, there is a granulomatous mass with a necrotic eosinophilic centre (left) enclosed by a thick wall of macrophages. Peripheral to the macrophages there are lymphocytes of the appendiceal lymphoid tissue. There is cell debris in the centre and a fragment of the threadworm is also visible (arrow). The lesion is a granuloma produced by a threadworm which has invaded the wall of the appendix, presumably from the lumen, which did contain a number of worms.   HE ×150

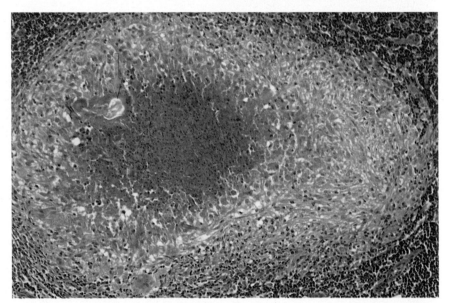

4.39 Acute appendicitis

Acute appendicitis is still common in Europe and the USA but uncommon in many other parts of the world, including Asia and Africa, probably for reasons of diet. The acute inflammatory process tends to spread along the muscular and serous coats, particularly if the lumen is obstructed. Suppuration and gangrene may develop distal to the obstruction, and perforation and peritonitis may follow. This shows the wall of the appendix, with the lumen at the top. The mucosa has been largely destroyed, only a few remnants of the glands remaining. The blood vessels are greatly dilated and the whole of the wall is infiltrated with polymorph leukocytes. The infiltrate is greatest in the submucosa where it amounts to pus formation. There is also an exudate of fibrin and polymorphs on the peritoneal surface (bottom).   HE ×60

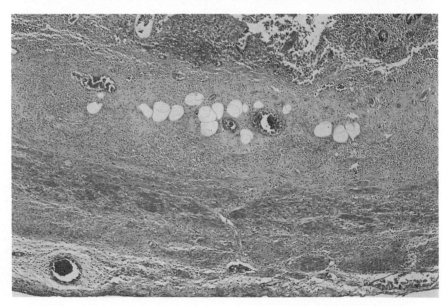

4.40 Acute appendicitis

This shows the muscle coats of the appendix at higher magnification. The muscle fibres cut in cross-section (left) are vacuolated but they are still viable, with well-stained nuclei as are the fibres cut longitudinally (right). These fibres however are separated by a pus-like inflammatory exudate which consists largely of polymorph leukocytes.

HE ×150

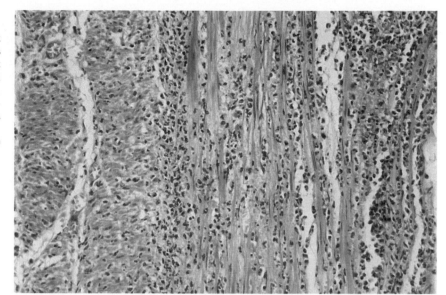

4.41 Acute appendicitis

This appendix was acutely inflamed near its tip, with fibrinous exudate on the peritoneal surface. This shows a focal lesion of the mucosa, with infiltration of the mucosa by large numbers of polymorph leukocytes and an eosinophilic inflammatory exudate. There is ulceration at the base of a gland (arrow). Small amounts of pus are present in the lumen of the affected gland (centre). Part of the lymphoid tissue of the appendix is visible at the bottom. Acute appendicitis probably starts in this way with a suppurative lesion at the bottom of a crypt but spread of the lesion to involve the whole appendix is greatly enhanced when the lumen is obstructed e.g. by a fecalith. HE ×150

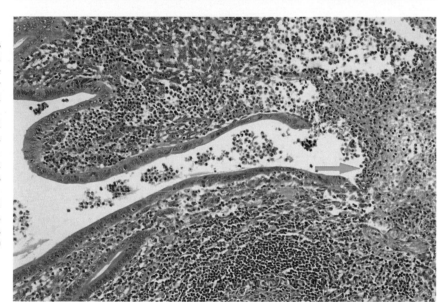

4.42 Chronic obliterative appendicitis

An appendix removed surgically for 'chronic appendicitis' may show, as here, obliteration of the lumen (centre and left) by fat and fibrous tissue. The mucosa and sub-mucosa have also disappeared, only the muscle coats (right) appearing relatively normal. These changes may have been produced by previous episodes of acute inflammation but they are much more likely to be the result of an ageing process in which the mucosal lymphoid tissue atrophies and fibrous replacement occurs. HE ×60

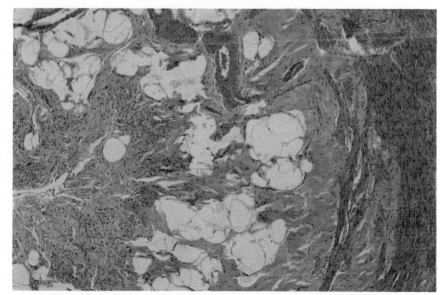

4.43 Hirschsprung's disease: colon

Hirschsprung's disease (idiopathic mega-colon) usually becomes evident soon after birth. The colon becomes distended and acute obstruction may occur. If the patient survives, hypertrophy of the muscle coats proximal to the abnormal part of the colon takes place. The defect is one of innervation with absence of parasympathetic ganglion cells (aganglionosis) from the intramural and submucosal plexuses of the distal colon. Unusually the whole colon may show the abnormality. Coordinated propulsive movement does not occur in the abnormal part of the colon. This shows the myenteric plexus. There are no ganglion cells but the nerve fibres appear hypertrophied.   HE ×360

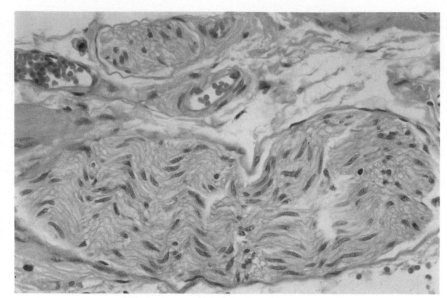

4.44 Pneumatosis cystoides intestinalis: colon

In this condition soft polypoid (grape-like) swellings project into the lumen of the colon. They consist of gas-filled cysts in the submucosa and may be misdiagnosed as neoplastic lesions. They may cause intestinal obstruction. The condition is sometimes associated with chronic lung disease. The mucosa of the colon is at the top and left. It appears normal, as does the muscularis mucosae. Beneath the latter, in the sub-mucosa, there is a large, round cystic space which is completely empty. The cyst is lined by large cells with eosinophilic cytoplasm (arrows) and scattered lymphocytes are present in the vicinity of the cyst. HE ×60

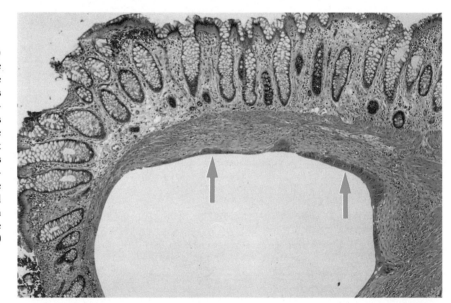

4.45 Pneumatosis cystoides intestinalis: colon

This shows part of the cyst (bottom) and its wall at higher magnification. The cyst is lined by multinucleated giant cells which have abundant eosinophilic cytoplasm (bottom). The surfaces of the giant cells facing the lumen of the cyst are flattened. The crypts of the colonic glands (top) are visible and the underlying muscularis mucosae is intact. Small numbers of lymphocytes and plasma cells are present in the connective tissues on both sides of the muscularis mucosae. The gas in the cysts is not air but is produced by bacterial action. The condition may also affect the small intestine.

HE ×235

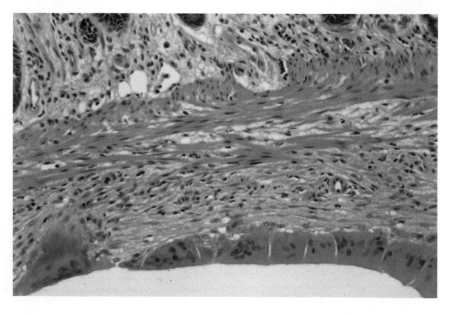

4.46 Pseudomembranous colitis

Pseudomembranous colitis is probably caused by a toxin produced by the anaerobic organism, *Clostridium difficile*. Treatment with the antibiotics lincomycin and clindamycin predisposes to its development. A biopsy of the mucosa may be required to establish the diagnosis. This is a specimen of rectal mucosa from a woman of 43 who developed post-operative diarrhea. The surface of the epithelium is covered (top) with a pseudomembrane which tends to assume a mushroom shape, the so-called 'summit' lesion. It consists of mucus (unstained), inflammatory cells (mainly polymorphs) and eosinophilic material (mainly fibrin and red cells). The mucosa beneath the pseudomembrane is ulcerated but the deeper parts of the colonic glands remain. The muscularis mucosae (bottom) appears normal.   HE ×60

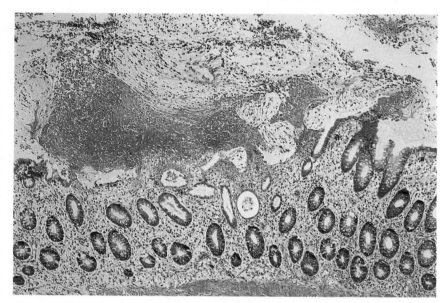

4.47 Pseudomembranous colitis

This shows the pseudomembrane and underlying mucosa at higher magnification. The pseudomembrane (left) consists of fibrin, mucus and inflammatory cells (trapped in the unstained mucus) and it is firmly adherent to the ulcerated mucosal surface (centre). The superficial parts of the mucus-secreting colonic glands have been destroyed but their somewhat distorted deeper parts have survived. The mucosa between the glands is infiltrated with inflammatory cells.   HE ×150

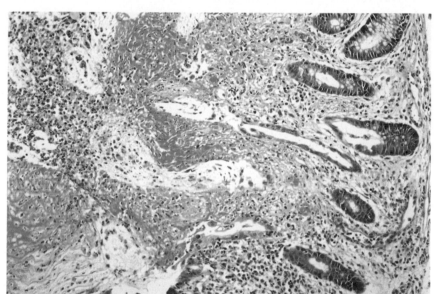

4.48 Epithelial dysplasia: colon

Patients with ulcerative colitis have a significantly increased risk of developing carcinoma of colon. Dysplastic changes precede the development of malignancy, and this shows dysplasia of the epithelial cells lining two glands in a woman of 62 with long-standing ulcerative colitis. The cells are pleomorphic, and only a minority show active secretion of mucin, most having eosinophilic cytoplasm. The nuclei in the smaller gland are round and fairly regular, but many of the nuclei of the closely-packed cells lining the larger gland are pleomorphic and hyperchromatic. There are several mitotic figures. Lymphocytes are present in the tissues around the glands. When dysplastic changes are being assessed it is important to avoid areas that are actively inflamed to prevent confusion with reactive changes. *See also 1.47, 1.48, 11.10–12.*

HE ×150

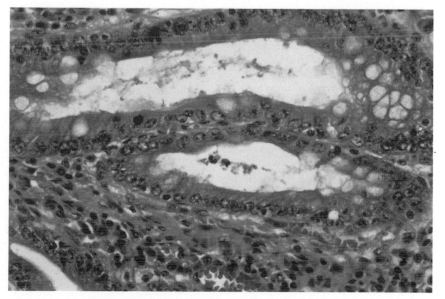

4.49 Villous adenoma: appendix

Villous adenoma (villous papilloma) is usually a single lesion and it tends to be located in the sigmoid colon or rectum. It is a lesion of older age groups. It tends to grow, encircling the lumen of the colon. The loss of fluid and electrolytes from a large lesion may be sufficient to cause hyponatremia, hypokalemia and dehydration. This specimen was unusual, in that it was a small lesion, located in the appendix, in a man of 54. It was blocking the lumen, producing a mucocele. It consists of slender papillae, each with a core of delicate connective tissue heavily infiltrated with chronic inflammatory cells and covered with very tall columnar epithelial cells. The nuclei of the epithelial cells are ovoid and show no pleomorphism.   HE ×150

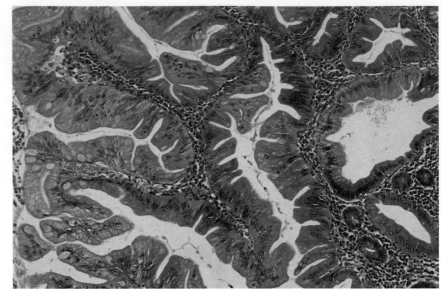

4.50 Villous adenoma: appendix

This shows the tip of one papillary process at higher magnification. The surface is covered with tall columnar cells, with abundant eosinophilic cytoplasm. The epithelial nuclei are ovoid and very regular in size and shape. There is no mitotic activity. There is some crowding of the nuclei which may be the result of oblique sectioning. The core of papilla is vascular and many lymphocytes and plasma cells are present.   HE ×360

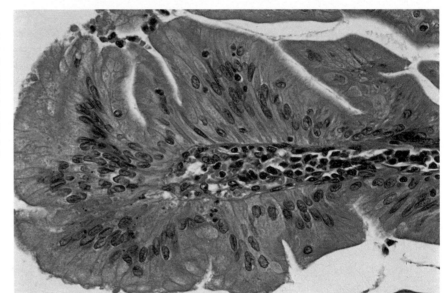

4.51 Tubulo-villous adenoma

Adenomatous polyps are common lesions in the colon. They are adenomatous neoplasms which occasionally undergo malignant change to adenocarcinoma. This is a longitudinal section of a polyp of colon. It is pedunculated, consisting of a mass of hyperchromatic epithelial cells (left) attached to the wall of the colon (right) by a slender stalk. The stalk consists of connective tissue and for most of its length it is covered with normal mucus-secreting colonic epithelium. This pale-staining epithelium gradually merges with the deeply staining epithelium of the polyp. Many tubules of irregular size and shape cut in longitudinal section and cross-section are visible within the polyp. There is no evidence of invasion of the stalk by the epithelium of the polyp.   HE ×12

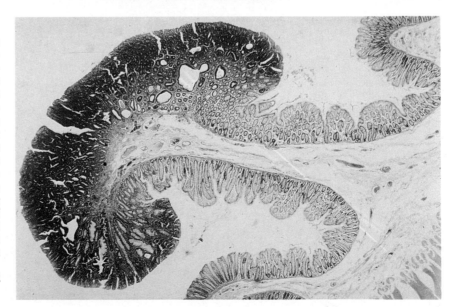

4.52 Villous adenoma: rectum

A man of 68 had a large (10 × 6cm) villous adenoma (villous papilloma) of the rectum removed surgically. Histologically it consisted of long papilliform processes growing more or less directly from the mucosal surface. This is the tip of one villus. The surface is covered with tall columnar cells with closely-packed large basophilic nuclei which show stratification (at higher magnification occasional mitotic figures were evident). There is no invasion of the stalk and the two glands cut in cross-section have a well-formed basement membrane. The epithelial cells secrete mucus but in greatly reduced amounts. The core of the villus is infiltrated with lymphocytes and plasma cells. Villous papillomas like this have a marked tendency to become malignant (carcinomatous). *See also 1.52.* HE ×235

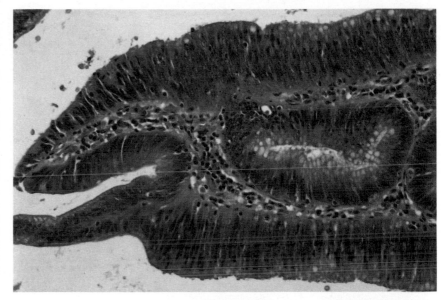

4.53 Adenocarcinoma: colon

The tumour took the form of an ulcer with a heaped-up ('rolled') edge. This shows the ulcer in cross-section. Normal colonic mucosa is just visible (centre right). It is continuous with a thicker and more basophilic layer of malignant epithelium (top right). This in turn merges with the raised mass of deeply staining tumour cells (thin arrow) which constitute the rolled edge of the ulcer. The ulcer crater is on the left (thick arrows). The floor of the ulcer consists of carcinomatous tissue, and malignant cells have invaded downwards and laterally, destroying the muscle coats and invading the pericolonic fat. Some muscle fibres survive (centre right). Lymphoid follicles are visible within the peri-colonic fat. Carcinoma of colon tends to spread not only by direct invasion and by the lymphatic channels but also by the portal vein to the liver. HE ×4

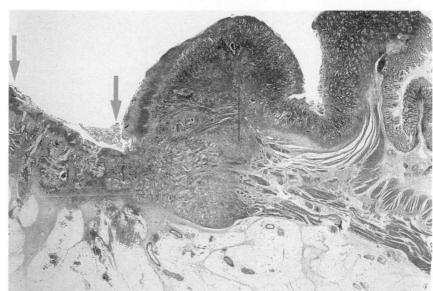

4.54 Adenocarcinoma: colon

This tumour has invaded the wall of the colon and penetrated to the peri-colonic tissues. It is a poorly differentiated tumour, consisting of sheets of cells with ovoid nuclei and eosinophilic cytoplasm. There is nuclear pleomorphism, and considerable mitotic activity was evident at higher magnification. There is an area of necrosis within the larger mass of cells. The tumour cells for the most part show little evidence of mucin secretion apart from the presence of several small acini (arrows). Special stains and electron microscopy usually reveal many minor 'micro-acini', however, even in a fairly anaplastic lesion like this one. The fibrous stroma between the groups of tumour cells contains many lymphocytes and plasma cells. HE ×200

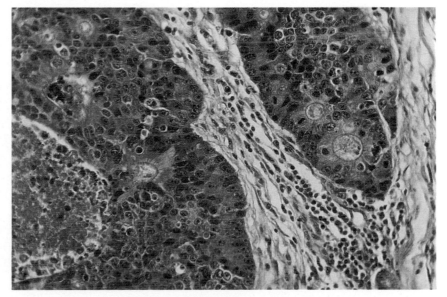

4.55 Benign lymphoid polyp: rectum

Lymphoid polyp (focal lymphoid hyperplasia) is a not uncommon lesion of the rectum, where it may cause bleeding or prolapse. It is located in the submucosa and forms a small soft polypoid mass covered with smooth intact mucosa and projecting into the rectum. Histologically it consists of mature lymphoid tissue resembling a lymph node. The lymphoid follicles beneath the capsule are atrophic, however, and the paracortical zones are enlarged. Although the lymphoid tissue is intersected by strands of connective tissue, there is no proper system of sinuses. Higher power examination confirms that the lymphoid tissue is mature and not lymphomatous. The lesion is probably hamartomatous in nature. Accurate diagnosis of this type of lesion is important, since lymphomas sometimes occur as primary lesions of the colon. HE ×40

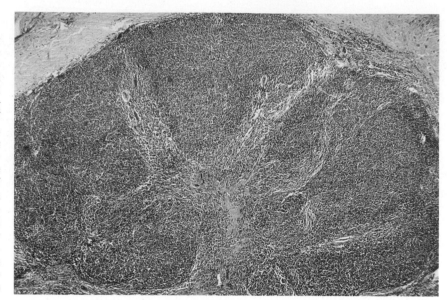

4.56 Lymphoplasmacytic lymphoma: colon

A woman of 81 had 30cm of a sigmoid colon and rectum removed surgically for diverticulosis. Multiple diverticula were present but in addition the mucosa of the whole of the resected bowel was heavily infiltrated with lymphomatous cells. The tumour also extended into the pericolonic fat, where it formed multiple nodules. This shows the colonic mucosa and submucosa with the mucus-secreting glands on the left. The tissues are diffusely infiltrated with small cells with basophilic nuclei, the infiltrate extending up to the surface epithelium. Numerous deposits of eosinophilic amorphous material (arrows) are present throughout the infiltrate. HE ×60

4.57 Lymphoplasmacytic lymphoma: colon

This shows the tumour cells at higher magnification. Many of them have small round basophilic nuclei and scanty cytoplasm (thin arrow). Others are obvious plasma cells, with abundant eosinophilic cytoplasm, the nucleus at one pole and a juxtanuclear 'halo'. There are numerous histiocytes with large vesicular nuclei (thick arrow) and small masses of eosinophilic amorphous material resembling amyloid (top). HE ×575

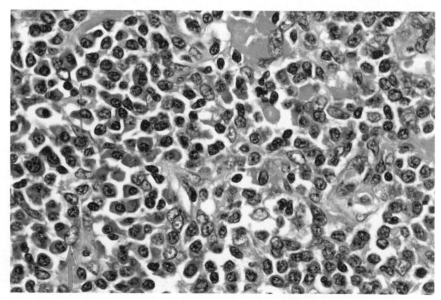

4.58 Malignant fibrous histiocytoma: retroperitoneum

Malignant fibrous histiocytoma tends to be located in deeper structures than the benign form. This one formed a large mass in the retroperitoneum. The tumour cells (arrows) have large vesicular round or ovoid nuclei and abundant eosinophilic cytoplasm. Each nucleus has a prominent central nucleolus. The cells were shown by histochemical methods to contain a number of enzymes characteristic of macrophage lysosomes: alpha-1-antichymotrypsin, lysozyme and cathepsin B. The small dark nuclei are mainly lymphocytic, although a few polymorph leukocytes are also present. Some lymphocytic lymphomas have large numbers of histiocytes and care must be taken to exclude this type of lesion.

HE ×470

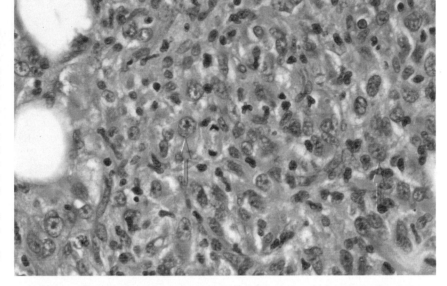

4.59 Mesothelioma: peritoneal cavity

There is a close link between inhalation of asbestos fibres and mesothelioma, and mesothelioma is usually found in the pleural cavity. The tumour may spread through the diaphragm to involve the peritoneal cavity. Primary mesothelioma of the peritoneal cavity does occur however albeit rarely. The tumour tends to spread widely over serosal surfaces including that of the intestine. This tumour appeared to be a primary peritoneal lesion. It had spread throughout the peritoneal cavity and this shows tumour growing on the surface of the small intestine (bottom). The tumour has a papillary structure, the small papillae consisting of a core of delicate connective tissue covered with closely packed cuboidal cells.　　HE ×235

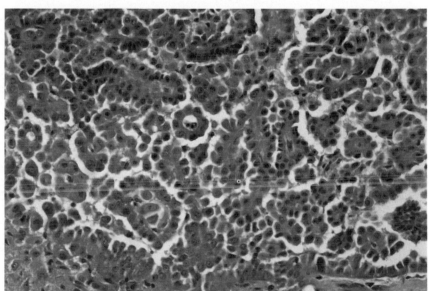

4.60 Mesothelioma: peritoneal cavity

The malignant mesothelial cells in this lesion, seen here invading the muscle coat of the small intestine (left), are elongated large cells with a relatively small nucleus and abundant eosinophilic cytoplasm with well-defined cell boundaries lying in a very loose pale-staining stroma. Vacuoles are present in the cytoplasm of many cells (arrows). Mesotheliomas secrete hyaluronic acid and histochemical tests for the presence of this in the stromal tissue can provide confirmatory evidence for the diagnosis.　　HE ×360

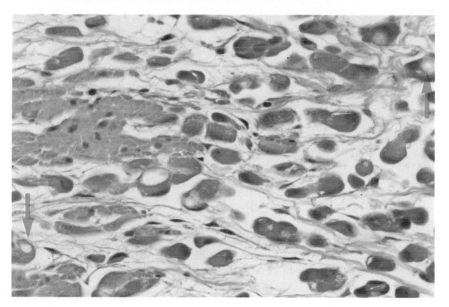

5.1 Congenital hepatic fibrosis: liver

In congenital hepatic fibrosis bands of dense fibrous tissue extend throughout the liver, enclosing islands of essentially normal parenchyma. This shows one such fibrous septum. Characteristically within it there are numerous mature bile ducts lined by cuboidal epithelium. Some of the bile ducts are dilated and some contain plugs of dense bile (centre right). A small nodule of liver cells (centre left) is also present within the fibrous tissue. In this case the patient was aged 15, the liver was enlarged and there was portal hypertension. The kidneys were polycystic, a condition known to be present in about half of the patients with congenital hepatic fibrosis.　　　　HE ×55

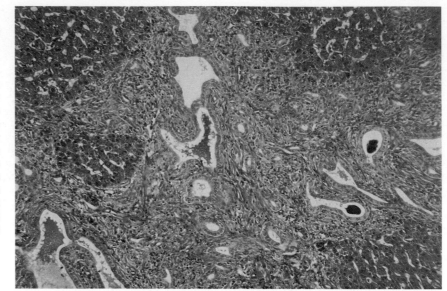

5.2 Dubin-Johnson syndrome: liver

The Dubin-Johnson syndrome is a rare benign condition in which there is a partial failure of the liver cells to secrete conjugated bilirubin into the bile canaliculi. Conjugated bilirubin is therefore released into the bloodstream. Intermittent jaundice results. The other constituents of the bile are unaffected. In the liver cells, as shown here, large coarse granules of a brown melanin-like pigment accumulate. The organ turns black but its architecture is undisturbed. The origin of the pigment is not known.

HE ×360

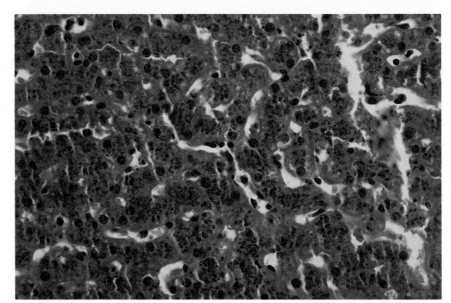

5.3 Alpha-1-antitrypsin deficiency: liver

Alpha-1-antitrypsin is synthesized in the liver, and its normal function is to inhibit the action of trypsin in the blood. In this condition it is retained within the liver, where it is well demonstrated by the periodic acid-Schiff (PAS) method as round bodies. This shows two lobules, on both sides of a portal tract containing a small bile duct running across the centre of the picture. Large numbers of small round bodies, stained purplish-red, are present within the hepatocytes. Most of the hepatocytes show severe fatty change. Fibrous tissue is increased in the portal tract and in severe cases macronodular cirrhosis develops. Pulmonary disease, usually emphysema, is often present and sometimes chronic pancreatitis. The staining reaction of the bodies is unaffected by pretreatment of the section with diastase.　　　　PAS ×360

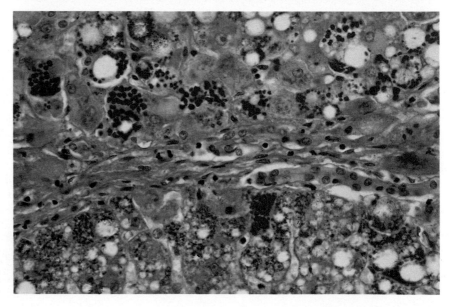

5.4   Hereditary hemorrhagic telangiectasia (Rendu-Osler-Weber syndrome): liver

In hereditary hemorrhagic telangiectasia, vascular hamartomas are present throughout the body. Most are in the skin and mucous membranes, particularly of nose and bowel. Generally the vessels are small thin-walled capillaries and venules and they may cause problems of bleeding. In the liver the lesions may take the form of discrete hemangiomas and this example has thick fibromuscular walls and the channels are filled with blood. It is located in the vicinity of a portal tract. The lesions may be extensive, however, and fibrosis may develop, to produce a form of cirrhosis.   HE ×60

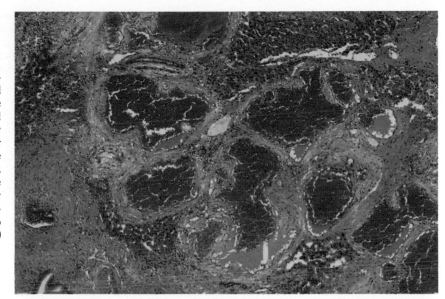

5.5   Cavernous hemangioma: liver

A not uncommon incidental finding during postmortem examinations is a spongy dark red nodule (sometimes wedge-shaped) which projects only slightly above the surface of the liver. It is a hemangioma of liver, and most are cavernous, composed of thick-walled vascular channels lined by flat endothelial cells. It is probably hamartomatous rather than neoplastic in nature. Rarely a hemangioma of liver bleeds, but as a rule the lesion is of no clinical significance. Some normal liver cells are visible on the left. *See also 1.49, 14.78.*   HE ×235

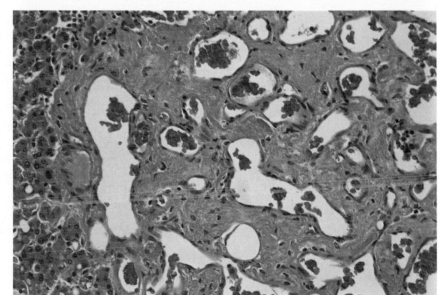

5.6   Chronic venous congestion and fatty change: liver

In chronic heart failure, when the right side of the heart fails to maintain its output, the inferior vena cava and hepatic veins become congested. In the centrilobular zone (right) and in the mid-zonal region (centre) the sinusoids are congested. The liver cells adjoining the portal tract (left) in which a branch of the portal vein and a small bile duct are visible, are relatively normal, but the centrilobular region shows severe fatty change and many hepatocytes are atrophic. Macroscopically, the contrast between the brown colour of the peripheral parts of the lobules and the yellow colour of the fatty central zones produces the 'nutmeg' pattern characteristic of chronic venous congestion. In more severe cases many hepatic cells disappear. Fibrous tissue appears to increase and cardiac cirrhosis results.   HE ×55

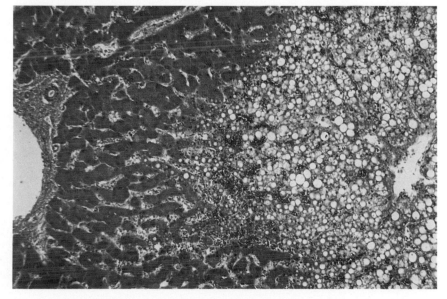

5.7 Budd-Chiari syndrome: liver

Narrowing or obstruction of the superior vena cava or of the main hepatic vein causes the Budd-Chiari syndrome. Intense congestion of the hepatic sinusoids results, and if the obstruction occurs suddenly (e.g. by thrombosis) the liver becomes swollen and acutely painful and ascites develops. Slowly-developing obstruction produces less acute clinical signs and symptoms. Atrophy of hepatocytes commonly occurs in the centres of the lobules. In this example the sinusoids in the centre of the lobule (lower right) are distended and full of blood, and the hepatocytes have largely disappeared, apart from a number of swollen degenerate bile-stained cells. In the less affected mid-zonal region the hepatocytes are atrophic but at the periphery of the lobule (left) they look comparatively normal.   HE ×150

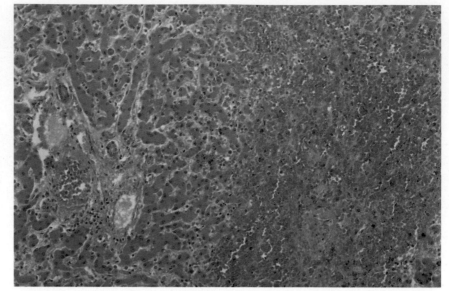

5.8 Peliosis hepatis: liver

Peliosis hepatis is an uncommon condition in which angiomatoid lesions develop throughout the liver. The lesions are associated with wasting disease such as advanced tuberculosis or malignancy and with the consumption of anabolic and contraceptive steroids. They take the form of blood-filled cysts up to 1cm in diameter. Several blood-filled cysts are shown here. They do not appear to have a distinct wall and are compressing the adjacent parenchymal cells. Hemosiderin-laden phagocytes are present in the adjacent liver (arrow). The cysts have no obvious connection with the hepatic sinusoids, which are not distended. The blood in the cysts is usually fluid but it may thrombose. Organization of thrombus in a cyst produces a stellate scar.   HE ×60

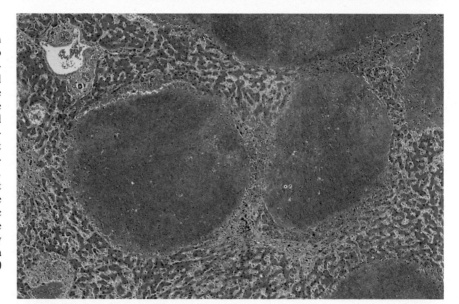

5.9 Nodular regenerative hyperplasia: liver

In nodular regenerative hyperplasia (nodular transformation of the liver) there are small nodules of regenerating hepatocytes throughout the liver. The condition can simulate micronodular cirrhosis fairly closely. Two nodules are shown here (left and right), with cords of liver cells between them which look relatively normal apart from the fact that the sinusoids are much wider than usual. There is no fibrosis in or around the nodules. Nodular regenerative hyperplasia may cause portal hypertension. The etiology is uncertain but there is an association with steroid therapy and chronic circulatory impairment.   HE ×60

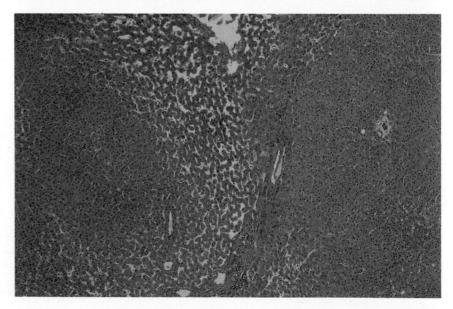

5.10 Amyloid: liver

Amyloid is a fibrillar material which is laid down in the tissues, usually extracellularly. It is associated with chronic inflammatory diseases such as rheumatoid arthritis and certain neoplastic conditions. In this case the patient had suffered from severe rheumatoid arthritis for many years. In the liver amyloid is generally deposited in the walls of the sinusoids, and in this example it appears as dense amorphous eosinophilic material (thin arrows) between and around the hepatocytes, separating them from the lumen of sinusoids (thick arrow). Some of the hepatocytes are atrophic, from pressure atrophy and ischemia. Many appear normal however and liver function may be well preserved in the presence of considerable amounts of amyloid. HE ×335

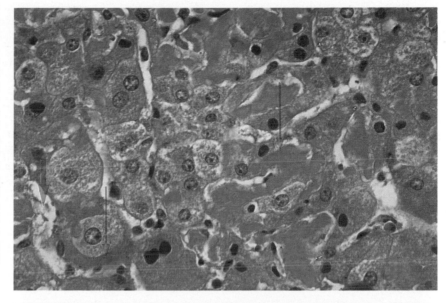

5.11 Alcoholic hepatitis: liver

Individuals vary considerably in their susceptibility to the toxic effects of alcohol, and alcoholic hepatitis develops in only about one-third of chronic alcoholics. The condition may progress to cirrhosis. This specimen was a wedge biopsy of liver removed at laparotomy from a man of 57 who was a known chronic alcoholic. There is ballooning degeneration of hepatocytes in the centrilobular region around the branch of the hepatic vein (arrow) and panlobular fatty change in the middle of the lobule. The hepatocytes at the periphery of the lobule, around the portal tract (left), are much less affected. It should be noted that ballooned hepatocytes have a central nucleus, whereas the nucleus in the fatty cells is located peripherally or is not visible. There is no evidence of cholestasis. HE ×60

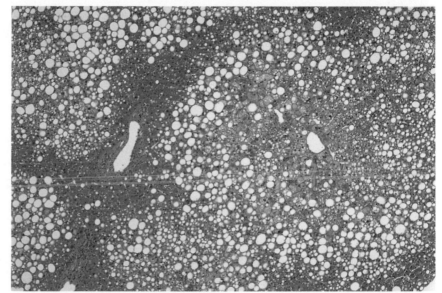

5.12 Alcoholic hepatitis: liver

At higher magnification the hepatocytes are swollen and contain fat droplets of widely varying size, some being very large. Some hepatocytes have pyknotic nuclei and are probably necrotic. Deeply eosinophilic alcoholic hyaline is present in several cells (arrows). A mild infiltrate of polymorph leukocytes is present. There is no evidence of cholestasis. Sometimes the cytoplasmic envelope around the larger droplets of fat ruptures and the droplets coalesce to form a large extracellular fat cyst which undergoes phagocytosis. Fibrosis usually follows. A stain for reticulin however showed no increase in reticulin fibres in this liver. There was also no increase in stainable iron. *See also 1.43.* HE ×360

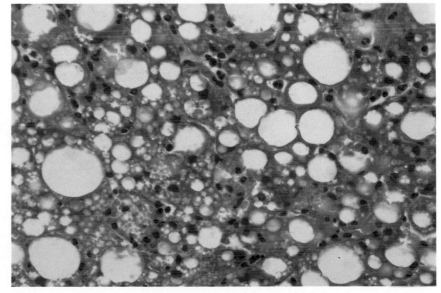

5.13 Gumma: liver

The gumma is a lesion of the tertiary stage of syphilis. There is usually extensive necrosis and destruction of parenchymal tissue, with subsequent formation of much scar tissue. This one formed a yellowish mass approx. 5cm dia in the liver. Most of it consists of amorphous necrotic material (left), enclosed in the capsule of dense fibrous tissue (centre) in which several small bile ducts (arrows) are trapped. Chronic inflammatory cells (mainly plasma cells) are also present in moderate numbers. Compressed and atrophic liver cells are visible on the right. Gummatous scarring of the liver may cause marked distortion of its structure (hepar lobatum). *See also 1.18, 9.14.*

HE ×60

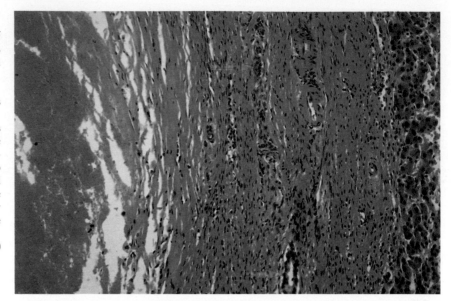

5.14 Visceral leishmaniasis: liver

Visceral leishmaniasis (kala-azar) is caused by the protozoon *Leishmania donovani*, which is 4–5μm dia and contains nuclei (1–2μm dia). The parasite is transmitted by the bite of the sandfly *Phlebotomus papatasii*. In a person with the illness, the affected organs (spleen, liver, kidneys etc) are swollen from the presence of large numbers of macrophages full of parasites. In this liver the cords of hepatocytes appear atrophic (some fixation shrinkage is present) whereas the phagocytic Kupffer cells lining the sinusoids are swollen, their cytoplasm containing large numbers of the protozoon, the nuclei of which appear as small blue dots (arrows). The fine structure of the *L. donovani* is not well demonstrated by histological sections, a smear stained by Giemsa or Leishman's stain being much more effective. HE ×575

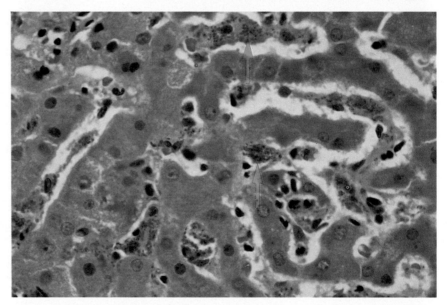

5.15 Cryptococcosis: liver

The causative organism, *Cryptococcus neoformans* (*Torula histolytica*) is a yeast 5–10μm dia which is difficult to stain because of the presence of a mucinous capsule. The organism evokes a chronic inflammatory reaction which is usually granulomatous but it is sometimes non-specific and may be very slight. A woman of 31 was found at laparotomy to have many small white nodules in her liver, and a portion of liver was removed for diagnostic purposes. Histologically, each nodule consists of necrotic liver surrounded by an inflammatory cellular infiltrate. The cells are epithelioid-type macrophages, multinucleated giant cells and large numbers of eosinophil polymorphs. The nuclei of the giant cells are large and vesicular (arrow). No microorganisms are visible. HE ×360

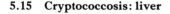
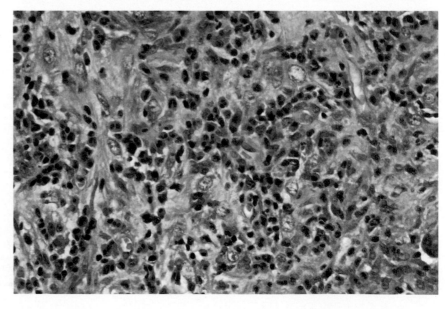

5.16 Cryptococcosis: liver

Cryptococci are difficult to detect in sections stained with hematoxylin and eosin because of the mucinous capsule around the organism. The capsule appears as a clear halo. The periodic acid-Schiff method for mucin however stains the capsule and so demonstrates clusters of the microorganism effectively; and in this section of the lesions shown in **5.15** the capsules appear as purplish-red circles (arrows). The protozoon is lying within vacuoles inside the giant cells.

PAS ×740

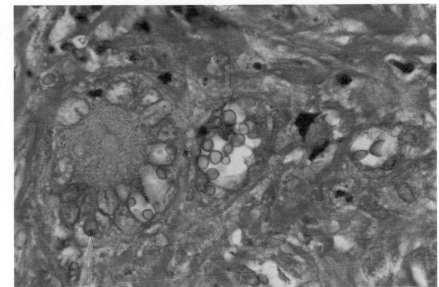

5.17 Cryptococcosis: liver

An even more effective method for detecting microorganisms such as fungi which have a mucinous capsule, is the Grocott method, which reveals the mucinous capsule by depositing silver within it. The silver appears black and in this section of the lesion shown in **5.15** the clusters of the microorganisms stand out against the blue-stained tissue.

Grocott method ×360

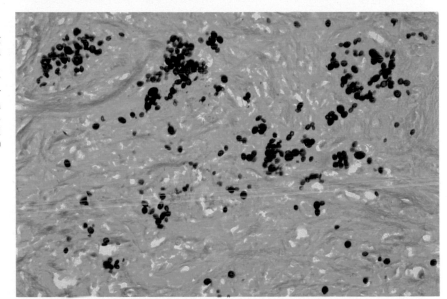

5.18 Schistosomiasis: liver

The flukes *Schistosoma mansoni* and *Schistosoma japonicum* inhabit the branches of the portal vein and their ova reach the liver where they excite an inflammatory reaction. The reaction is initially acute but becomes granulomatous. This shows a degenerate ovum enclosed in a capsule of phagocytes and cellular connective tissue. Many of the adjacent hepatocytes are atrophic. Schistosomiasis does not lead to cirrhosis but may cause extensive fibrosis of the liver (pipe-stem fibrosis) and portal hypertension. *See also 1.39, 10.54.*

HE ×360

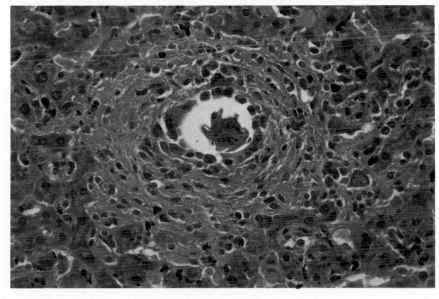

5.19 Viral hepatitis: liver

There are three common types of viral hepatitis: hepatitis A, hepatitis B and non A- non B hepatitis. It is not possible to distinguish between them histologically. In acute viral hepatitis there is diffuse inflammation of the liver and widespread necrosis of parenchymal cells. This patient died several weeks after the onset of hepatitis B. There is extensive destruction of liver cells (left) and infiltration by lymphocytes and plasma cells which is widespread but most intense in the portal tracts (left). Many small bile ducts have survived in the portal tracts. The hepatic cell plates have been destroyed (lobular disarray) and the surviving hepatocytes form clusters of various sizes. Many hepatocytes are atrophic. Bile capillaries are blocked by bile (arrows), evidence of severe cholestasis. HE ×150

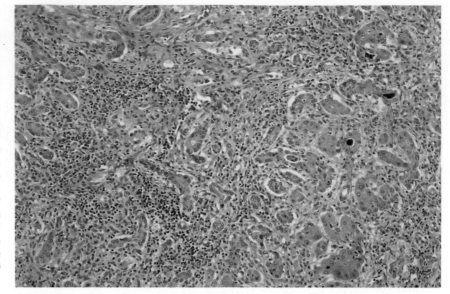

5.20 Viral hepatitis: liver

In viral hepatitis severity of infection varies considerably from a mild pyrexial upset illness without jaundice to a fulminant hepatic lesion with diffuse necrosis and hepatic failure. This is a later stage in a less severe illness than that shown in **5.19**. The remaining hepatocytes are very swollen and hydropic (ballooning degeneration). The infiltrate of chronic inflammatory cells (lymphocytes and plasma cells) is concentrated to a large extent in the portal tracts which are considerably expanded. Several degenerate hepatocytes in the form of acidophil bodies are evident. HE ×150

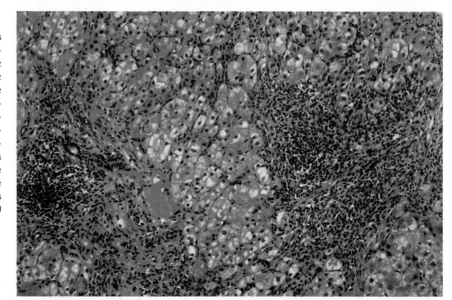

5.21 Viral hepatitis: liver

This is a case of hepatitis A. In HE sections of the liver there was little evidence of hepatocyte damage and only a moderate infiltrate of inflammatory cells in the connective tissue. A silver stain for fine connective tissue fibres (reticulin) has been used, to demonstrate the delicate reticulin network and lobular structure of the liver, and it reveals broad bands of dark-staining fibres linking several portal tracts. The degree of bridging fibrosis seen here, some of which is probably attributable to condensation of the stroma following destruction of liver cells, is probably reversible. It may however be a precursor of cirrhosis. There is no increase of fibres around the central veins (arrows). Reticulin stain ×55

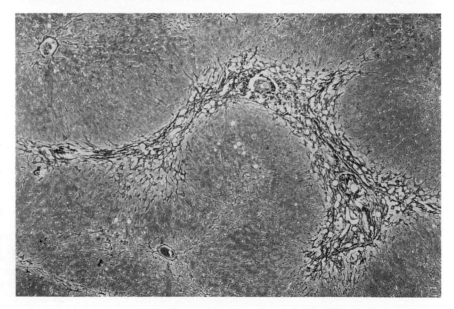

5.22 HBsAg in viral hepatitis: liver

The patient suffered from chronic hepatitis and the portal tract (left) is infiltrated by chronic inflammatory cells. The section has been reacted with an antibody to the surface antigen of the B virus using the indirect immunoperoxidase method, and in the cytoplasm of many of the hepatocytes there is a dark brown mass which represents surface antigen of the hepatitis B virus (HBsAg). Electron microscopy confirms that cells giving this reaction contain HBsAg in the smooth endoplasmic reticulum of the cell. Cytoplasm containing HBsAg in high concentrations have a ground-glass appearance in HE sections and also stain with orcein. In some countries a significant minority of the population are carriers of HBsAg.

Indirect immunoperoxidase method for HBsAg ×235

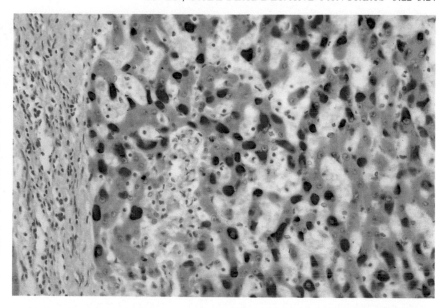

5.23 Subacute viral hepatitis: liver

This was a fatal case of hepatitis B. There has been extensive destruction of hepatocytes with condensation of stroma (centre and bottom) and the connective tissue is infiltrated by chronic inflammatory cells. The surviving liver cells are regenerating, with formation of rounded hyperplastic nodules (left and upper right). There is evidence of stasis of bile, however, in one of the nodules (upper right), with plugging of the canaliculi by bile (thin arrows). A small bile duct is visible (thick arrow).   HE ×150

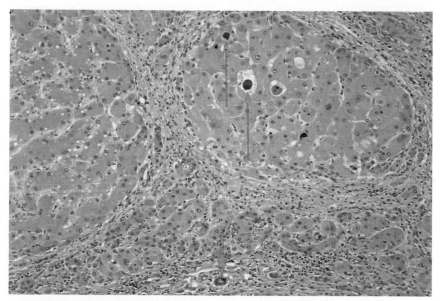

5.24 Chronic active hepatitis: liver

Some patients with acute B-viral hepatitis progress to chronic active hepatitis. Others present with chronic active hepatitis without a history of an acute phase. This patient had had acute hepatitis. The portal tract is enlarged and infiltrated with chronic inflammatory cells which have eroded the limiting plate and extend into the lobule, with necrosis of individual hepatocytes (piecemeal necrosis). Connective tissue stains demonstrated an increase of fibrous tissue in the portal tract. Most of the hepatocytes (right) are swollen (ballooning degeneration) but some individual cells are shrunken and necrotic (acidophil bodies).   HE ×150

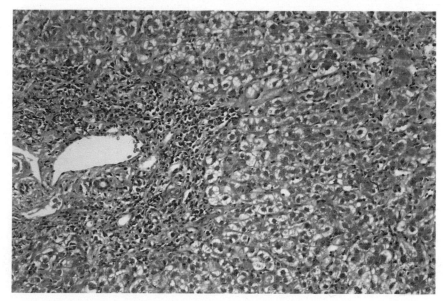

5.25 Chronic active hepatitis: liver

This is the same case as that shown in **5.24**. The hepatocytes adjacent to the portal tract (right) are very swollen and hydropic (severe ballooning degeneration) with only wisps of cytoplasm around the nucleus. A bile canaliculus (thin arrow) is plugged with bile. The number of chronic inflammatory cells in this part of the portal tract with its small bile duct (thick arrow) is comparatively small but fibroblastic activity and an increase in fibrous tissue are evident. The latter features are suggestive of progression to cirrhosis and a poorer prognosis.

HE ×360

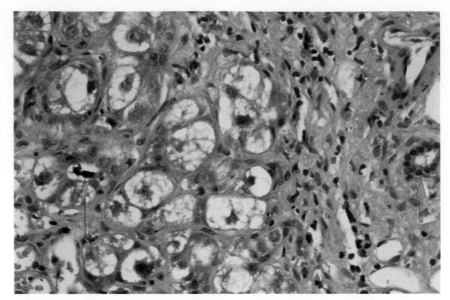

5.26 Early macronodular cirrhosis: liver

The margins of two hyperplastic nodules of liver cells (regenerative nodules) are shown. The cords of hepatocytes are very irregular and some of the hepatocytes are large and contain two or more nuclei with prominent nucleoli. Multinucleated hepatocytes are evidence of regeneration. Large fat droplets are present in some hepatocytes and others show ballooning degeneration. A reticulin stain showed that the liver cell plates were two or more cells thick. A band of mature fibrous tissue (arrow) extends between the nodules from a portal tract (top right). Comparatively few lymphocytes are present in it. The junction between the fibrous septum and the regenerating nodules is well-defined.   HE ×235

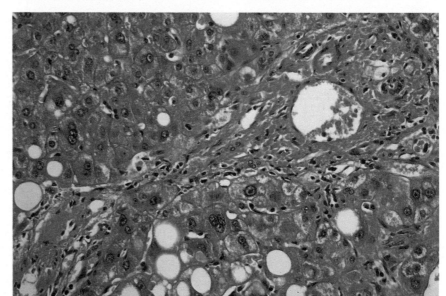

5.27 Ischemia of cirrhotic nodules: liver

In a cirrhotic liver, the blood supply to the nodules of regenerating hepatocytes is often precarious and ischemia may develop, e.g. after a severe hemorrhage from esophageal varices. When the ischemia is less severe, instead of necrosis, fatty change occurs. In this nodule most of the hepatocytes in the centre contain large fat droplets. The cells at the periphery of the nodule are affected only slightly or not at all whereas in alcoholic fatty cirrhosis the whole of the nodule shows fatty change.   HE ×150

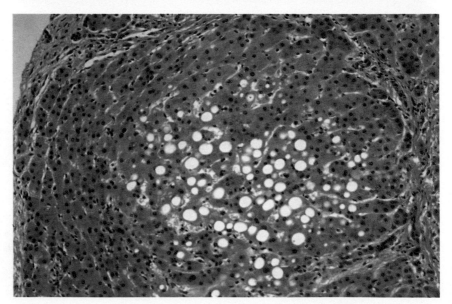

5.28 Cryptogenic cirrhosis: liver

Not infrequently cirrhosis develops for which there is no obvious cause. This is termed cryptogenic cirrhosis and it is often macronodular. This is a section of the liver stained for reticulin from a man who had had portal hypertension for many years before dying of liver failure. The cause of his liver disease was not known. Large pale, sharply-defined nodules of parenchymal cells are enclosed within broad bands of darkly-staining reticulin. The unstained areas in the reticulin are vascular channels. The reticulin framework within the nodules is also well demonstrated and in places the width of the cords of hepatocytes shows that they are two cells thick.

Reticulin stain ×60

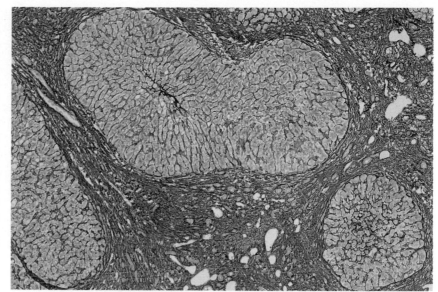

5.29 Focal nodular hyperplasia: liver

A woman of 48 was found, at laparotomy for resection of a carcinoma of colon, to have a whitish lesion 4mm dia in the liver. The lesion was resected. It consists of nodules of hepatocytes of normal appearance separated by the bands of connective tissue. There is no lobular arrangement within the nodules. Bile ductules are present in the connective tissue and also a fairly intense infiltrate of lymphocytes. There is also a collection of lymphocytes in the centre of one of the nodules (right). The nodules merged with normal-looking liver at the periphery and there was no fibrous capsule. Focal nodular hyperplasia may be mistaken for cirrhosis but it is a localized lesion, probably of a hamartomatous nature. It is liable to bleed.

HE ×150

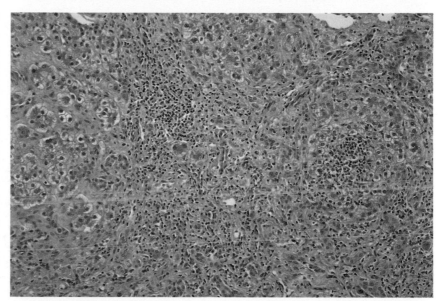

5.30 Idiopathic hemochromatosis: liver

In idiopathic hemochromatosis excess of iron is absorbed from the gut. Much of it is stored in the liver in hepatocytes and bile duct epithelium as hemosiderin. Hepatocytes are destroyed and fibrosis develops which eventually leads to cirrhosis, generally micronodular in type. This specimen has been stained by Perls' method which stains the iron blue (the Prussian blue reaction). Three nodules of regenerating liver cells are shown, separated by broad bands of fibrous tissue. Nearly all the liver cells contain iron in high concentration but the Kupffer cells contain comparatively little. In the bands of fibrous tissue the small bile ducts (arrow) contain iron, and there are collections of iron-laden phagocytes.

Perls' method ×55

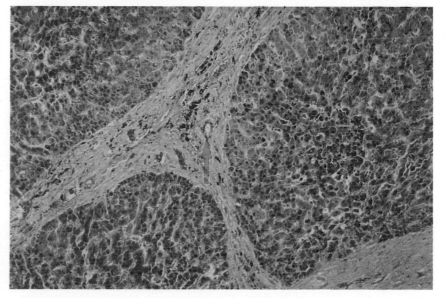

5.31 Drug-induced intrahepatic cholestatis: liver

This shows the centre of a hepatic lobule with the central vein on the lower right. There is very marked centrilobular cholestasis, with many bile canaliculi plugged with inspissated bile. The liver cells appear comparatively normal and there is a notable lack of inflammatory cell infiltrate. There is no evidence of necrosis or hepatitis elsewhere. Centrilobular cholestasis is very characteristic of the jaundice induced by anabolic steroids and oral contraceptives. The cause in this case, a woman of 52, was thought to be chlorpromazine, even though there was no evidence of necrosis of liver cells or of a cellular inflammatory infiltrate in the portal tracts. The condition usually resolves completely when the hormone or drug is stopped.   HE ×360

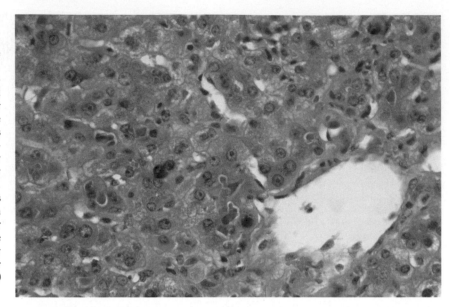

5.32 Intrahepatic biliary atresia: liver

This is not a single entity, since damage to the intrahepatic bile ducts can occur in a number of different conditions including viral infections of the liver and cholangitis. The lesion also occurs in association with a variety of chromosomal defects. This is the liver of a child who had long-standing cholestatic jaundice. Inspissated bile plugs the bile canaliculi (arrows). There is very extensive fibrosis which involves portal tracts. However, no interlobular bile ducts are visible in the fibrous tissue and very few inflammatory cells are present.   HE ×60

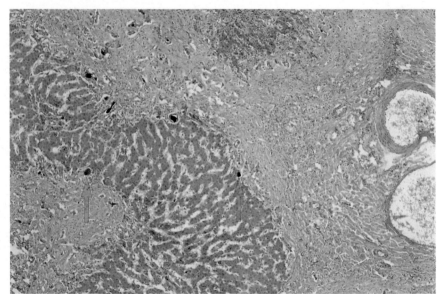

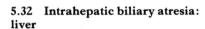

5.33 Extrahepatic bile duct obstruction: liver

In extrahepatic bile duct obstruction there is often evidence of cholangitis and portal inflammation. Secondary infection, usually low grade, tends to occur. The portal tract is wider than normal, with loose edematous connective tissue arranged typically in concentric layers around an interlobular bile duct filled with inspissated bile (left of centre). The epithelial lining of the duct is flattened and deficient in places. The number of inflammatory cells in the connective tissue is fairly small, but at one point where there is loss of the limiting plate, they are fairly numerous. Proliferation of small ducts is evident (arrows).   HE ×75

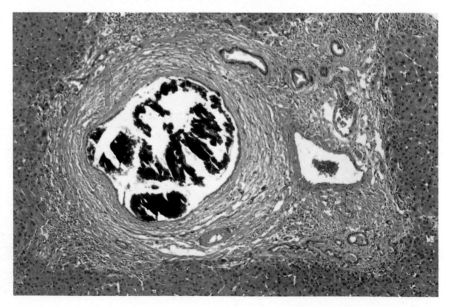

5.34 Extrahepatic bile duct obstruction: liver

Many of the bile canaliculi (arrows) are plugged with bile, and there is a so-called bile lake, an area of necrotic liver (bile infarct) into which extravasated bile has seeped. The cytoplasm of macrophages and some atrophic hepatocytes adjacent to the bile lake also contain bile. Like bile infarcts, bile lakes are usually located near portal tracts. They tend to be found in large duct obstruction but they may be found also in primary biliary cirrhosis and alcoholic cirrhosis. HE ×360

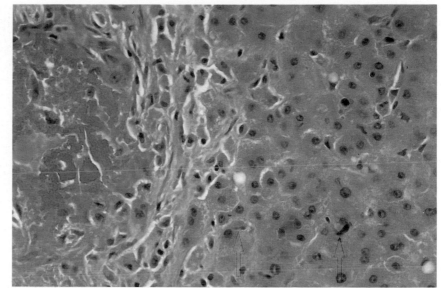

5.35 Extrahepatic bile duct obstruction: liver

This patient had long-standing obstruction of the common bile duct. Chronic cholangiohepatitis has resulted in the formation of edematous fibrous tissue in the portal tracts, which are accordingly increased in width. The periphery of the lobules is clearly delineated but proliferation of the bile ductules has occurred. Characteristically the centrilobular zones are unaffected. The surviving liver has the appearance of a piece of a jigsaw. There is a tendency for the widened portal tracts to coalesce, to produce secondary biliary cirrhosis of a monolobular type. HE ×60

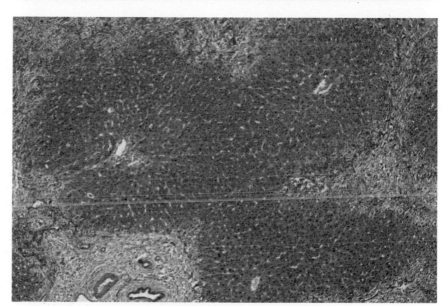

5.36 Extrahepatic bile duct obstruction: liver

This is the lesion (secondary biliary cirrhosis) illustrated in **5.35**. It shows the junction of one of the greatly widened portal tracts and the liver parenchyma. Within the edematous cellular connective tissue (left) in the portal tract there are many basophilic proliferating bile ductules and small numbers of chronic inflammatory cells. The centrilobular zone (arrow) is unaffected. HE ×150

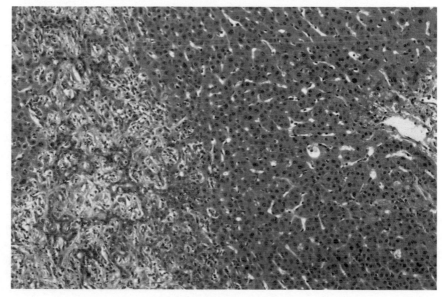

5.37 Primary biliary cirrhosis: liver

In primary biliary cirrhosis an inflammatory process causes degeneration of the epithelium of the smaller (interlobular and septal) intrahepatic bile ducts. There is evidence of an autoimmune reaction and the serum generally contains an anti-mitochondrial antibody. Later, necrosis of liver cells occurs, probably as a result of cholestasis, and cirrhosis eventually develops. There is an infiltrate of chronic inflammatory cells in the portal tracts which are broadened. This varies in intensity from one tract to another. Small bile ducts and ductules are absent but there is neither cholestasis nor significant damage to the liver parenchyma apart from a few acidophil cells (necrotic hepatocytes, arrows). There is piecemeal necrosis in the most severely affected portal tract (right).　　HE ×60

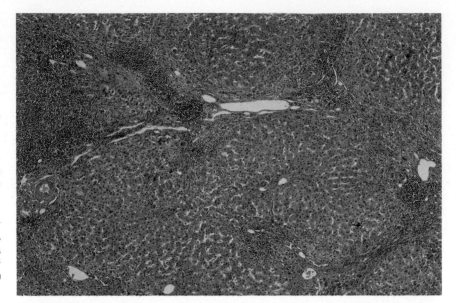

5.38 Primary biliary cirrhosis: liver

This shows one portal tract at higher magnification. It is infiltrated with lymphocytes and plasma cells. One interlobular duct is seen in cross-section (arrows). Its epithelial cells are very swollen and the wall is infiltrated with lymphocytes. Some cells are lying in the lumen. There is ballooning degeneration of some hepatocytes (top right) and some erosion of the adjacent limiting plate. The distribution of the duct lesions is irregular in primary biliary cirrhosis, and the large intrahepatic and extrahepatic ducts are unaffected.　　HE ×360

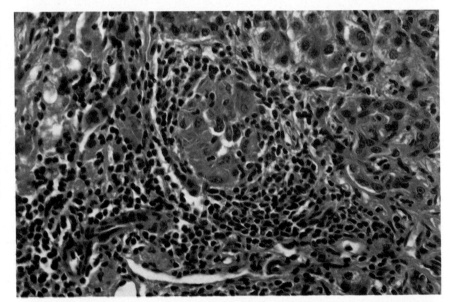

5.39 Primary biliary cirrhosis: liver

In addition to the many lymphocytes and plasma cells in this portal tract there is a small round granuloma composed of large macrophages and occasional lymphocytes (left). There is no necrosis in the granuloma which has developed in relation to a damaged bile duct. The hepatocytes (top) adjoining the portal tract show ballooning degeneration and piecemeal necrosis.
　　HE ×360

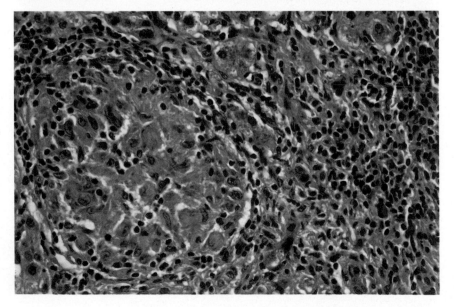

5.40 Primary biliary cirrhosis: liver

As the disease progresses, proliferation of bile ductules is often seen. In this broadened portal tract no bile ducts remain, but large numbers of basophilic proliferating bile ductules are present in the loose connective tissues, along with many lymphocytes and plasma cells. There is some piecemeal necrosis (arrow).   HE ×235

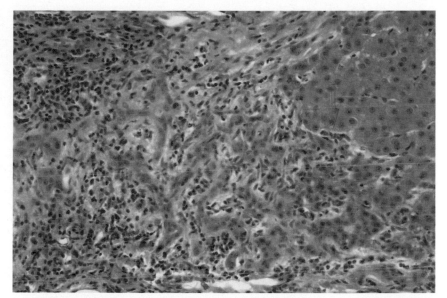

5.41 Primary biliary cirrhosis: liver

This is a more advanced stage of the disease. Fibrous septa have formed, joining up adjacent portal tracts but leaving the centrilobular regions unaffected, to produce a pattern of monolobular fibrosis similar to that seen in extrahepatic bile duct obstruction. There is patchy lymphocytic infiltration of the portal tracts. Proliferation of bile ductules is not a feature. Primary biliary cirrhosis can usually be distinguished from extrahepatic bile duct obstruction by the absence of bile ducts and the presence of a more intense infiltrate of lymphocytes in the portal tracts.
          HE ×60

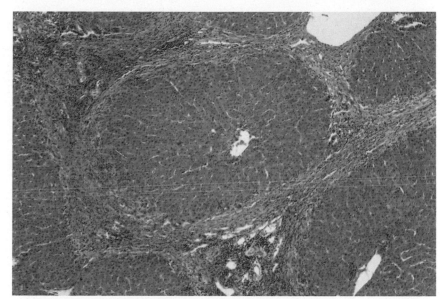

5.42 Ulcerative colitis: liver

Lesions are not infrequently seen in the liver in patients with chronic inflammatory bowel disease. A man of 60, with severe long-standing ulcerative colitis affecting the whole colon and rectum but now inactive, developed sclerosing cholangitis which caused fibrosis and narrowing of the common bile duct. There is also marked increase in fibrous tissue within the liver in the portal tracts, around the interlobular bile ducts (left). The fibrous tissue is fairly heavily infiltrated with lymphocytes and these are extending into the adjacent nodule of hepatic parenchyma. There is severe cholestasis, with plugging of bile canaliculi by bile and also bile staining of the hepatocytes, many of which show ballooning degeneration. Some of the hepatocytes have two or more nuclei. The changes in the liver were diagnosed as secondary biliary cirrhosis.   HE ×150

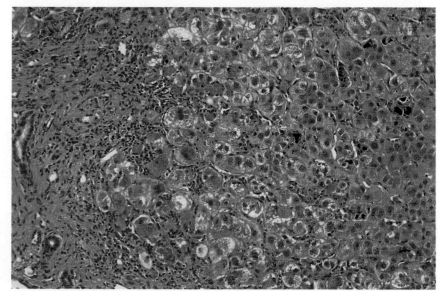

5.43 Liver cell adenoma: liver

Liver cell adenoma (benign hepatoma) is very uncommon but it is increasing in incidence, probably because of the use of the contraceptive pill and sex hormone therapy. The cells of the adenoma closely resemble normal hepatocytes but tend to be slightly larger. They form trabeculae two or three cells thick. There is little or no mitotic activity. There are no bile ducts or portal tracts but bile canaliculi can be detected. A fibrous capsule is usually present but it is generally incomplete, and tumour cells do not invade the adjacent parenchyma. A notable feature of the lesion is its vascularity which may lead to serious intraperitoneal hemorrhage. Two areas of hemorrhage (bottom) are seen here.   HE ×150

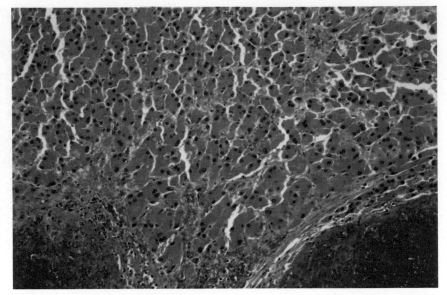

5.44 Hepatocellular carcinoma: liver

The incidence of hepatocellular carcinoma (hepatoma) varies markedly in different parts of the world, being relatively much more common in parts of Africa and Asia. The great majority of cases arise in cirrhotic livers, and men are affected more often than women. This lesion was in the cirrhotic liver of a man of 67. It consists of trabeculae of large polyhedral cells between which lie thin-walled sinusoids (centre left). The cells have round vesicular nuclei in which there is a very prominent central nucleolus. The cytoplasm is eosinophilic and vacuolated. The trabeculae are broad and several cells thick in places. There is some acinar formation, the acini containing bile-stained debris (arrows). Glycogen could be demonstrated in the cells but an orcein stain for HBsAg was negative.   HE ×360

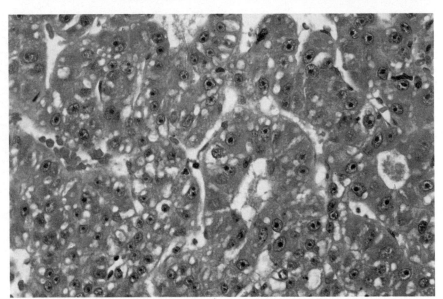

5.45 Hepatocellular carcinoma: liver

This tumour was in a woman of 41. The tumour caused enormous enlargement of the liver, which weighed 6,120g. There was no evidence of cirrhosis. It consisted of multiple rounded masses, several of which showed extensive necrosis and hemorrhage. Histologically there was a wide range of structure. Much of the tumour consisted of solid trabeculae but acinar formation was common; and in this part of the tumour cells form irregular groups. The cytoplasm of the cells contains large vacuoles which had contained fat. It was also possible to demonstrate abundant glycogen in them. Thin-walled vascular channels are present. This type of structure is reminiscent of renal carcinoma (hypernephroma).   HE ×360

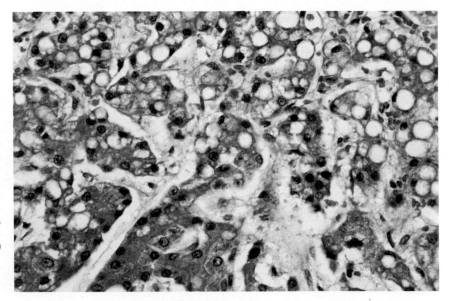

5.46 Hepatocellular carcinoma: liver

A woman of 72 with no history of ethanol abuse and no evidence of jaundice was admitted to hospital because of the recent onset of ascites and persistent diarrhea. She died 6 weeks later, and post-mortem examination revealed cirrhosis and a hepatocellular carcinoma. The tumour was bile-stained in places. It is a poorly differentiated carcinoma, composed of cells showing marked nuclear pleomorphism, with frequent mitoses. The cytoplasm is relatively scanty. In this area there is a tendency to acinar formation, the centres of some acini containing cells or cell debris. HBsAg was not detectable in the tumour cells. The tissue at the bottom is compressed and rather atrophic liver.   HE ×235

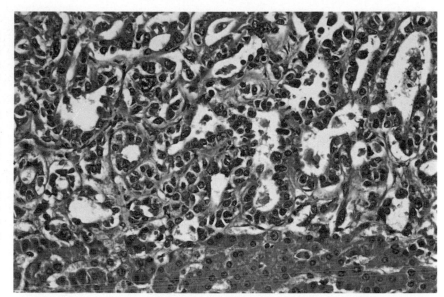

5.47 Cholangiocarcinoma: common bile duct

Cholangiocarcinoma most often arises in the porta hepatis at the junction of the right and left hepatic ducts. The presenting sign is usually jaundice. This tumour, however, was a small mass within the common bile duct near the ampulla of Vater. It is a papilliferous adenocarcinoma, the deeply basophilic columnar epithelial cells of the tumour forming small and large ductular structures into which long finger-like papilliform processes project. Some of the ductular structures contain considerable amounts of cell debris. Cholangiocarcinomas are generally sclerosing (desmoplastic) tumours and characteristically there is abundant fibrous stroma between the duct-like structures of this tumour. The stroma is also heavily infiltrated with lymphocytes.   HE ×120

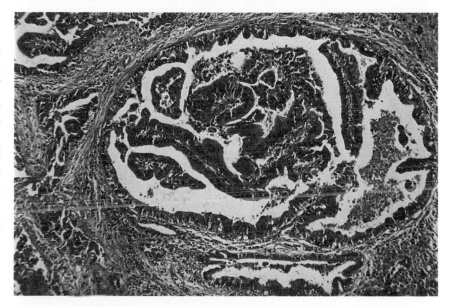

5.48 Secondary melanoma: liver

The liver is a common site for secondary tumours. The metastases are usually multiple and often large. Sometimes the malignant cells infiltrate the liver diffusely, without forming large masses. In this case the tumour cells are lying within the sinusoids. They have only a moderate amount of cytoplasm and nuclei (thin arrows) which are generally larger than the hepatocyte nuclei (thick arrows). There is a prominent central nucleolus in the tumour cell nuclei. Nuclear pleomorphism is comparatively slight, however, and no mitotic figures are present in this field. The brown pigment (top right) in the cytoplasm of the hepatocytes is hemosiderin. A special stain did reveal small quantities of melanin within the tumour cells, confirming that the deposit was metastatic malignant melanoma.

HE ×360

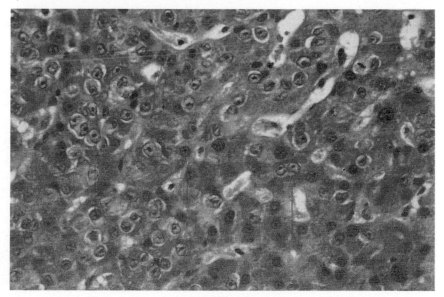

5.49 Cholesterolosis: gall bladder

The lining of this gall bladder from a man of 72 was yellow from the presence of small bright yellow deposits in the folds of the mucosa (strawberry gall bladder). This shows one of the folds in cross-section. The surface is covered with columnar epithelium but the fold is greatly swollen from the presence within it of large macrophages with abundant pale finely-granular ('frosted-glass') cytoplasm. The frosted-glass appearance is caused by the presence of lipid, and it is the lipid which gives the mucosal folds their yellow colour. Cholesterolosis precedes formation of gallstones and gallstones are usually present in gall bladders showing cholesterolosis.   HE ×320

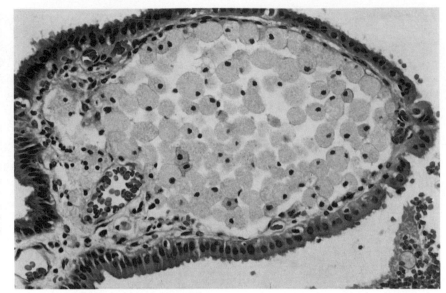

5.50 Chronic cholecystitis: gall bladder

The gall bladder was small, with a thickened opaque fibrous wall, and contained many 'mixed' gallstones. The mucosa (left) is chronically inflamed and thickened. The epithelial cells are full of mucin, the normal mucosal folds are lacking and there is a dense infiltrate of inflammatory cells (lymphocytes, plasma cells and eosinophil leukocytes). The muscle coat (right) is hypertrophied and infiltrated by small numbers of chronic inflammatory cells. There is however little increase in fibrous tissue.

HE ×120

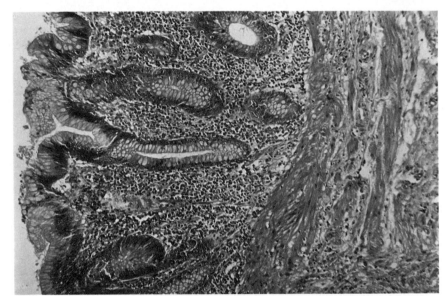

5.51 Acute hemorrhagic pancreatitis: pancreas

The etiology and pathogenesis of acute pancreatitis are still uncertain but most cases are associated with disease of the biliary tract, trauma or alcoholism. In the hemorrhagic form there is extensive necrosis of the organ, and pancreatic enzymes, particularly lipase and trypsin, are released in an active form into the pancreas itself and into adjacent tissues such as the omentum, where they cause necrosis. In this focus the exocrine tissue of the pancreas (left) has been spared to a large extent whereas the adjacent fatty tissue (centre and right) has disintegrated, leaving only cell debris and hemorrhage. Necrotic fat has a high affinity for calcium which quickly deposits in it, and the basophilia of the necrotic tissue (bottom) is attributable to its presence.   HE ×135

5.52 Chronic sclerosing pancreatitis: pancreas

In this form of chronic pancreatitis, which is associated with alcoholism, the pancreas becomes hard and nodular. It is sometimes larger than normal but is more often reduced in size. Histologically there is marked increase in fibrous tissue, with bands of collagenous tissue around and within the pancreatic lobules (interlobular and intralobular fibrosis). The fibrous tissue contains a mixture of lymphocytes, plasma cells and polymorph leukocytes. The exocrine tissue is abnormal and the acinar epithelium is flat or cuboidal instead of columnar. The lumen of the glands is much increased. It may be difficult to differentiate chronic pancreatitis from carcinoma of the pancreas but the normal appearance of the nuclei of the epithelial cells in pancreatitis is a helpful feature.

HE ×150

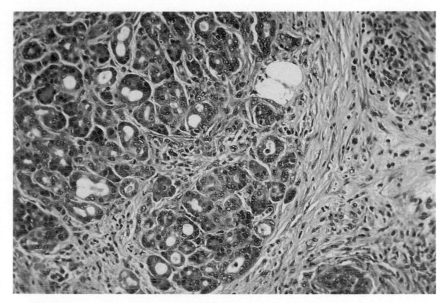

5.53 Cystic fibrosis: pancreas

In cystic fibrosis (fibrocystic disease) there is an abnormality of the secretory activity of exocrine glands in many organs, including pancreas, liver, alimentary tract, lungs and sweat glands. The pancreas becomes small, firm and gritty. This shows one lobule of pancreas. The ductules and acini are lined by atrophic flat cells and the lumen distended with dense laminated secretion. Dense fibrous tissue has formed around the lobule and between the individual glands (interlobular and intralobular fibrosis). Similar secretion filled the ducts. Loss of the exocrine secretion of the pancreas produces a malabsorption syndrome. HE ×160

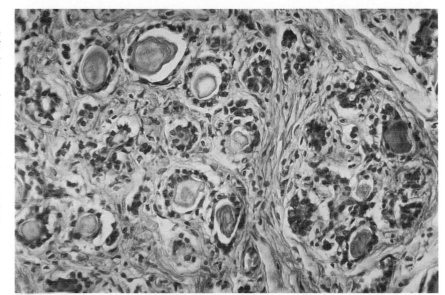

5.54 Carcinoma: pancreas

Carcinoma of pancreas is a not uncommon tumour which arises most frequently in the head of the pancreas and less often in the body. It rarely occurs in the tail of the organ. It usually obstructs the main ducts of the pancreas, so that the pancreatic secretions fail to reach the duodenum and malabsorption results. The common bile duct also is obstructed and jaundice develops. It is usually a scirrhous tumour, histologically an adenocarcinoma. In this example the tumour cells are pleomorphic and form cords and acini, lying in abundant densely fibrous stroma. The acini, though rudimentary, are reminiscent of the acini of the exocrine tissue of the pancreas. HE ×135

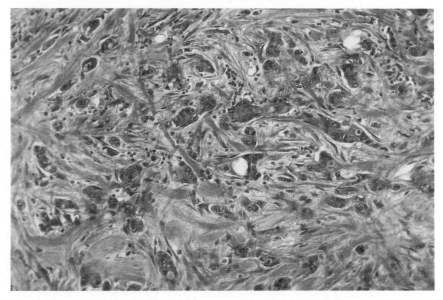

6.1 Atheroma: coronary artery

Atheroma is a patchy thickening of the intima of arteries, caused mainly by the deposition of lipid and fibrous tissue. This is a coronary artery from a middle-aged man who suffered from severe angina pectoris. The lumen is reduced to less than half normal diameter by a greatly thickened intima. The intima nearest the lumen consists of dense fibrous tissue, apart from one area rich in lipid-filled macrophages (thin arrow). Deep in the intima on the right and pressing on an atrophic media (thick arrow) are a crescent-shaped mass of eosinophilic material and large numbers of small clefts which contained cholesterol crystals. The gap deep in the intima on the left contained similar material. The dense fibrous tissue beneath this focus is patchily calcified (double arrow). The media is breached under the calcified areas.   HE ×15.

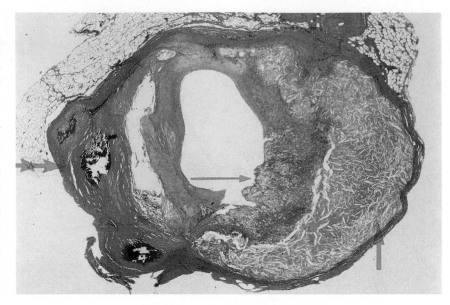

6.2 Atheroma: coronary artery

This is the same coronary artery as shown in **6.1**. The lumen of the artery is at the top right corner, and the band of smooth muscle at the bottom is the atrophic media. The intima is enormously thickened, by the presence deep in it (centre and left) of amorphous material containing large numbers of cholesterol crystals (the unstained clefts). There are many foamy (lipid-filled) macrophages and chronic inflammatory cells in this zone and also in the thick layer of dense fibrous tissue layer (arrow) which separates it from the lumen.   HE ×60

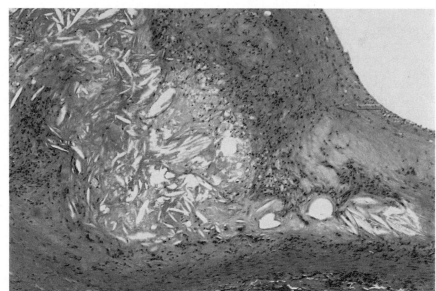

6.3 Atheroma: coronary artery

The lumen of the artery is on the right. The tissue lining the artery was for the most part dense fibrous tissue (**6.1**) but in this area it consists of large macrophages with abundant finely-vacuolated (foamy) cytoplasm, accompanied by strands of fibrous tissue. Similar cells were present in the cholesterol-rich material deep in the intima. The origin of the lipids of atheromatous plaques and their mode of entry into the intima are uncertain but their composition suggests an origin from the plasma.   HE ×150

6.4 Atheroma: coronary artery

This shows the junction of the intimal plaque (left) and the media (right). The internal elastic lamina (arrows) is wrinkled and although broken in many places elsewhere is intact. It has reduplicated to form a second thinner elastic lamina. Fibrous tissue has replaced much of the muscle in the media (right). The plaque (left) consists of dense fibrous tissue, much of it acellular and hyaline. Special stains would reveal the presence of both fibrin and lipid. A notable feature is the presence of small blood vessels. Similar vessels are present in the media, and it is likely that the intimal vessels are extensions of these. Hemorrhage from small blood vessels in intimal plaques can cause a rapid increase in the size of the plaque and consequent reduction in the lumen of the vessel, thereby simulating thrombotic occlusion.   IIE ×235

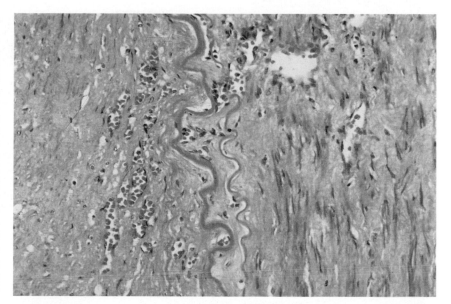

6.5 Aneurysm: femoral artery

The media beneath an atheromatous plaque tends to atrophy (**6.1, 6.2**) and a severely atheromatous artery may be so weakened as to undergo aneurysmal dilatation. Atheroma is now the commonest cause of aneurysm of aorta in many countries. The aneurysm is usually fusiform. A similar lesion may affect other large arteries and this one was in the femoral artery near its origin. The muscle fibres of the medial coat (right half) have been completely replaced by dense fibrous tissue. Small blood vessels are present, as well as many small lymphocytes. The lumen of the aneurysm (left half) contains thrombus which consists largely of red-stained fibrin (left). The lipid-rich zone, with many cholesterol clefts, between the fibrin and the fibrous wall of the aneurysm is probably the remnant of a large atheromatous plaque.   HE ×135

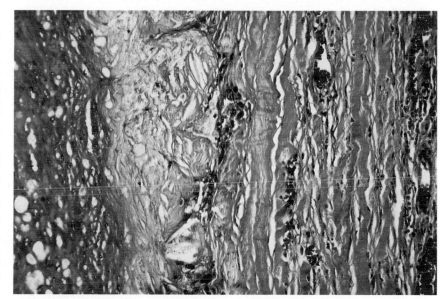

6.6 Congenital ('berry') aneurysm: cerebral artery

'Berry' aneurysms are not congenital but there is probably a defect in the medial coat of the artery at the site which leads to aneurysm formation if the internal elastic lamina degenerates. This aneurysm and the associated vessels were removed surgically from a man of 23. It was a saccular aneurysm 2cm dia. The wall consists of an inner layer of dense hyaline collagen (left of centre) and an outer layer (right) of loose vascular connective tissue. Chronic inflammatory cells are also present in the latter and hemosiderin was demonstrated. There are deposits of calcium salts in the hyaline layer. No medial muscle fibres or elastic laminae remain. The lumen of the aneurysm (left) contains thrombus which is adherent to the hyaline fibrous tissue. There is no endothelial lining.

HE ×150

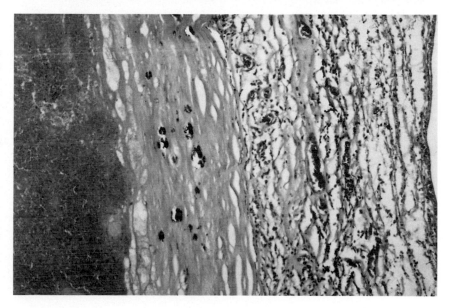

6.7 Arteriolosclerosis: spleen

Arteriolosclerosis tends to affect older people but develops earlier and is more severe when systemic hypertension is present. It is often particularly severe in individuals with diabetes mellitus. The organs most involved are kidney, pancreas, liver and spleen, but with increasing age the arterioles of the spleen and to a lesser extent of the renal glomeruli are liable to undergo hyaline thickening of their walls even in the absence of systemic hypertension. Arteriolosclerosis reduces the blood flow to the tissues. In this spleen, acellular hyaline material has been deposited beneath the endothelium of several arterioles (possibly one tortuous vessel cut three times). The lumen of the vessel is greatly reduced. The muscle fibres of the media are atrophic. The surrounding tissue consists of the small lymphocytes of the white pulp.　　HE ×250

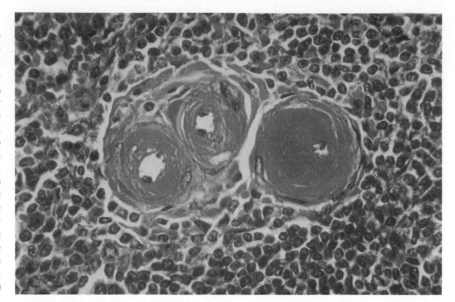

6.8 Giant cell arteritis

Giant cell arteritis is focal granulomatous inflammation of the wall of the artery, with destruction of the muscular and elastic tissues. The temporal arteries are the commonest site. The patient is usually over 50 years of age. This is the temporal artery from a man of 65 who complained of headaches and tenderness in the temporal region. The lumen was full of thrombus which was being organized. This shows the tunica media and intima (the lumen is out of the picture at the top). The internal elastic lamina and most of the muscle of the media have been destroyed, only a few muscle fibres (cut in longitudinal section) remaining (bottom). In their place is a cellular granulomatous tissue consisting of histiocytes, multinucleated giant cells, lymphocytes and plasma cells. At higher powers fragments of elastic laminae could be detected.　　HE ×235

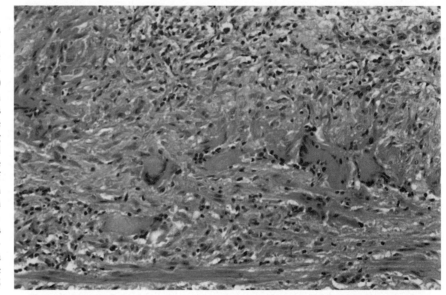

6.9 Giant cell arteritis: femoral artery

This was an endartectomy specimen 17cm long from a woman of 63. The lumen of the artery was full of thrombus. This shows the inner two-thirds of the wall of the artery. The intima (right) is thick and fibrous. The media has been destroyed and replaced by a granulomatous tissue which contains large numbers of lymphocytes, plasma cells, histiocytes and multinucleated giant cells, many of them of Langhans type. Surprisingly the internal elastic lamina was intact.　　HE ×150

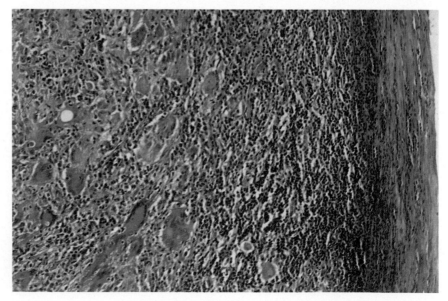

6.10 Polyarteritis nodosa: artery

Polyarteritis nodosa affects medium-sized muscular arteries and arterioles. Foci of necrosis develop in the wall accompanied by an acute inflammatory response. Thrombosis and vascular occlusion often develop, with ischemia and frequently infarction of the tissues supplied. The vessel may rupture, with severe hemorrhage. This artery is in the healing phase. Inflammatory cells, mostly lymphocytes, are still numerous in the adventitia but the necrotic remnants of the arterial wall have been removed by phagocytes. Parts of the muscular media (thin arrow) and of the internal elastic lamina (thick arrow) have survived. The vessel has thrombosed, and organization of the thrombus has led to the formation of connective tissue which fills the original lumen. The small channel in the centre is the result of partial recanalization.   HE ×270

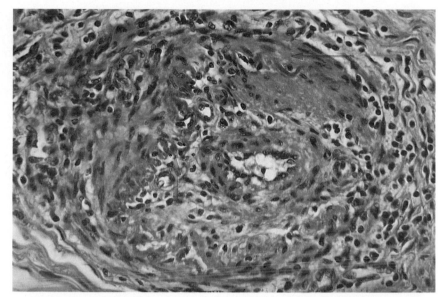

6.11 Thrombotic thrombocytopenic purpura (Moschowitz's syndrome): heart

A girl of 16 suffered from fever, purpura, hemolytic anemia and vague neurological signs. This is the myocardium. Hemorrhage (top left) has occurred into the edematous interstitial tissues of the heart. The most notable feature, however, is the presence of eosinophilic material within the small blood vessels. There is no inflammatory reaction in the vicinity. These are fibrin thrombi and similar microthrombi were present in small vessels in other organs but not in large vessels. The thrombi resulted from intravascular activation of the blood-clotting mechanism or to vascular damage (thrombotic microangiopathy); and fragmentation of red cells by the fibrin strands caused the hemolytic anemia (microangiopathic hemolytic anemia). HE ×135

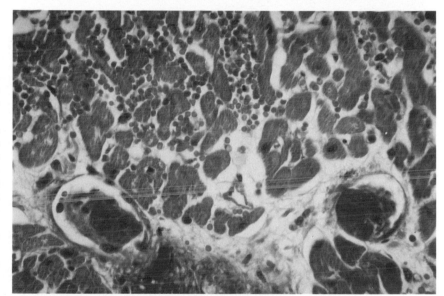

6.12 Progressive systemic sclerosis: artery

Progressive systemic sclerosis, a rare disease of unknown etiology, invariably involves small arteries and arterioles. There is intimal thickening and reduction of blood flow, with atrophy and loss of specialized tissues and replacement by fibrous tissue. This is a digital artery. The finger was ulcerated and gangrenous at its tip. The lumen is blocked by dense hyaline collagenous tissue. The internal elastic lamina is very convoluted, suggesting that the artery is contracted. It is also broken at several points. The media is atrophic and lightly infiltrated by lymphocytes. The adventitia is very thick and fibrous and there is a moderate infiltrate of chronic inflammatory cells near the media.
HE ×120

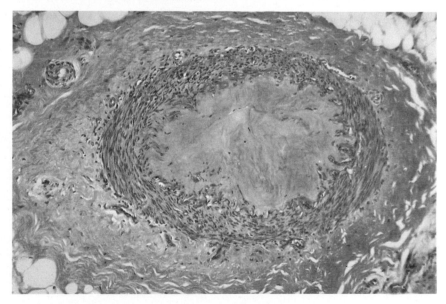

6.13 Thrombosis: coronary artery

The right main coronary artery of this patient was suddenly blocked by thrombosis. He died 4 hours later. The intima of the artery (bottom) is atheromatous and the part adjoining the lumen (top) is thick and fibrous. Thrombus occupies the lumen of the artery (centre and top). It is adherent to the intima at one point (thin arrows). It consists of pale-staining finely granular masses of fused platelets (thick arrows) and eosinophilic strands of fibrin. Trapped leukocytes are present throughout the thrombus, along with collections of erythrocytes.

HE ×150

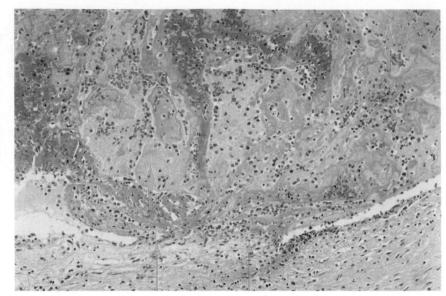

6.14 Thrombosis: coronary artery

Platelets are a major component of thrombus that forms in an artery and they initiate the process of thrombus formation. Fibrin strands however usually form alongside the aggregates of platelets. The resulting thrombus is pale. Only when the blood flow slows does the content of red cells increase to significant proportions. This is the same vessel as in **6.13**, showing part of the thrombus occupying the lumen at higher magnification. The pale-staining granular material (thin arrow) consists of fused platelets, and the tortuous fibrillary material is fibrin (thick arrows). The leukocytes include large mononuclear cells (monocytes, lymphocytes and polymorph leukocytes). Red cells are present, the densely eosinophilic material on the left consisting of fused red cells.

HE ×360

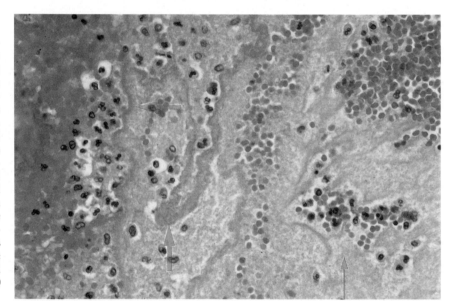

6.15 Infarct: heart

Infarction of the myocardium is caused by acute ischemia, following sudden narrowing of a coronary artery. In most cases the artery is blocked by occlusive thrombus which has formed over an atheromatous plaque. In this case the infarct was one week old, as estimated from the clinical history. This shows the edge of the infarct. The necrotic muscle fibres (left half of picture) stain a deeper red than the bands of living muscle on the right. The necrotic fibres retain their shape but the nuclei no longer stain. Between the living and dead muscle there is a richly cellular and vascular tissue containing many polymorphs, macrophages and fibroblasts. If the patient had survived, the dead muscle would have been digested by the phagocytes and the fibroblasts would have formed a fibrous scar.

HE ×55

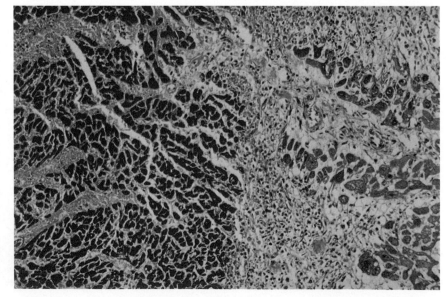

6.16 Infarct: heart

This is the same lesion as that in **6.15**, showing the necrotic muscle fibres of the infarct at higher magnification. There is no nuclear staining, the nuclei having disappeared by karyolysis. The fibres themselves are more deeply eosinophilic than normal fibres but the staining is patchy, with areas of pallor. The striations, however, are still detectable. In the interstitial tissue there are, in addition to some nuclear fragments, macrophages which have migrated into the dead muscle. If the patient survives long enough, all the necrotic muscle even in a large infarct is removed by phagocytosis.　　HE ×135

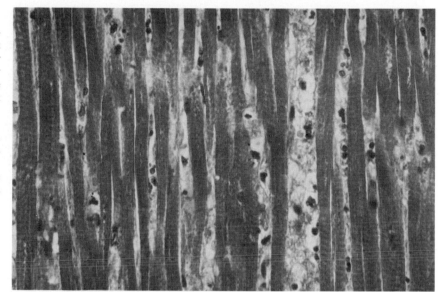

6.17 Chronic myocardial ischemia: heart

Not uncommonly, parts of the myocardium are found in which the muscle fibres are partly replaced by fibrous tissue. Characteristically, as shown here, the surviving muscle fibres are located mainly around blood vessels and whilst some of the fibres are thin and atrophic, others appear hypertrophied. The fibrous tissue (lower left and upper right) is fairly cellular. In cases like this the coronary arteries are invariably severely atheromatous, with the lumen much reduced in size. It is assumed, therefore, that the changes in the myocardium are the result of chronic ischemia. However it is difficult, if not impossible, to exclude the possibility of a previous undetected infarction as the cause of the changes.　HE ×150

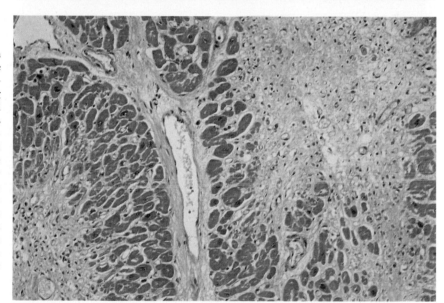

6.18 Aneurysm: heart

When an infarct of myocardium heals, the necrotic muscle is replaced by fibrous tissue. Sometimes the fibrous tissue fails to withstand the high pressure of the blood in the left ventricle and stretches. An aneurysm of the ventricle results. The patient, a man of 42, developed an aneurysm of the left ventricle 6 months after coronary thrombosis and infarction of the myocardium. The aneurysm was resected surgically. Two elliptical pieces of the wall of the aneurysm were removed, one 7 × 3.5cm and the other 7 × 2cm. The endocardial surface of both pieces appeared white and fibrosed. The cavity of the ventricle is on the left. The wall of the aneurysm consists largely of dense collagenous fibrous tissue, with only a few scattered muscle fibres remaining.　　HE ×60

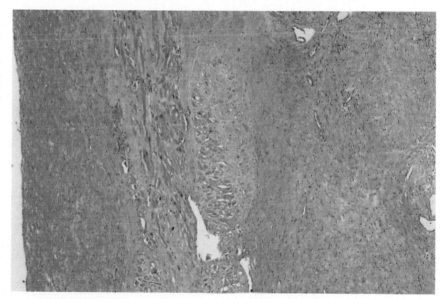

6.19 Organizing thrombus: renal artery

Thrombosis of a renal artery is uncommon. Atheroma is an important predisposing factor and in diabetes mellitus atheroma of the renal arteries and their main branches is often severe. A man of 53 with systemic hypertension had an atrophic right kidney removed surgically. This is the renal artery. The wrinkled internal elastic lamina and the inner part of the media are visible on the right. They show no abnormality. No intima or endothelium is visible, and the lumen (centre and left) is occupied by a loose, rather myxoid, connective tissue. There are small fragments of eosinophilic material which are probably fibrin (arrow).

HE ×150

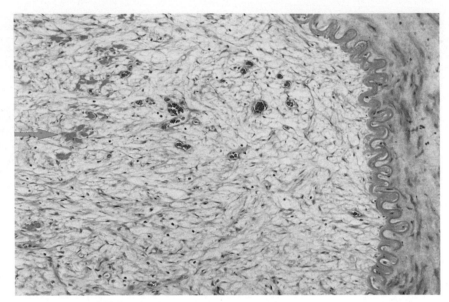

6.20 Organizing thrombus: renal artery

At higher magnification the internal elastic lamina is seen to be reduplicated (arrow) but not broken. There is no endothelium or distinct intima. The connective tissue in the lumen (centre and left) is pale-staining and myxoid, from the presence of much connective tissue mucin. Special stains would show abundant reticulin but relatively few mature collagen fibres. The blood vessels are thin-walled and widely patent. A small amount of hemosiderin is present (top left). The cells are elongated fibroblasts and round macrophages. This renal artery had thrombosed some months prior to the removal of the kidney and the connective tissue in the lumen has been produced by organization of the thrombus. *See also 1.28.*

HE ×235

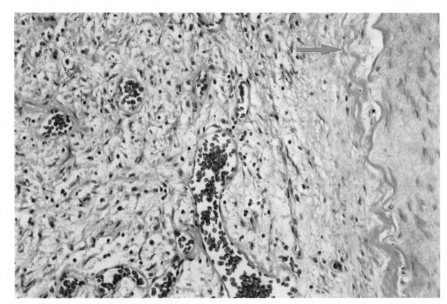

6.21 Syphilitic mesaortitis: aorta

The aorta is often affected in tertiary syphilis. The disease attacks the thoracic aorta, causing inflammation of the small blood vessels (vasa vasorum) in the adventitia which supply the media of the vessel. This shows the outer half of the aortic wall, the adventitia being just out of sight on the right. The vasa vasorum are dilated and surrounded by a cellular infiltrate consisting almost entirely of plasma cells. The musculo-elastic laminae in the vicinity of the vessels have been disrupted by the inflammatory reaction. The reduction in blood supply to the media as a result of obliterative endarteritis of the vasa vasorum is also destructive of the media and further weakens the wall of the aorta.

HE ×235

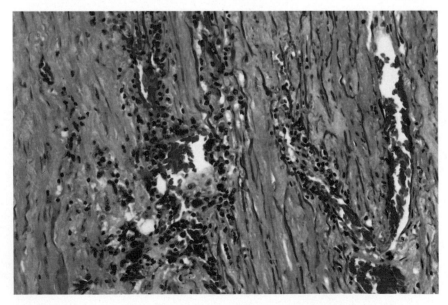

6.22 Rheumatoid arthritis: aorta

Rheumatoid arthritis is a systemic disease, one of the group of connective tissue diseases. An acute necrotizing arteritis which tends to cause thrombosis of the vessel and possibly infarction, occurs frequently in some members of this group. A man of 38 with a history of rheumatoid arthritis had his aortic valve removed surgically for aortic incompetence. This shows the inner media of the aorta and part of the intima on the left. The intima is thick and fibrous, and a small blood vessel and chronic inflammatory cells are present in it. There is an infiltrate of chronic inflammatory cells in the media, surrounding the vasa vasorum (arrow), and marked destruction of the eosinophilic musculo-elastic laminae in the vicinity of the inflamed vessels. Similar changes were present in the outer half of the media.

HE ×150

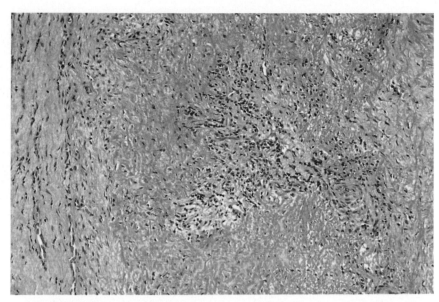

6.23 Rheumatoid arthritis: aorta

This shows the inflammatory reaction around the vasa vasorum (periarteritis) in the media at higher magnification. The reaction is granulomatous, consisting of mononuclear cells of various types, including histiocytes, lymphocytes and small numbers of plasma cells. The vessel at the centre of this reaction is difficult to detect, since it has been cut longitudinally and its lumen has been practically occluded by endarteritis. Fragments of elastic tissue from the disrupted medial laminae are visible (arrows).

HE ×360

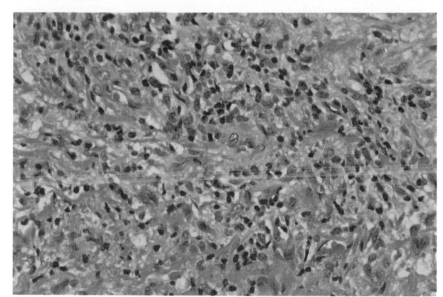

6.24 Rheumatoid arthritis: aorta

The inflammatory reaction centred on the vasa vasorum and causing periarteritis and endarteritis has a destructive effect on the musculo-elastic laminae of the medial coat of the aorta, similar to that in syphilitic mesaortitis. This is made more obvious by stains for elastic tissue. This is the outer two-thirds of the aorta, stained to show elastic fibres (black) and collagen fibres (purple-red). All the elastic laminae are disrupted and broken into small fragments, the gaps being filled by collagenous connective tissue.

Elastic-van Gieson ×150

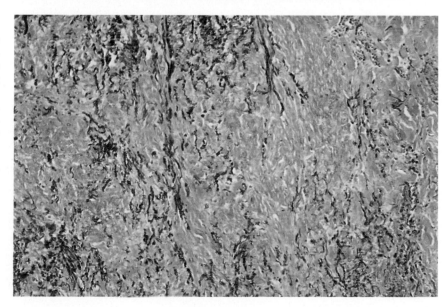

6.25 Medial degeneration and dissecting aneurysm: aorta

Degenerative changes are often present in the media of the aorta in middle-aged and elderly people, with loss of elastic tissue and muscle. An aorta affected in this way is liable to rupture, allowing blood to enter the media and track within it, splitting it into an outer and inner layer: a dissecting aneurysm. A man of 50 developed a dissecting aneurysm of the thoracic aorta. The blood had entered the media through a slit-like opening 5cm above the aortic valve. The adventitia is at the top and the inner media at the bottom. The media has been split by the aneurysm, the lumen of which is visible (centre left). Blood is tracking beyond the lumen, separating the elastic laminae (thin arrows). There is marked increase in basophilic connective tissue mucin in the inner media (thick arrows).   HE ×60

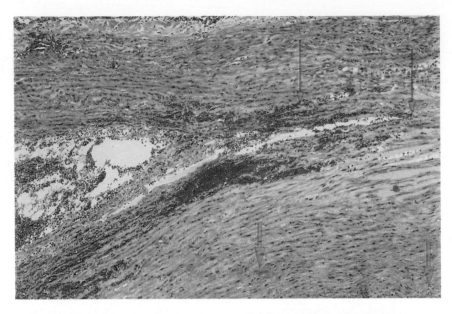

6.26 Medial degeneration and dissecting aneurysm: aorta

The lumen of the aneurysm is on the right. It is lined by a layer of thrombus consisting of platelets, fibrin and red cells which is firmly adherent to the media. The inner half of the media (centre and left) shows reduced cellularity, with fewer muscle cells, and there is some disruption and loss of the musculo-elastic laminae, with formation of small pools of basophilic connective tissue mucin (arrows). This is mucoid medial degeneration.   HE ×150

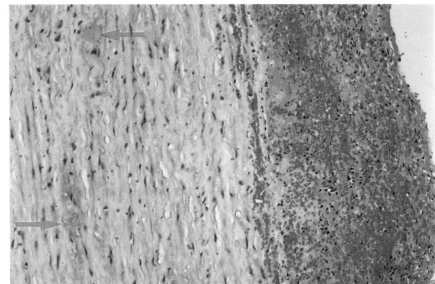

6.27 Medial degeneration and dissecting aneurysm: aorta

This is the inner media of the aorta. The eosinophilic muscle fibres and their associated laminae of elastic tissue are reduced in number and some are broken. The muscle fibres are atrophic and thin. Between the remaining laminae there are considerably increased amounts of connective tissue mucin. Most of the mucin is basophilic but some of it is almost unstained (arrow). Loss of elastica and smooth muscle fibres from the media and increase of connective tissue mucin is sometimes called Erdheim's medial degeneration. The term medionecrosis is also applied occasionally but true necrosis is rarely seen.   HE ×235

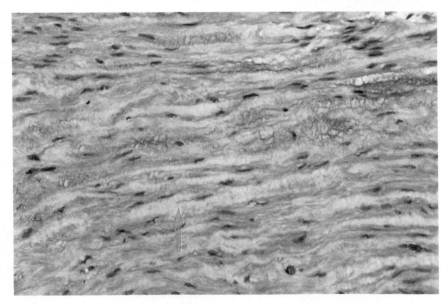

6.28 Medial degeneration and dissecting aneurysm: aorta

The loss of elastic tissue from an aorta showing medial degeneration is made more obvious by the use of a specific stain for elastic tissue. The loss is invariably greater than suspected from HE sections. This section of the inner half of the aortic wall and part of the aneurysm has been stained to show elastic fibres (black). The lumen of the aorta is at the bottom. The dissecting aneurysm (top) contains yellow-stained blood and thrombus. The media shows extensive disruption and loss of the elastic laminae. The unstained parts of the media from which the elastic laminae have disappeared contained mucinous material.

Elastic-van Gieson ×60

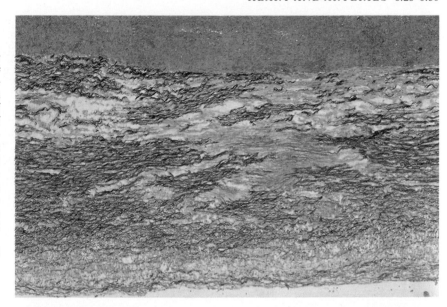

6.29 Myxoid degeneration: mitral valve

Myxoid degeneration occasionally occurs in the cusps of the mitral valve. The cusps are liable to stretch and become 'floppy'. They may prolapse during ventricular systole, rendering the mitral valve incompetent. Older people are generally affected but younger individuals with a defect in collagen synthesis such as Marfan's syndrome may suffer from the condition. A man of 57 was diagnosed as having the floppy valve syndrome affecting his mitral valve. The valve was removed surgically. The surface of a cusp is on the left. The cusp is thickened and its centre consists of myxoid connective tissue. There is separation of the cells and fibres of the connective tissue, with abundant pale-staining connective tissue mucin between the fibres. The chordae showed a similar increase in myxoid tissue. HE ×150

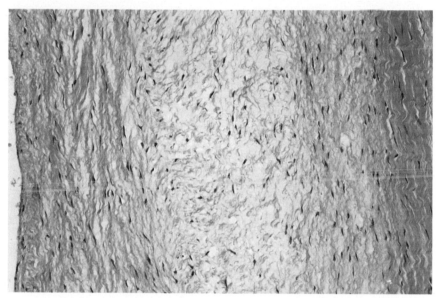

6.30 Mucopolysaccharidosis (Hurler's syndrome): mitral valve

In this condition, also known as gargoylism, there is a defect of mucopolysaccharide metabolism: there is a deficiency of the lysosomal enzyme alpha-1-iduronidase and a consequent failure in the breakdown of mucopolysaccharides. Excess mucopolysaccharides accumulate in various types of cell in many tissues including the valves of the heart. This shows a mitral valve cusp which was greatly thickened. The connective tissue cells are characteristically swollen and vacuolated, and in the connective tissues of the deeper parts of the cusp (right) there is a massive accumulation of purplish-red mucopolysaccharide.

PAS ×120

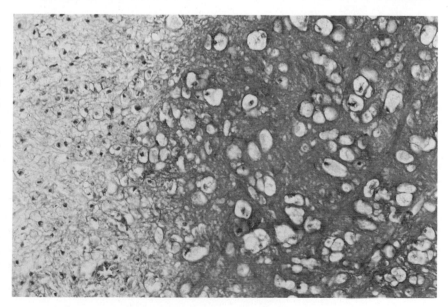

6.31 Brown atrophy: heart

With increasing age the yellowish-brown pigment lipofuscin tends to accumulate in many tissues. The lipofuscin granules are residual bodies; that is, the indigestible remnants of cell organelles and cytoplasmic materials. When the parenchymal cells of an organ have atrophied, because of increasing age or the presence of a wasting disease, the condition of 'brown atrophy' results. In this case death followed prolonged cachexia caused by malignant disease. At necropsy the heart was small and lacked epicardial fat, and it looked browner than normal. The muscle fibres are atrophic and narrow (the fragmentation is artefactual) and collections of lipofuscin granules are present at the poles of the nuclei in several muscle fibres.

HE ×850

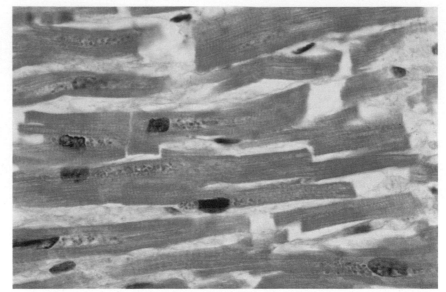

6.32 Viral myocarditis: heart

A number of different viruses can cause inflammatory changes in the myocardium. Coxsackie viruses (groups A and B) are particularly important in this context, mainly affecting children (including infants) and young adults. They produce a diffuse inflammatory reaction in the interstitial tissues throughout the heart. In this example the interstitial tissues of the myocardium are heavily infiltrated with lymphocytes, plasma cells and macrophages. A few eosinophil leukocytes are also present. The muscle fibres are separated by the cellular infiltrate and inflammatory edema. There is also destruction of muscle fibres, many of which are necrotic and undergoing fragmentation. The inflammatory changes in Coxsackie myocarditis are considered to be reversible, but it is possible that occasionally diffuse interstitial fibrosis results. HE ×360

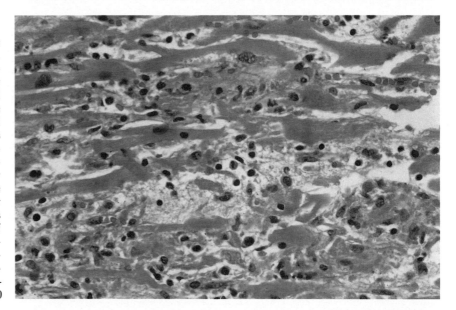

6.33 Toxic (diphtheritic) myocarditis: heart

The diphtheria bacillus does not invade the tissues but forms a toxin which is absorbed into the bloodstream. The toxin is injurious to the nervous system and the cardiovascular system. The myocardium and particularly the conducting system are often affected. In this heart the muscle fibres show varying degrees of damage, ranging from loss of striations to complete necrosis and fragmentation (arrow). Diphtheria toxin can also cause fatty change in the heart muscle. There is marked edema of the interstitial tissue, with separation of the muscle fibres, but the infiltrate of inflammatory cells is fairly slight, the cells being mostly lymphocytes. HE ×200

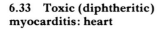
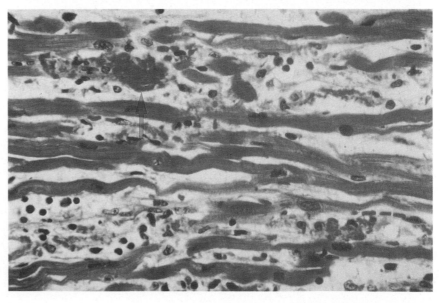

6.34 Acute rheumatism: heart

There is a pancarditis in acute rheumatism. In the myocardium the pathognomonic lesion is the Aschoff body, one of which is shown here. It is a collection of pleomorphic histiocytes with large basophilic nuclei each of which contains a very prominent nucleolus which gives the cell an 'owl-eye' appearance. Several of the cells are binucleate. The cytoplasm is fairly abundant and slightly basophilic. In the centre of an Aschoff body there is often necrotic collagenous tissue. This Aschoff body was characteristically located near a small blood vessel (not visible in this field). The large cells in the Aschoff body could be myogenic but are more likely to be histiocytic in origin. Aschoff bodies heal by fibrosis but the function of the myocardium is not generally permanently affected.　　　HE ×580

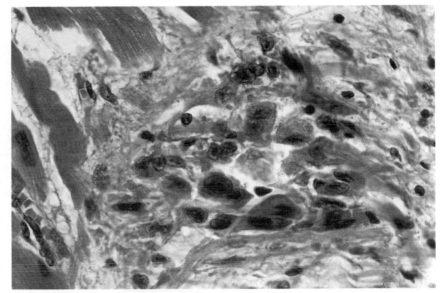

6.35 Acute rheumatism: heart

In rheumatic fever, the endocardium is inflamed and small thrombi form along the lines where the valve cusps come into contact during closure. These 'vegetations' consist mainly of fused platelets initially but fibrin usually participates later in their formation. A mitral cusp is shown here, with a deeply eosinophilic vegetation (left) closely adherent to its surface. The cusp is thicker and more fibrous than normal and macrophages, lymphocytes and plasma cells are present along with small blood vessels (arrow). The boundary between the cusp and the vegetation is indistinct, the base of the vegetation (centre) being phagocytosed by macrophages. Fibroblasts are also present and organization of the vegetation produces a fibrous scar which deforms the cusp and tends to render the valve incompetent and/or stenotic.　　　HE ×235

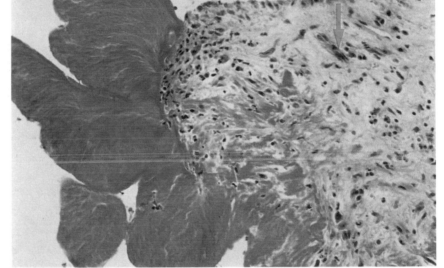

6.36 Rheumatoid arthritis: aortic valve

Necrosis of connective tissue occurs frequently in rheumatoid arthritis and is particularly common in the skin in the form of rheumatoid nodules. However rheumatoid nodules occasionally occur in other tissues and this one was in the aortic valve of a man of 43. The valve was removed surgically and a prosthesis implanted. The centre of the nodule consists of amorphous necrotic material (bottom), enclosed by a wall of elongated histiocytes with ovoid nuclei. The histiocytes tend to be aligned side-by-side (palisading). Outside the histiocytic zone there are round macrophages and lymphocytes. The presence of the destructive necrotizing lesion shown here had rendered the aortic valve incompetent and led to its removal. *See also 1.17, 14.29.*　　HE ×360

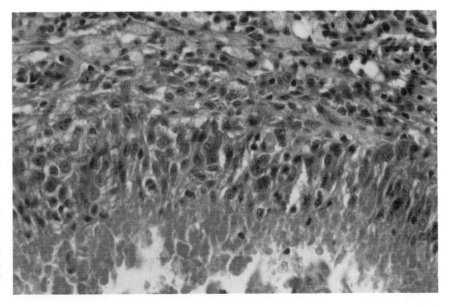

7.1 Laryngeal nodule: vocal cord

Laryngeal nodules (singer's node) arise in adults, and in men more often than in women, probably as a result of chronic 'irritation', heavy smoking and singing being predisposing factors. They are generally located on the true vocal cords and are not neoplastic. This lesion was a smooth swelling 10 × 5mm on the right vocal cord of a woman of 68. It consists typically of eosinophilic sparsely cellular connective tissue in which there are dilated thin-walled blood vessels (lower right). The stratified squamous epithelium of the vocal cord (left) is intact but mildly dysplastic with vacuoles between the epithelial cells.   HE ×150

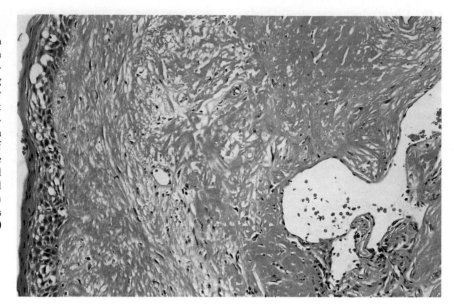

7.2 Laryngeal nodule: vocal cord

This shows the stroma and epithelium of the nodule at higher magnification. The epithelium is extensively vacuolated but the dysplastic changes are slight and of no significance. Hyperkeratosis is occasionally found over the line of contact of the nodule with the other vocal cord. The stroma contains comparatively few cells and a large quantity of amyloid-like material. Although the material looks like amyloid it is however hyalinized collagen. In some nodules the stroma is vascular and almost hemangiomatous but in others it is more dense and fibrous.   HE ×360

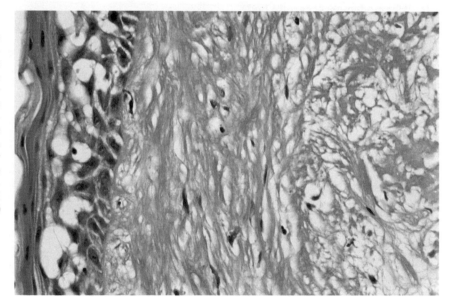

7.3 Myxedema: vocal cord

In myxedema there is increase in the connective tissue mucins throughout the body which accounts, for example, for the thickening of the skin. The larynx is also involved so that the voice becomes husky and low-pitched. This shows part of a vocal cord. The surface of the cord is just out of the picture on the left and some fibres of the laryngeal muscles are visible on the right. In the connective tissue between them there is a wide zone (centre) which has a greatly increased content of basophilic connective tissue mucins. The many clefts in this zone are occupied by connective tissue cells.   HE ×120

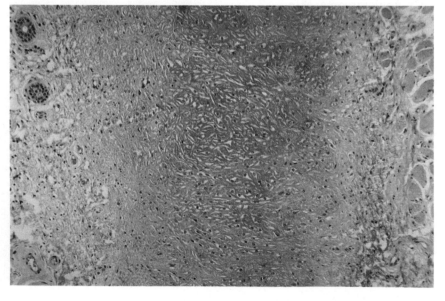

7.4 Sarcoidosis: larynx

Sarcoidosis is characterized by the presence of follicles composed of epithelioid cells. Virtually any tissue may be involved. This is the anterior commissure of the larynx of a 48-year-old woman. The keratinized squamous epithelium (left) is attenuated but intact. Lymphocytes are present in the basal layers. The underlying connective tissue is heavily infiltrated by chronic inflammatory cells and epithelioid cells. The epithelioid cells form several round follicles (thin arrows). A multinucleated giant cell of Langhans type is present in one follicle (thick arrow). Langhans-type giant cells are often present in considerable numbers and may contain asteroid bodies in their cytoplasm. Healing of sarcoid lesions tends to produce much fibrous tissue which can interfere with the functions of a tissue. *See also 1.23, 2.32, 7.48, 14.10.* HE ×190

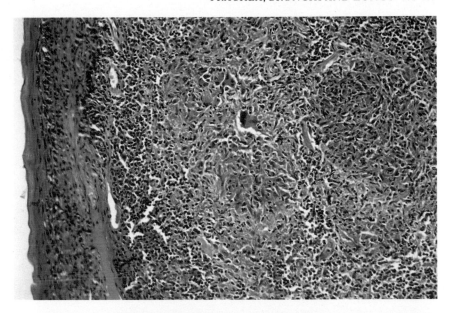

7.5 Juvenile papilloma: larynx

Juvenile papilloma is probably caused by a virus. It is often multiple (papillomatosis) and may grow to a considerable size, blocking the larynx. It occurs on the true vocal cord but may also involve the false cord and subglottic region. The tumour is composed of multiple papillary processes, each consisting of vascular core of connective tissue (thin arrow) covered with a thick layer of well-differentiated squamous epithelium. The underlying respiratory-type epithelium on the surface of the larynx (thick arrow) has been compressed by the tumour and is atrophic; and the submucosa (bottom) is infiltrated with chronic inflammatory cells. HE ×60

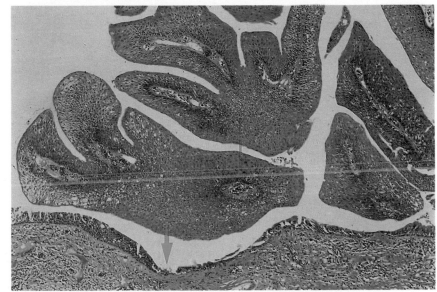

7.6 Juvenile papilloma: larynx

This shows the squamous epithelium of the papilloma at higher magnification. The maturation pattern of the epithelial cells is orderly. Many of the cells in the most superficial layers (left) are vacuolated. The nuclei in the basal layers are basophilic and there is considerable mitotic activity. Papillomas of the larynx in children tend to recur over a long period but remain well-differentiated. They sometimes disappear at puberty. *See also 1.51, 3.10, 3.11.* HE ×350

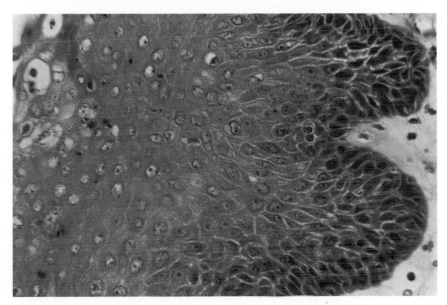

7.7 Squamous carcinoma: larynx

Epidermoid carcinoma of the larynx occurs most often in older men. Most arise in the vocal cords but they may also be located above or below the cords or on the aryepiglottic folds. Almost all arise from the stratified squamous epithelium and they are usually well-differentiated. This lesion was an ulcerated mass 1cm dia on the left vocal cord of a man of 60 who had suffered for some months from progressive hoarseness and difficulty in swallowing. The surface epithelium (top) is unbroken. The tumour in the underlying tissues is a squamous cell carcinoma, consisting of sheets of eosinophilic cells. In several of these there are foci of keratinization (arrows). Small collections of dark-staining lymphocytes are present beneath the surface epithelium.   HE ×60

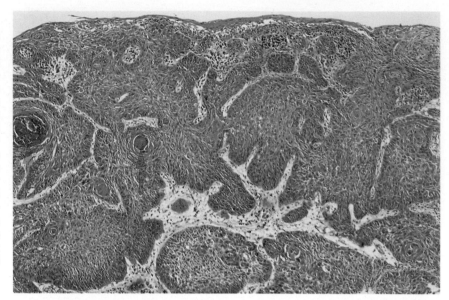

7.8 Squamous carcinoma: larynx

The surface epithelium (left) is extremely attenuated but not ulcerated. The tumour consists of well-differentiated squamous epithelial cells with abundant eosinophilic cytoplasm. There are foci of keratin formation (arrow). The nuclei of the tumour cells show only slight dysplasia and the number of mitotic figures is small. There is slight lymphocytic infiltration of the connective tissue stroma near the surface. Carcinoma of the larynx tends to remain localized. It is locally destructive, however, and ulceration may lead to infection and eventually to bronchopneumonia through inhalation of infected material.   HE ×150

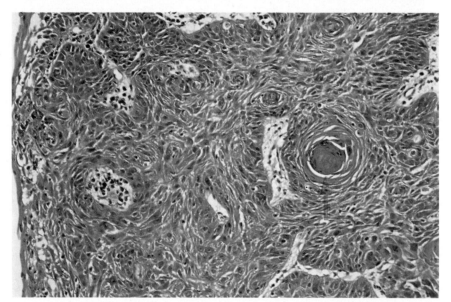

7.9 Aspiration of amniotic fluid: lung

When there has been some difficulty during childbirth, the fetus may draw excessive amounts of amniotic fluid into its lungs and subsequently fail to expel it completely. This is more liable to occur in premature infants with respiratory difficulties. This is the lung of a newborn infant. Dense granular material fills the air spaces and within it there are many keratotic epithelial squames (arrows). The granular material is coagulated protein. Special stains would also demonstrate the presence of mucus and lipid. Small numbers of polymorph leukocytes are also present and the capillaries in the walls of the alveoli are congested, suggesting the onset of pneumonia.   HE ×200

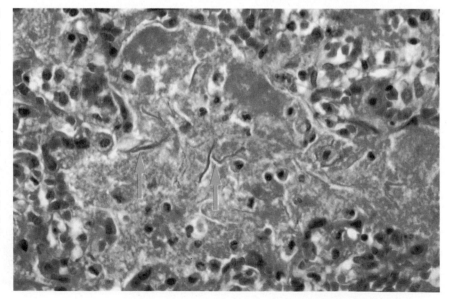

7.10 Atelectasis neonatorum: lung

Failure of the lungs of the newborn infant to expand fully is an important cause of hypoxia and death. It is particularly common in infants born prematurely and in those suffering from birth trauma. This specimen is from an infant who lived for only a few hours. The lungs were small and rubbery. The structure of the infant's lung is reminiscent of fetal lung, with large numbers of unexpanded and virtually airless alveoli (arrows). The capillaries in the walls of the alveoli are engorged. The wall of a bronchiole, with its characteristic lining of columnar epithelium, is visible at the bottom.
HE ×200

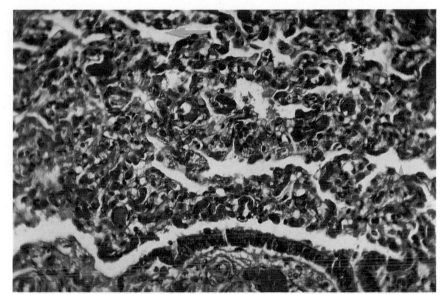

7.11 Respiratory distress syndrome (hyaline membrane disease): lung

Respiratory distress syndrome is particularly liable to affect premature infants. The child develops dyspnea and cyanosis, and its blood pressure falls. The cause is a deficiency of surfactant secretion by the type II cells in the pulmonary alveoli. The child's respiratory movements then fail to keep the alveoli fully expanded, with subsequent exudation of plasma proteins and particularly fibrin into the air spaces. In the lung of this infant a band of amorphous eosinophilic material, probably consisting largely of fibrin, lines the surface of the alveolar duct (part of a bronchiole is just visible on the left). Some of the associated alveoli are collapsed. The capillaries are extremely congested in the walls of the alveoli, and there is some infiltration by inflammatory cells. HE ×280

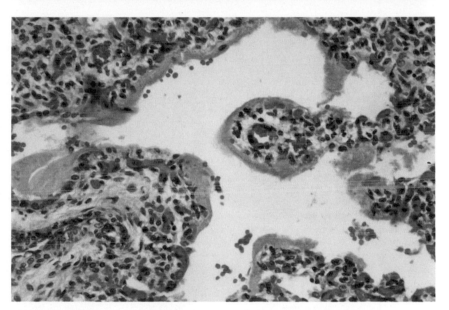

7.12 'Shock' lung

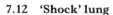

A man of 57 died 21 days after gastrectomy for a carcinoma of stomach. For two weeks prior to his death he had been ventilated, latterly with a high concentration of oxygen, in order to maintain an adequate PO_2. At postmortem examination the lungs were extensively consolidated. The alveoli are filled with amorphous proteinaceous material and bands of hyaline eosinophilic material line the thickened alveolar walls. Mononuclear cells of various types including many macrophages are present in the lumen. There is also early organization and fibrosis, with elongated fibroblasts in the alveolar walls and within the alveoli. High concentrations of oxygen are toxic to alveolar epithelium and capillary endothelium, and this toxic effect is probably responsible for many of the changes previously attributed to 'shock'.
HE ×235

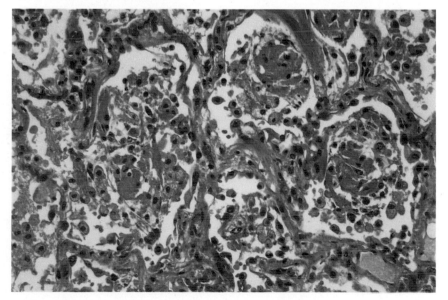

7.13 Congestion and edema: lung

This patient died of acute left ventricular failure. The lungs were heavy and edematous and frothy fluid was present in the air passages. The failure of the left ventricle had caused the blood pressure to rise abruptly in the pulmonary veins and in the capillaries in the walls of the alveoli in the lungs. The pulmonary alveoli are filled with eosinophilic fluid which has leaked from the congested capillaries. A transudate of this type is usually cell-free but the polymorph leukocytes in the alveoli suggest that bronchopneumonia was starting to develop.
HE ×150

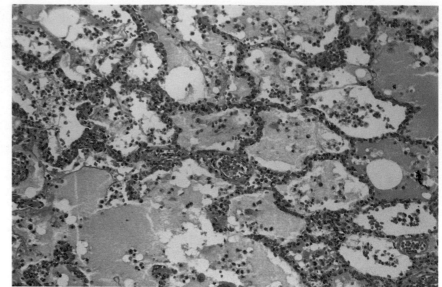

7.14 Chronic venous congestion: lung

The lung is from a person with mitral stenosis and long-standing pulmonary venous hypertension. The raised blood pressure in the pulmonary capillaries has caused them to become swollen and tortuous. Red cells have leaked into the lumen of the alveoli, and macrophages also present in the alveoli have ingested them. Some of the macrophages are very large and distended with hemosiderin derived from the ingested red cells: pulmonary hemosiderosis. Hemosiderin-laden macrophages tend to congregate around the respiratory bronchioles and may release iron-containing salts which are deposited in the connective tissue fibres of the lung. Increase in connective tissue then causes brown induration of the lungs.
HE ×235

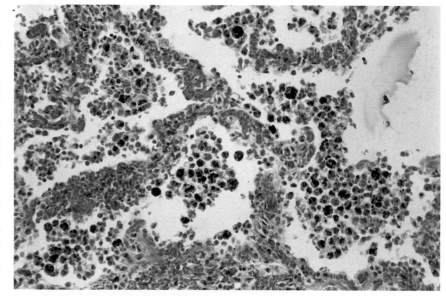

7.15 Pulmonary (arterial) hypertension: lung

Many diseases can cause pulmonary arterial hypertension and pulmonary vascular disease. This patient had very severe chronic bronchitis and emphysema, with consequent chronic hypoxia. In this pulmonary arteriole (60μm dia) the elastic fibres appear black, collagen red and muscle orange-yellow. There is localized thickening of the intima (thin arrow) caused by bands of longitudinally-orientated muscle fibres; and a well-developed muscular media is present (thick arrow), similar to that found in systemic arterioles. 'Muscularization' of the terminal parts of the pulmonary vascular tree as shown here is caused by the chronic hypoxia and it leads to increased resistance in the pulmonary vascular (arterial) system and to pulmonary hypertension.
Elastic-van Gieson ×580

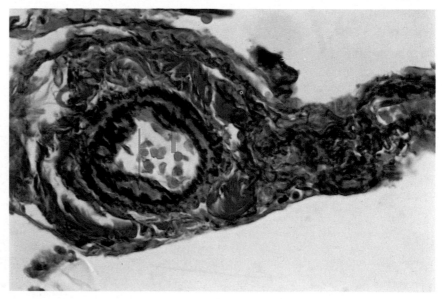

7.16 Pulmonary (arterial) hypertension: lung

This patient had a patent ductus arteriosus. The flow of blood from the aorta to the pulmonary artery had caused severe pulmonary hypertension which had been present for many years. The elastic tissue is stained black and the blood is orange-yellow. The branch of the pulmonary artery (thin arrow) has a thick hypertrophied media. Closely associated with this artery are numerous small abnormal blood vessels (top centre and bottom left) which constitute the so-called angiomatoid lesion. These thin-walled vessels are formed by dilatation of small branches of the pulmonary artery proximal to sites of occlusion of the artery. One dilated vessel (thick arrow) cut longitudinally has a thin wall and resembles a vein. It is however a grossly dilated arteriole.　　　　Elastic-van Gieson ×70

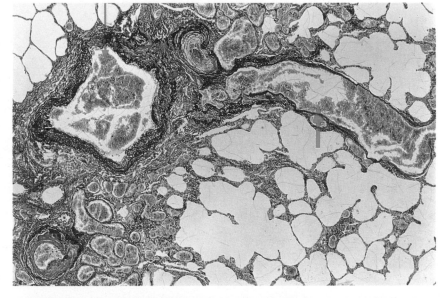

7.17 Infarct: lung

An infarct of lung is usually a dark red hemorrhagic wedge-shaped mass, with its 'base' on the pleural surface. Occlusion of a branch of the pulmonary artery is not sufficient by itself to cause infarction, and increased pulmonary venous pressure is usually also present. It presumably acts by impairing pulmonary blood flow but other factors may also play a part. Mitral stenosis was present in this case. The walls of the alveoli are necrotic. The capillaries are full of blood but endothelial and alveolar epithelial cells cannot be distinguished. The alveolar spaces contain necrotic cells and cell debris, much of it derived from red cells. There is also abundant fibrin.　　HE ×150

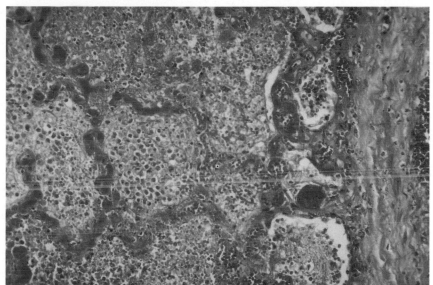

7.18 Obstruction of bronchus: lung

A carcinoma of bronchus was obstructing the bronchus. The lung distal to the obstruction collapsed and the bronchial secretions and desquamated cells were retained in the affected lung. The alveoli are full of macrophages with abundant foamy (lipid-laden) cytoplasm accompanied by smaller numbers of plasma cells and lymphocytes. Eosinophilic exudate (arrow) is also present in the alveoli. The lipid within the macrophages gave the affected segment of lung a definite yellow colour macroscopically, and this type of lesion is sometimes referred to as endogenous lipid pneumonia. Exogenous lipid pneumonia is caused by the inhalation of lipoid material.
HE ×160

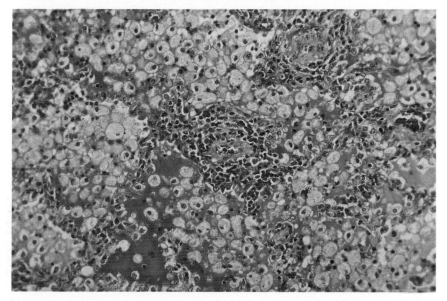

7.19 Lipid (aspiration) pneumonia: lung

Lipid may be aspirated from a variety of sources. These include drops or sprays used for nasal conditions, or mineral oil (liquid paraffin) taken as a laxative; and radio-opaque material used for bronchography may also remain in the lungs. The lipid tends to be irritant and the inflammatory reaction it produces may lead to considerable fibrosis. In this case the patient had used nasal drops containing mineral oil, and a mass was found in the right lower lobe. In the alveoli there are many macrophages with large clear vacuoles in their cytoplasm; and there is evidence of a more acute inflammatory reaction in the form of polymorph leukocytes (e.g. bottom left) and dilated capillaries in the alveolar walls. A fat stain confirmed that the vacuoles in the macrophages contained lipid.   HE ×200

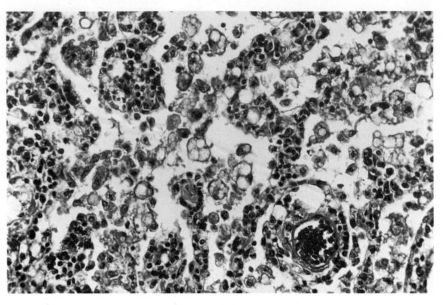

7.20 Lipid (aspiration) pneumonia: lung

In this case also the aspirated lipid was almost certainly mineral oil. The lung parenchyma has been destroyed and replaced by dense fibrous tissue in which there are large spaces. These spaces contained lipid which was lost during processing of the tissues. It was demonstrated however in frozen sections. The lesion caused by aspirated lipid may be symptomless and only discovered incidentally, e.g. in an X-ray of chest.   HE ×135

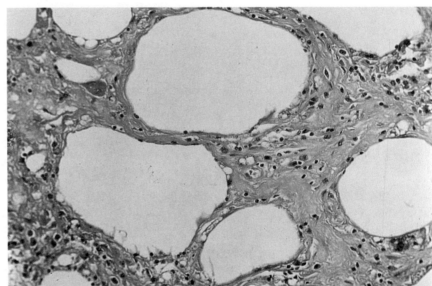

7.21 Asthma: lung

In an asthmatic attack an exudate containing mucus and a serous component is secreted into the bronchi. A woman aged 27 years who had suffered from asthma for 4 years died in status asthmaticus. At postmortem the small bronchi and bronchioles were obstructed by plugs of viscid secretion and there was patchy collapse of both lungs. The right ventricle of the heart was neither hypertrophied nor dilated. The lumen (top) of this small bronchus contains exudate in which there are many cells, mostly eosinophil leukocytes. The epithelial lining consists of large mucus-secreting cells. The epithelial basement membrane is thickened and the smooth muscle of the wall is hypertrophied. Eosinophil leukocytes and lymphocytes are also present in the wall. The adjacent alveoli appear normal and emphysema was not present.   HE ×150

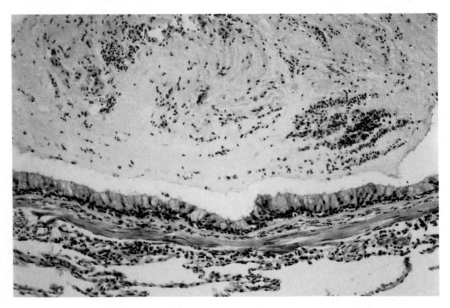

7.22 Bronchiectasis and emphysema: lung

One of the most effective ways of studying bronchiectasis and emphysema is a thin (300μm) slice of the whole lung (a Gough-Wentworth section). This is part of such a section showing the lower lobe and part of the upper lobe. Practically all the bronchi in the lower lobe are dilated (ectatic), the changes extending as far as the pleura. The ectatic bronchi were filled with greenish mucopus which was lost during processing of the tissue. Their walls are thin and no cartilage is visible. Fibrosis is not evident, however, around them. Several bronchi in the upper lobe are similarly affected. Much carbon is present in the centres of many lobules and dilatation of the respiratory bronchioles has produced focal dust emphysema (coalworker's pneumoconiosis).   IIE ×1

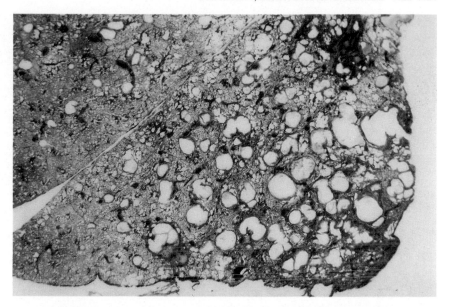

7.23 Cystic fibrosis: lung

The exact nature of the defect in cystic fibrosis is not known but many secretions, including sweat secretion are abnormal. In the lungs there is plugging of the bronchi and bronchioles with tenacious mucoid secretion and saccular bronchiectasis tends to develop. This shows a bronchiole filled with basophilic mucus. The epithelial lining is intact and of normal structure. Mucus is present also in many of the associated alveoli (arrows). There are many inflammatory cells within the mucus and in the walls of the alveoli. Many of the small blood vessels are dilated. Secondary infection and severe inflammation are liable to develop in mucus-filled bronchioles and lead to destruction and obliteration of the lumen by fibrous tissue (obliterative bronchiolitis).

HE ×120

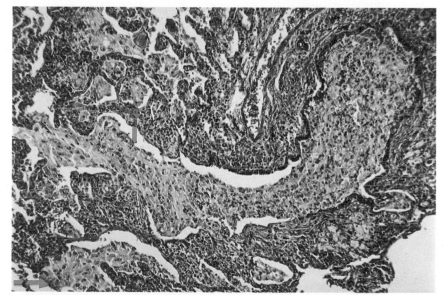

7.24 Cystic fibrosis: lung

This is a small bronchus from another case in cross-section. It is ectatic and the lumen is filled with pale-staining secretion in which there are many polymorphs and large macrophages. The even dispersion of the cells throughout the mucus suggests that its consistency is firmer than that of normal mucus. The respiratory mucosa is extensively ulcerated with most of the wall of the bronchus lacking an epithelial lining. The remaining epithelium (left) consists of cuboidal cells which do not appear to secrete mucin. The wall of the bronchus and surrounding tissues are hyperemic and infiltrated by polymorphs, lymphocytes and plasma cells.   HE ×150

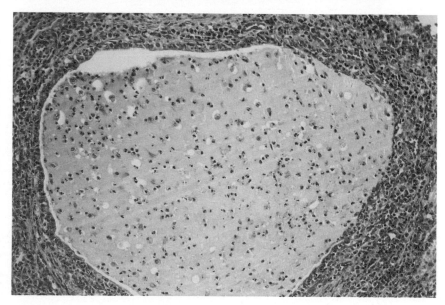

7.25 Bronchiectasis: lung

Inflammatory changes are often seen in ectatic airways, and in many instances lymphoid tissue including mature lymphoid follicles is present. The term follicular bronchiectasis is sometimes applied when lymphoid follicles are a prominent feature. This shows a terminal bronchiole (left and top) lined by cuboidal and columnar epithelial cells and with many polymorphs and macrophages in the lumen. Adjacent to the bronchiole there is a collection of mature lymphocytes (right). Venules with thick hyalinized walls are prominent in the follicle but there is no germinal centre. HE ×135

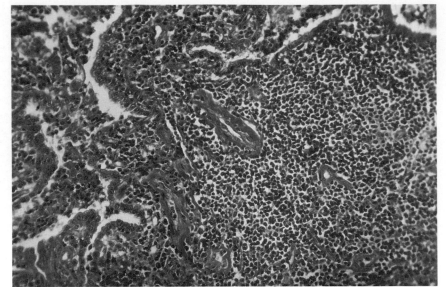

7.26 Centrilobular emphysema: lung

In centrilobular emphysema the abnormal air spaces are dilated respiratory bronchioles in the centres of the acini (centriacinar or centrilobular emphysema). Carbonaceous dust is often present in town-dwellers in the same site as the emphysematous lesion but its presence is not invariable. This is part (9 × 6cm) of a whole lung section prepared by the Gough-Wentworth technique. A linear deposit of carbonaceous dust is present beneath the pleura (bottom). Large abnormal air spaces (up to 2cm dia) are present in the centres of almost all the lobules and within the emphysematous spaces there are carbon-laden septa. The alveoli in the peripheral parts of the lobule appear relatively normal, as do the branches of the pulmonary artery (arrows).

Whole lung section, unstained ×1.3

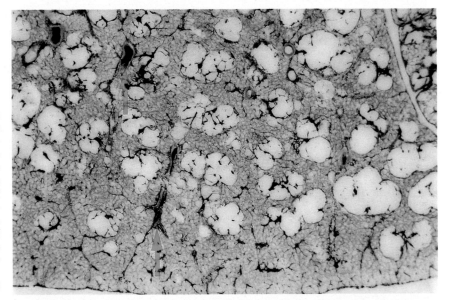

7.27 Panacinar emphysema: lung

In panacinar (panlobular) emphysema the whole of each pulmonary acinus is disorganized, the normal alveoli being replaced by large thin-walled spaces. This is part (9 × 6cm) of a whole lung section prepared by the Gough-Wentworth technique. Apart from the large blood vessels, the normal architecture of the pulmonary acini has been largely destroyed and the lung now consists of thin-walled air spaces of variable size and shape. There is no evidence of inflammation or of fibrosis. In panacinar emphysema the emphysematous spaces are formed by the loss of alveolar walls and coalescence of adjacent alveoli. Alveolar ducts and respiratory bronchioles are also involved in the more severe forms. The branches of the pulmonary artery appear normal.

Whole lung section, unstained ×1.3

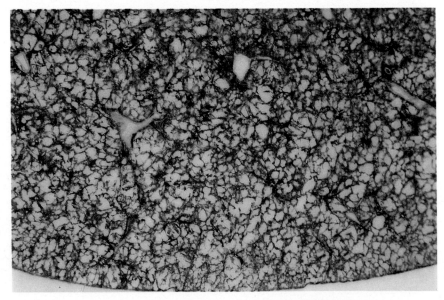

7.28 Panacinar emphysema: lung

This is a histological section of a lesion similar to that shown in **7.27**. The pleura is on the right. Beneath it there are deposits of carbonaceous dust (thin arrow). The alveolar walls are atrophic and thin and in places they have broken down to form spaces considerably larger than normal alveoli. Several collections of pigment-laden macrophages are present (thick arrow).　　　HE ×110

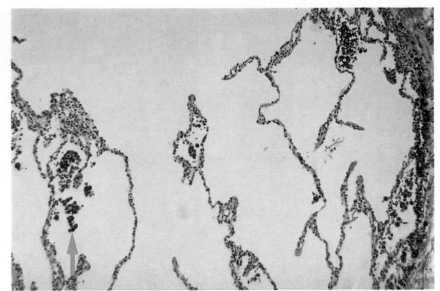

7.29 Acute bronchitis: lung

Acute inflammation of the bronchi is probably initiated in most cases by viruses or mycoplasmas, with subsequent invasion by pyogenic bacteria such as *Haemophilus influenzae* or *Staphylococcus aureus*. The inflamed respiratory epithelium secretes increased quantities of mucus (catarrhal bronchitis) but if the inflammation becomes more intense, polymorphs appear in increasing numbers and the mucus becomes mucopurulent or even purulent. This is a very acute lesion. The lumen of the bronchus (left half) is occupied by pus-like secretion consisting largely of neutrophil polymorphs. Polymorphs are also present within the epithelium, and the blood vessels in the submucosa (right) are greatly dilated. The epithelium is ulcerated at one point (thin arrow). Fibrin is present beneath the epithelium (thick arrow).　　　HE ×200

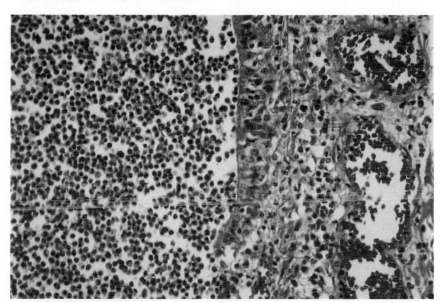

7.30 Influenzal bronchitis: lung

The influenza virus attacks the respiratory epithelium and may cause extensive necrosis. The mucociliary defence mechanism of the respiratory tract is weakened and bacterial infection, particularly with *Staphylococcus aureus*, and bronchopneumonia may follow with great rapidity. This shows a small bronchus from a fatal case. The lumen is on the left. The epithelial lining has been completely destroyed by the viral infection, leaving only a layer of necrotic cell debris. The submucosa is intensely hyperemic with dilatation of the small blood vessels. Fibrin strands are present among the eosinophilic strands of fibrin but the cellular infiltrate consists almost entirely of lymphocytes. In a less severe case bacterial infection would tend to increase the number of polymorph leukocytes.

　　　HE ×235

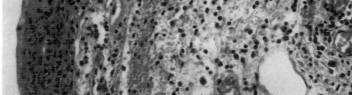

7.31 Bronchopneumonia: lung

In pneumonia the lung is inflamed, the causative agent (usually a microorganism) being in the alveoli, whereas in alveolitis the inflammatory reaction (allergic in nature) is in the walls of the alveoli. In bronchopneumonia, the agent reaches the alveoli via the air passages and bronchioles, and evokes an acute inflammatory reaction in the associated alveoli. In this example polymorph leukocytes, accompanied by smaller numbers of red cells, fill the alveolar ducts and alveoli. The alveolar capillaries are dilated and polymorphs are present also in the walls of the alveoli. Special stains would show the presence of fibrin and bacteria in the inflammatory exudate in the lung. The inflammatory exudate in the air spaces solidifies the lung (consolidation).   HE ×150

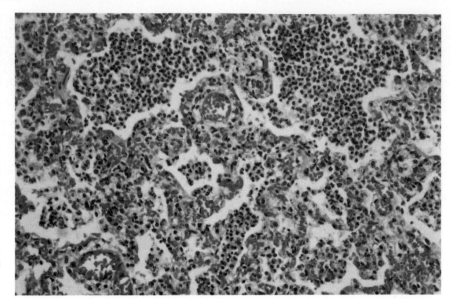

7.32 Lobar pneumonia: lung

As in bronchopneumonia the route of infection in lobar pneumonia is by the bronchi and bronchioles, but within the lung the reaction is more acute and the inflammatory exudate spreads the infection directly through the lung tissue, until it reaches the pleural surface. In this way a whole lobe or more than one lobe rapidly becomes uniformly consolidated. This shows lobar pneumonia at a fairly early stage. The alveoli are filled with inflammatory exudate which contains large numbers of polymorph leukocytes, along with red cells, degenerate macrophages and fibrin. A Gram stain would reveal large numbers of microorganisms, usually *Streptococcus pneumoniae*. The capillaries in the alveolar walls are greatly dilated. The affected lobe would appear brownish-red and solid macroscopically (red hepatization).   HE ×335

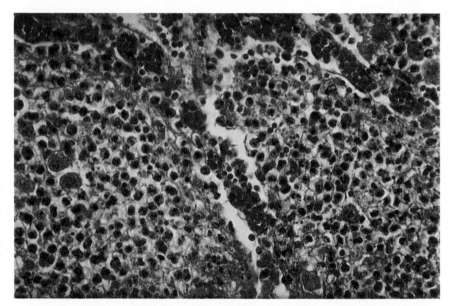

7.33 Lobar pneumonia: lung

Postmortem, the consolidated lobes in this case were not red and hyperemic macroscopically as in **7.32** but grey (grey hepatization). Histologically the alveolar walls (arrrow) are thin and the alveolar capillaries inconspicuous, perhaps as a result of compression by the large amount of inflammatory exudate which fills the alveoli. The exudate in some alveoli is largely composed of fibrin whereas in others many more polymorph leukocytes are present. It is the reduction in the vascularity of the consolidated lung that makes it appear grey to the naked eye.   HE ×135

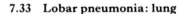
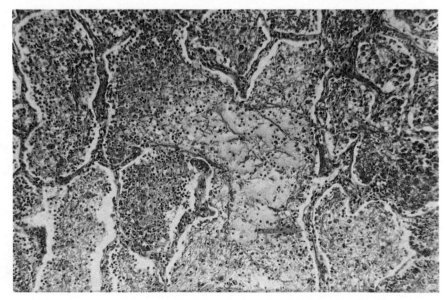

7.34 Fibrinous pleurisy: lung

In acute inflammatory conditions of the lung such as lobar pneumonia, an inflammatory exudate may form in the pleural cavity, with deposition of fibrin on the visceral pleura. After the acute phase of the illness, fibrin on the pleural surface undergoes organization. Organization is already well advanced in this case. The fibrin is at the top and the pleural surface is at the bottom. Beneath the elastic tissue of the pleura there are carbon-laden macrophages in the subpleural tissues. There are no serosal cells on the pleura which is covered with granulation tissue consisting of large numbers of dilated blood vessels, macrophages and fibroblasts. In time this tissue would organize the remaining fibrin and convert it into fibrous tissue. HE ×235

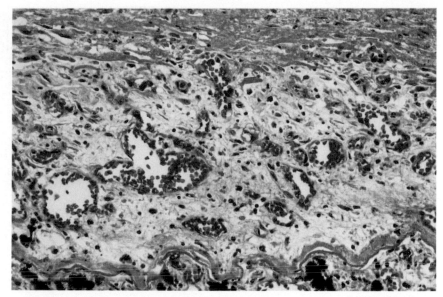

7.35 Fibrinous pleurisy: lungs

The process of organization of the layer of fibrin on the surface of the visceral pleura has reached a more advanced stage than that shown in **7.34**. The elastic tissue of the pleura and the subserosal connective tissue are just visible on the right. All the fibrin has been removed by the macrophages, and the fibroblasts in the granulation tissue have formed considerable amounts of collagen, thereby converting the granulation tissue into cellular but still fairly vascular fibrous tissue. The end-result will be a layer of fibrous tissue which if abundant may greatly restrict the respiratory movements of the lung. HE ×134

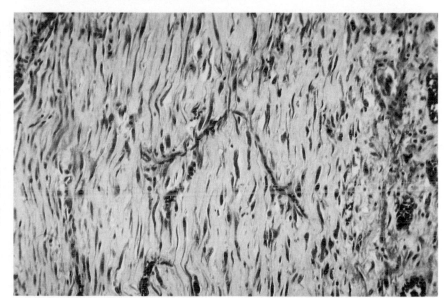

7.36 Pleural plaque: lung

In people exposed to asbestos, distinctive plaques not infrequently form on the parietal pleura. They are well-defined, with a smooth shiny surface, and on cutting have a firm consistence not unlike that of cartilage. This plaque is from a man of 41 with a history of occupational exposure to asbestos. It consists of very cellular connective tissue. The cells are large fibroblasts and fibrocytes. The nuclei of the fibroblasts are vesicular and pleomorphic and an abnormal (tripolar) mitotic figure is visible (arrow). Other parts of the plaque were less cellular, with more abundant hyaline collagen. No asbestos (ferruginous) bodies are found in these plaques as a rule. HE ×360

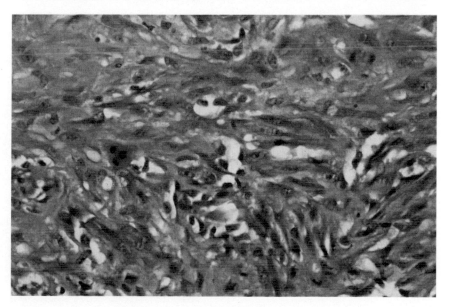

7.37 Tuberculosis: lung

Tuberculosis is caused by *Mycobacterium tuberculosis*, and the route of infection is generally by inhalation of droplets from other individuals with the disease. The bacillus evokes a mononuclear cell response in the tissues although polymorphs sometimes appear initially. The macrophages evolve into epithelioid cells which adopt a follicular arrangement. This, however, is a miliary tubercle in the lung, the infection being blood-borne. Centrally it consists of follicular collections of epithelioid histiocytes and two multinucleated giant cells. The peripheral part of the follicles consists of small lymphocytes and plasma cells. In a section stained by the Ziehl-Neelsen method tubercle bacilli would be detectable. Several pulmonary alveoli are visible around the tubercle.   HE ×235

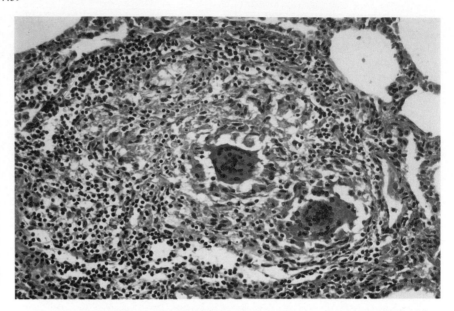

7.38 Tuberculous bronchopneumonia: lung

Large numbers of tubercle bacilli may be discharged into a bronchus, from a caseous lymph node or from a caseous focus in the lung, and inhaled. Tuberculous bronchopneumonia results. An inflammatory exudate forms in the infected alveoli and the lung undergoes consolidation. The alveoli (right) are filled with abundant eosinophil exudate containing small numbers of macrophages. On the left the consolidated lung has become necrotic (caseation). The necrotic (caseous) material has disintegrated and been coughed up, leaving a cavity. At the edge of the caseous area (centre) there is much blue-staining nuclear debris. A Ziehl-Neelsen stain would reveal large numbers of tubercle bacilli throughout the focus.   HE ×70

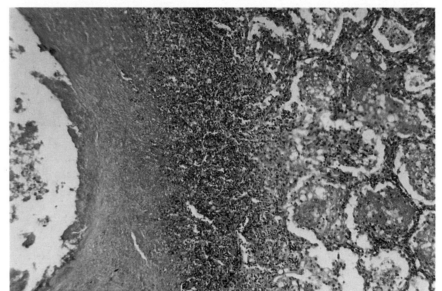

7.39 Tuberculous bronchopneumonia: lung

This shows consolidated pulmonary alveoli at higher magnification. The alveolar capillaries (bottom) are dilated and thrombosis may have occurred (arrow). The inflammatory exudate filling the alveoli consists largely of macrophages many of which are degenerate and beginning to disintegrate. Small numbers of polymorph leukocytes are also present. Tuberculous lesions tend to undergo necrosis on a large scale and the whole of the consolidated lung may become caseous.   HE ×335

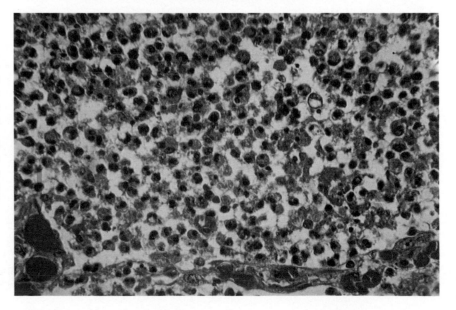

7.40 Tuberculous bronchopneumonia: lung

Tubercle bacilli are very difficult to stain and are not rendered visible for example by an ordinary Gram stain. This section has been stained however by the Ziehl-Neelsen method for staining acid-fast bacilli which uses a powerful stain (carbol fuchsin), heat, and a mordant. The nuclei of the macrophages which fill the consolidated alveoli are coloured blue and large numbers of red tubercle bacilli (many of them curved rods) are congregated around them in the cytoplasm.　　　Ziehl-Neelsen ×580

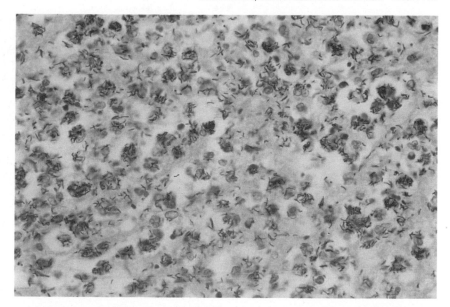

7.41 Silicosis: lung

Small particles (less than 5μm dia) of free silica (silicon dioxide) are very fibrogenic. They are toxic to the macrophages that phagocytose them and the cells are killed. The cycle is repeated and eventually fibroblasts envelop the particles in dense fibrous tissue. In man the nodules are several mm dia. In this case there were many hard silicotic nodules scattered throughout both lungs. Histologically a nodule consists of virtually acellular (hyalinized) fibrous tissue containing deposits of brown particles. The brown particles are non-fibrogenic coal dust, the silica particles being colourless and not visible. There are very few blood vessels in the nodule. The cracks in the tissue were caused by shrinkage during processing.　　　HE ×135

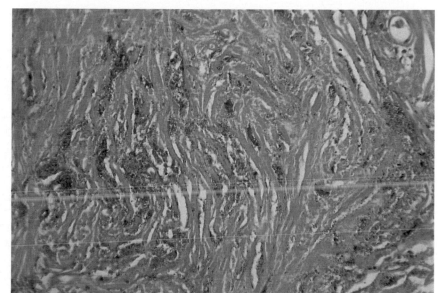

7.42 Asbestos (ferruginous) bodies: lung

Asbestos fibres in the lung are too large to be phagocytosed and often acquire a proteinaceous golden-yellow 'coat' which is rich in hemosiderin. Several clusters of asbestos bodies of various lengths are present in the bronchiole and alveolus. The coat of most of the bodies is highly segmented and most have bulbous swellings, usually at the ends of the fibre. An eosinophilic exudate and macrophages are also present in the air spaces. With sensitive methods asbestos bodies are detectable in the lungs of a large proportion of the population of many countries. Their significance is uncertain but their presence in the lung or in the sputum does not signify asbestosis.　　　HE ×360

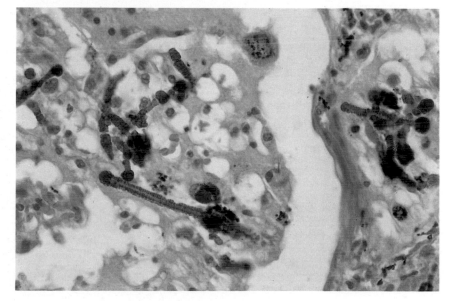

7.43 Pulmonary alveolar proteinosis: lung

Pulmonary alveolar proteinosis is a rare disease of unknown etiology. About one-third of those affected die. The symptoms are cough and shortness of breath, with loss of weight and increasing weakness. Postmortem there are areas of consolidation in the lungs. This shows one such area. The alveoli are full of eosinophilic amorphous exudate in which there are many clefts which contained cholesterol crystals. The alveolar walls are compressed and thin apparently from compression by the exudate in the alveoli. The alveolar capillaries are inconspicuous. There is a moderate infiltration of lymphocytes. There is no fibrosis. The eosinophilic exudate is rich in protein and lipid and probably comes from degenerate granular pneumocytes, some of which are visible (arrow). HE ×150

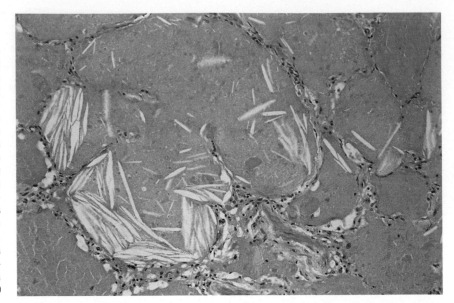

7.44 Pulmonary alveolar proteinosis: lung

This is part of the lesion shown in **7.43**. Present along with the eosinophilic amorphous material in the alveoli are large round cells with abundant granular cytoplasm (arrow). These are granular pneumocytes. They are the probable source of the amorphous material, and many of the cells in the alveoli have small pyknotic nuclei and are probably degenerate. The walls of the alveoli are thickened and infiltrated with lymphocytes. HE ×235

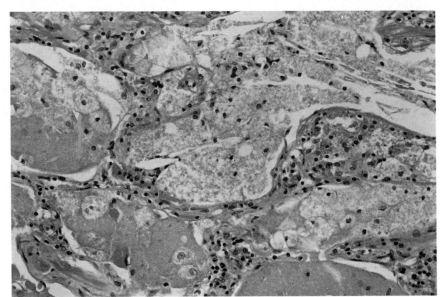

7.45 Extrinsic allergic alveolitis: lung

Organic dusts can induce an Arthus-type reaction in the walls of the pulmonary alveoli, by reacting with circulating precipitating antibodies. The dust may be derived from moulds (mouldy hay in farmer's lung) or bird droppings (bird-fancier's lung). This is a diagnostic specimen from a 36-year-old woman. The alveolar walls are thickened from the presence of lymphocytes, plasma cells and one large multinucleated giant cell of the Langhans type (arrow). The epithelial cells lining the alveoli are more prominent than normal and a few desquamated cells lie within the alveoli. Multinucleated giant cells are characteristic of extrinsic allergic alveolitis, in which an identifiable organic antigen has been inhaled. The source of the antigen in this case was a budgerigar (bird-fancier's lung). HE ×235

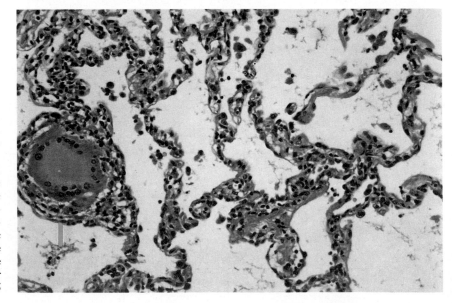

7.46   Cryptogenic fibrosing alveolitis: lung

In cryptogenic fibrosing alveolitis an auto-immune mechanism may operate, since there is a not infrequent association with connective tissue diseases. Inhalation of an external agent of an unknown nature may also play a part. Macrophages and granular (Type II) pneumocytes collect in the alveolar spaces and the walls of the alveoli are thickened. This is a diagnostic specimen from a man of 52. It consists of irregular air spaces with thick walls of cellular connective tissue infiltrated by lymphocytes and plasma cells. Some of the air spaces are still recognizably alveolar. They are lined by prominent epithelial cells and contain many mononuclear cells with abundant eosinophilic cytoplasm (e.g. top centre). As with extrinsic allergic alveolitis, progressive fibrosis may end in honeycomb lung. HE ×235

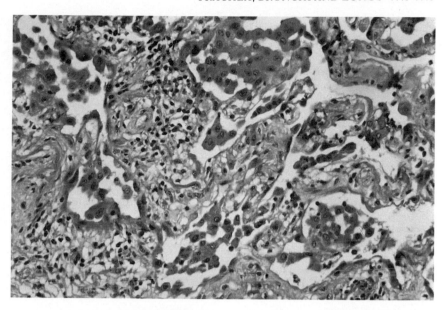

7.47   Honeycomb lung

Fibrosis of the interstitial tissues of the lung (fibrosing alveolitis) may progress until considerable parts of the lungs are converted into small cystic spaces (1–2cm dia) with thick fibrous walls. The affected parts of the lung have a honeycomb appearance. The lesion is associated with a variety of diseases, particularly the connective tissue group, and a number of drugs and toxic substances (including beryllium and cadmium). In this case the underlying disease was progressive systemic sclerosis. The normal structure of the lung has been destroyed and in its place there are large air-filled spaces with thick walls of vascular fibrous tissue. High-power examination showed that the spaces were lined with cuboidal epithelium of the bronchiolar type and there were many lymphocytes and plasma cells in the fibrous tissue.
HE ×8

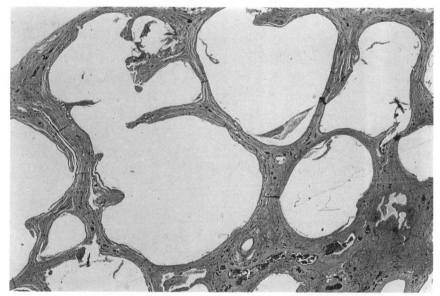

7.48   Sarcoidosis: lung

Sarcoidosis is a granulomatous (mononuclear) inflammatory disease characterized by the formation in many tissues, most notably lymphoid tissue, of collections (follicles) of epithelioid cells similar to those found in miliary tubercles. The lungs are not infrequently involved. This is a diagnostic biopsy specimen from the lung of a woman of 30 suspected clinically of having asbestosis. The alveolar walls are thickened and infiltrated by lymphocytes and macrophages. Several epithelioid follicles (thin arrows) are also present. These contain multinucleated giant cells (Langhans cells) and one follicle is enclosed in fibrous tissue (thick arrow). There is no necrosis. Macrophages are present in some alveoli. A minority of cases of sarcoidosis progress to honeycomb lung. *See also 1.23, 2.32, 7.4, 14.10.*
HE ×150

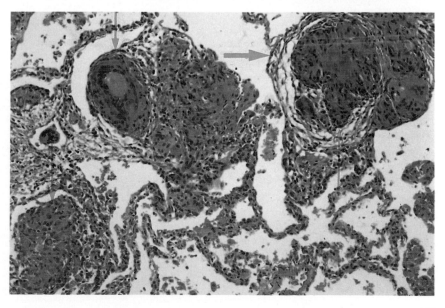

7.49 Wegener's granulomatosis: lung

Wegener's granulomatosis often affects not just the lung but also the upper respiratory tract. The basic lesion is an angiitis, with necrosis and granulomatous inflammation. Vascular lesions are often present in other organs and tissues, notably the kidneys. The affected parts of the lung are consolidated and reddish-grey but necrosis is often extensive and accompanied by cavitation. In this lesion the lung tissue has been replaced by cellular granulomatous tissue consisting of macrophages, lymphocytes, plasma cells and loose connective tissue. Occasional giant cells of the Langhans type (thin arrow) are present at the periphery of the granulomatous tissue. The deeply eosinophilic material on both sides of the granulomatous tissue is necrotic lung and granulatomous tissue. Small amounts of carbon dust are visible (thick arrow).　　　　　HE ×55

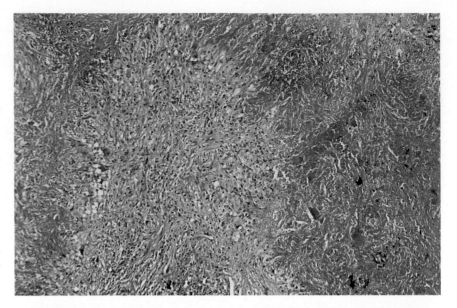

7.50 Kaolin pneumoconiosis: lung

Kaolin (china clay) is a silicate of aluminium, and inhaled particles are not fibrogenic. The dust is not harmless, however, and kaolin pneumoconiosis occasionally occurs if exposure to the dust is very severe. The patient in this case was a china clay worker who had been heavily exposed to kaolin dust over many years. Both lungs felt nodular at postmortem examination. The visceral pleura is at the bottom. The adjacent alveoli are lined by cuboidal epithelial cells (arrow) and their lumen contains many large macrophages with abundant brown cytoplasm. Kaolin dust is practically colourless but at higher magnification the cytoplasm of the macrophages was seen to be full of dust particles. The macrophages show no signs of degeneration. There is no evidence of fibrosis.　　　　　HE ×360

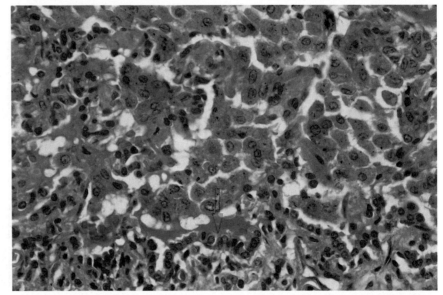

7.51 Hydatid disease: lung

Hydatid disease is caused by the tapeworm *Echinococcus granulosus*. The adult worm is only 3–5mm long and has only a few segments. It is a common parasite of dogs, and hydatid disease is produced in another animal when it acts as intermediate host, as a result of swallowing ova from a dog's feces. Hydatid cyst forms a slowly-growing space-occupying mass up to 20cm dia. It is multilocular, containing many daughter cysts. On the left is the cuticular layer of the cyst wall (the lumen and the innermost germinative layer are out of the picture on the left): amorphous densely-staining laminated chitinous material (arrow) which is being digested by macrophages. The cuticular layer is enclosed by the adventitial layer (centre and right) which consists of fibrous tissue containing many eosinophil leukocytes.　　　　　HE ×120

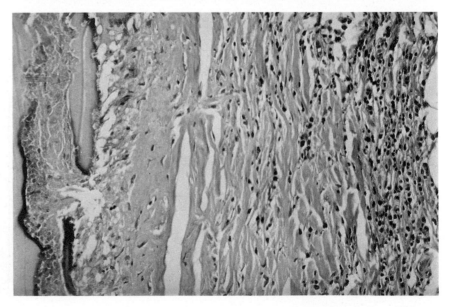

7.52 Adenochondroma: lung

Adenochondroma (hamartoma) of lung is uncommon, usually occurring in adults and often found by chance, e.g. on X-ray of the chest. It is a firm, well-demarcated benign lesion, generally loculated. Macroscopically the cut surface is firm and glistening from the presence of nodules of cartilage. This example was an incidental finding in a man of 52 years. It was removed surgically to exclude the possibility of a malignant tumour. It consists of cartilage (top), fibrous tissue, fat and lymphoid tissue. These tissues are intersected by numerous clefts and gland-like spaces lined by epithelium. The epithelial cells are most cuboidal or flattened, although in some areas it is respiratory (bronchial) epithelium.

HE ×60

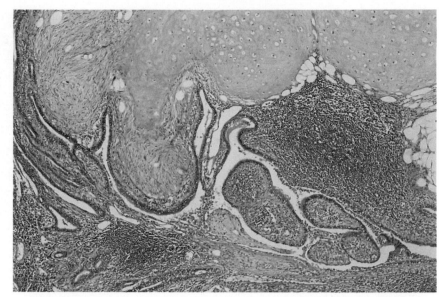

7.53 Adenochondroma: lung

This shows the various tissues (cartilage, lymphoid tissue and epithelium) of the adenochondroma at higher magnification. They are fully mature, with no nuclear pleomorphism or mitotic activity. In this part of the mass the epithelium is flattened or cuboidal rather than respiratory in type. Some cell debris is present in the cleft-like spaces. The precise nature of adenochondroma of lung is uncertain but it is probably primarily a hamartomatous overgrowth of the connective tissue elements of the wall of a bronchus with secondary involvement of bronchial epithelium.

HE ×150

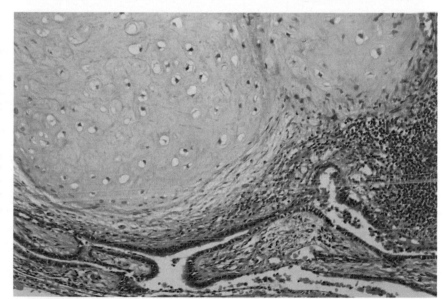

7.54 Adenoma: bronchus

Adenoma of bronchus forms about 1% of neoplasms of lung. It is not a single well-defined entity but includes carcinoid and cylindroma (adenocystic carcinoma). This tumour is a carcinoid. It may present with chronic cough or hemoptysis or it may obstruct a bronchus and cause collapse of the lung and 'pneumonia'. The cells, arranged in closely-packed cords, are very regular in form, with round or ovoid nuclei containing one or more nucleoli. No mitoses are present. Their cytoplasm is fairly abundant. The stroma is scanty, consisting of delicate connective tissue (thin arrow) and small blood vessels which are collapsed and inconspicuous (thick arrow). Bronchial carcinoid cells are argyrophil-positive but argentaffin-negative. Electron microscopy shows neurosecretory granules in the cytoplasm.

HE ×360

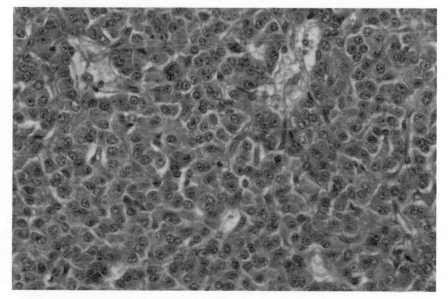

7.55 Carcinoma (oat-cell): bronchus

Oat-cell carcinoma of bronchus is closely linked with cigarette smoking. Like carcinoid tumour of bronchus it arises from Kultchitsky cells in the bronchial epithelium. Neurosecretory granules can be demonstrated in the cytoplasm of the tumour cells and account for the positive argyrophil reaction they sometimes give. The bronchial epithelium (top) is intact but atrophic and stretched over the tumour. It rests on a thickened basement membrane. The tumour consists of closely-packed small cells of uniform shape and size. They have ovoid basophilic nuclei and cells which appear round have been cut transversely. They have very little cytoplasm. There are numerous mitoses (thin arrow). Pyknotic nuclei and nuclear fragments are fairly numerous and an area of necrosis (thick arrow) is present. HE ×235

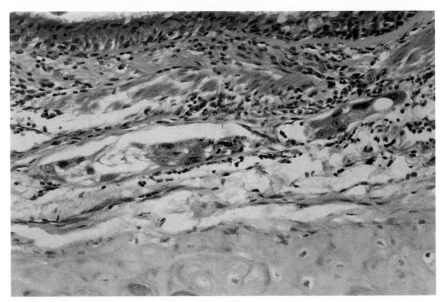

7.56 Carcinoma: bronchus

This is the bronchial mucosa several cm from a primary carcinoma of bronchus. The lumen of the bronchus and its lining of ciliated respiratory epithelium are visible at the top and the bronchial cartilage at the bottom. A lymphatic channel (arrow) in the submucosa has been cut longitudinally, and in its lumen there are groups of pleomorphic tumour cells. The cells practically fill the lumen and are probably part of a finger-like extension of the primary tumour along the lymphatic channel. Permeation of lymphatic channels as well as embolic spread to the lymph nodes are a well-recognized mode of spread of highly malignant tumours. There is some lymphocytic infiltration in the vicinity of the channel. *See also 1.54.*
HE ×235

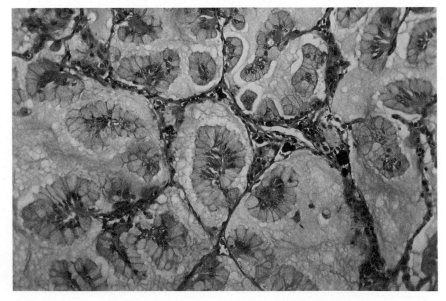

7.57 Alveolar carcinoma: lung

Alveolar carcinoma arises in the periphery of the lung, apparently distal to the bronchial epithelium. The cells grow within the alveoli, using the alveolar walls for support. Possible cells of origin are the Clara cell and the Type II (granular) pneumocyte. This tumour is composed of tall mucin-secreting columnar cells which are lining the alveolar walls and also forming papillary growths into the lumen. Mucin is also present within the alveoli. The walls of the alveoli are thin but apparently intact. Secondary adenocarcinoma can produce a similar lesion in the lungs and a diagnosis of alveolar carcinoma is valid only after the presence of a primary carcinoma in another organ such as the alimentary tract has been excluded. This is generally possible only by postmortem examination, which was performed in this case. HE ×135

7.58 Pulmonary blastoma: lung

Pulmonary blastoma is a rare lesion which may present with hemoptysis but is sometimes detected on routine X-ray examination of the chest. It is a well-circumscribed, lobulated firm mass, which histology shows to contain both epithelial and mesenchymal elements. It is probably best regarded as a 'mixed' tumour rather than a true blastoma arising from a pluripotential cell. It is a highly cellular lesion, consisting of pale-staining round epithelial cells (thin arrow) and strands of basophilic elongated mesenchymal elements (thick arrow). There is no necrosis. Two large vessels are lymphatics. HE ×150

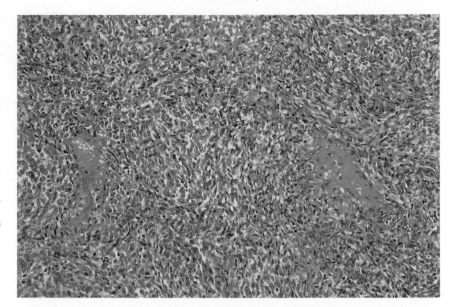

7.59 Pulmonary blastoma: lung

Epithelial elements predominate in this field: the cells are of regular size and shape and tend to form loose clusters (e.g. left of centre). They have round vesicular nuclei and a moderate amount of cytoplasm. A nucleolus is visible in some of the nuclei. No mitotic activity is evident and there is no necrosis. The mesenchymal tissue consists of strands of elongated basophilic connective tissue cells (arrow) and dilated thin-walled blood vessels (centre right). The appearances in pulmonary blastoma have been compared to fetal lung. The lesion is not highly malignant and although metastasis may occur, some cases have survived for long periods. HE ×235

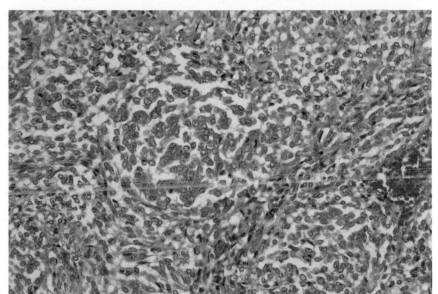

7.60 Secondary sarcoma: lung

The capillary bed of the lungs forms an effective filter and malignant cells in the bloodstream are frequently retained. The lungs are therefore a common site for metastatic tumour. This usually takes the form of multiple discrete masses but a solitary metastasis may be found. This metastasis was a hemorrhagic necrotic mass (4 × 4 × 3cm) in the right lower lobe of a man of 76. It appeared to be solitary and was removed surgically. It is a sarcoma composed of large cells showing considerable pleomorphism of nucleus and cytoplasm. Many are elongated and strap-like and others are more round and multinucleated. The appearances suggest that it is a rhabomyosarcoma but cross-striations could not be demonstrated in the cytoplasm. The site of the primary was unknown. HE ×360

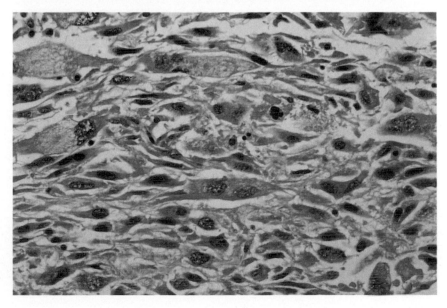

7.61 Secondary carcinoma: lung

A 'shadow' was found on X-ray examination in the upper lobe of the right lung of a man of 51. The lobe was excised surgically and found to contain a firm pale mass 3cm dia close to the main bronchus. It consists of cords of large cells separated by a delicate stroma of thin-walled blood vessels lined by flat endothelial cells (arrows). The cells have abundant pale granular cytoplasm. The vesicular nuclei show very little pleomorphism. No mitoses are evident. Despite its well-differentiated appearance, the tumour had infiltrated the adjacent lung deeply. The appearance of the cells and the sinusoidal structure of the stroma suggested that it was probably a metastasis from a primary renal carcinoma. Further clinical investigation confirmed this.   HE ×360

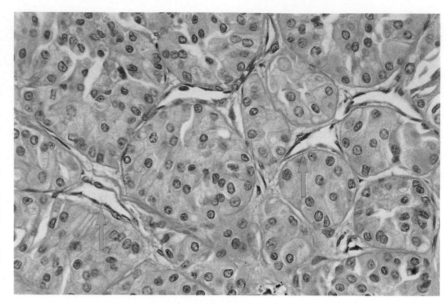

7.62 Lymphocytic lymphoma: lung

In the lung lymphocytic lymphoma may take the form of a well-defined mass or it may produce consolidation of part or all of a lobe. This is a lobectomy specimen from a man of 42 with an undiagnosed lesion in the left lower lobe. It formed a soft, creamy-white, wedge-shaped mass with its apex at the main bronchus and extending 5cm to the pleura where it was 8cm dia. Multiple fine white nodules were detectable in the surrounding lung. The lung parenchyma is extensively infiltrated with small lymphocytes. The infiltrate is greatest in the walls of the bronchioles and the larger blood vessels but extends into the alveolar walls. No germinal centres are present. Large numbers of macrophages and eosinophilic fluid are present in some of the alveoli.   HE ×60

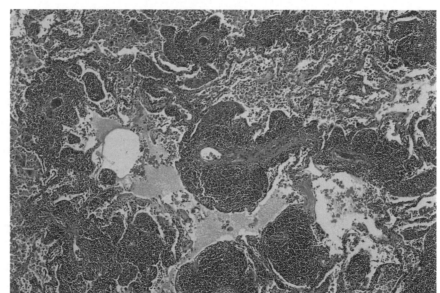

7.63 Lymphocytic lymphoma: lung

At higher magnification the lymphomatous infiltrate consists of a uniform population of small cells with round basophilic nuclei and only a thin rim of cytoplasm. They resemble mature small lymphocytes. Occasional histiocytes are present among the lymphocytes. The tumour was diagnosed as a low-grade lymphocytic lymphoma. There are hemosiderin-containing macrophages and lymphocytes in the alveoli.   HE ×150

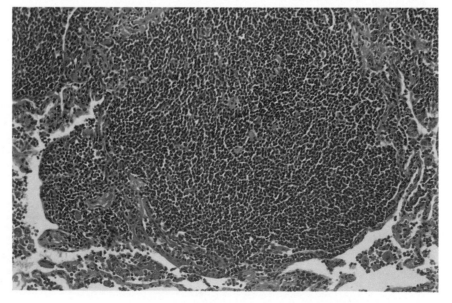

7.64 Mesothelioma: pleura

Mesothelioma is an increasingly common neoplasm of pleura and peritoneum, closely associated with asbestos. Exposure to the dust may be relatively slight and many years may elapse between exposure and development of the tumour. The tumour forms a continuous layer over the pleural surfaces, encasing the lung and obliterating the pleural cavity. It does not as a rule penetrate deeply into lung tissue. This tumour has a sarcomatous structure, consisting of elongated cells resembling fibroblasts or smooth muscle cells, with ovoid basophilic nuclei and eosinophilic cytoplasm. No connective tissue fibrils are evident. There is considerable nuclear pleomorphism but no mitoses are visible in this field. At higher magnifications small vacuoles were detectable in the cytoplasm of some of the cells.　HE ×235

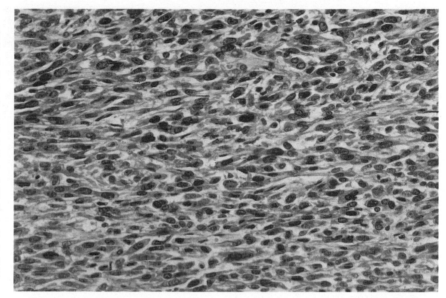

7.65 Mesothelioma: pleura

In this mesothelioma the malignant cells are forming tubules or gland-like spaces (pseudoacini). A mesothelioma which forms pseudoacini is termed an epithelial mesothelioma. In this case however the tumour cells lining the pseudoacini are unusually flat and attenuated and more like endothelium (or mesothelium) than epithelium, and only a minority of the spaces are lined by cuboidal cells. No mitoses are evident among the tumour. The tubules are separated by abundant spindle-cell stroma.

HE ×150

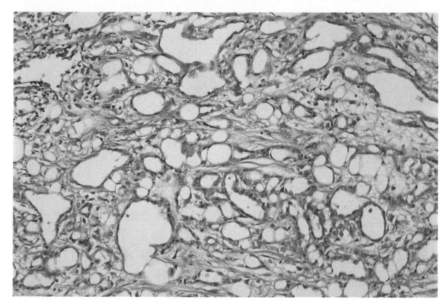

7.66 Mesothelioma: pleura

Mesothelial cells secrete hyaluronic acid and a positive result with stains for connective tissue mucin helps in making the diagnosis. The periodic acid-Schiff method, which stains epithelial mucins purplish-red, is also helpful; and in this section the vacuoles within the individual cells are unstained (hyaluronic acid is Schiff-negative), and there is no Schiff-positive secretion within the tubules (the strongly-staining material is cell debris). The absence of Schiff-positive epithelial mucin helps to exclude adenocarcinoma. The numerous delicate (reticulin) fibrils between the tumour cells which form a kind of basement membrane to the tubules, are Schiff-positive. The nuclear structure of the malignant cells is well demonstrated.　PAS ×360

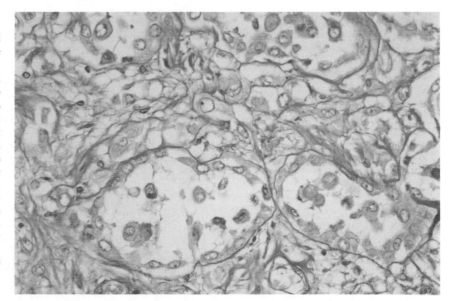

8.1 Infarct: pituitary

Ischemic necrosis of the pituitary is sometimes associated with childbirth. The pituitary enlarges during pregnancy and if the trauma and hemorrhage of childbirth cause shock and hypotension, the pituitary's blood supply is liable to fail. If this leads to extensive necrosis of the pituitary, a characteristic clinical syndrome develops (Simmond's disease or Sheehan's syndrome). Viable pituitary tissue consisting of pituicytes with well-stained nuclei remains (left). The other tissue (centre and right) is necrotic: the sinusoids are dilated and only the cytoplasm of the pituitary cells stains. The small deeply-staining nuclei are leukocytic. HE ×150

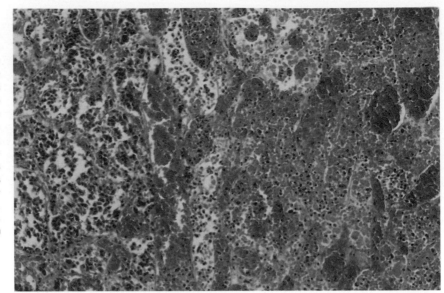

8.2 Chromophobe adenoma: pituitary

Chromophobe adenomas usually produce symptoms and signs by pressure rather than by hormone secretion. An 81-year-old man died of coronary artery disease and heart failure and a lobulated mass (3 × 3 × 4cm) was found protruding from the pituitary fossa. It had compressed the anterior cerebral artery and caused infarction of part of the right frontal lobe of the brain. A pituitary tumour had been resected 24 years previously. Histologically the tumour consists of groups of closely-packed polygonal cells separated by a delicate vascular stroma. The cells have small central nuclei which vary in size. There are no mitoses. The cytoplasm is eosinophilic, but special stains showed that it was a chromophobe adenoma. HE ×235

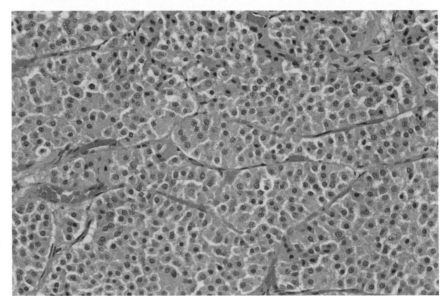

8.3 Acidophil (eosinophil) adenoma: pituitary

Acidophil adenomas causing acromegaly and gigantism are usually large and may cause extensive destruction of tissue around the pituitary fossa. This is an acidophil adenoma of pituitary, removed surgically from a man of 43 with acromegaly. It consists of cords of cells of uniform appearance: each has a round nucleus basally situated in the abundant cytoplasm. There is no nuclear pleomorphism and mitoses are not present. The abundant stroma between the cords of tumour cells is very vascular with many dilated thin-walled blood vessels. Special stains showed sparse granules in the cytoplasm of the tumour cells and immunohistochemistry confirmed the presence in them of growth hormone. HE ×360

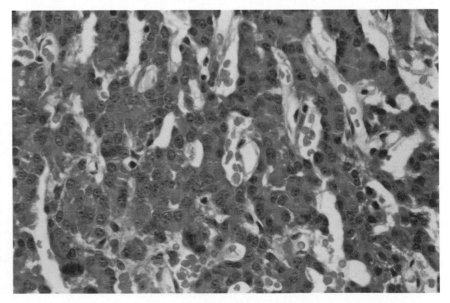

8.4 Acidophil (eosinophil) adenoma: pituitary

An acidophil adenoma was removed surgically from a 44-year-old man who had suffered from acromegaly for many years. The tumour consists of sheets of round cells with abundant eosinophilic cytoplasm and vesicular nuclei which are moderately pleomorphic. Many of the nuclei contain a prominent nucleolus. There is no mitotic activity. The cytoplasm is finely granular and in some cells a pale-staining juxtanuclear halo (arrow) is visible. The stroma consists of small thin-walled blood vessels.
HE ×860

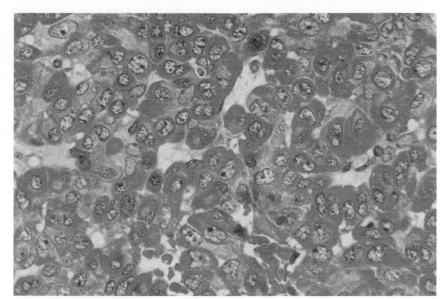

8.5 Acidophil (eosinophil) adenoma: pituitary

Sections stained by hematoxylin and eosin are usually of limited value in distinguishing the various types of pituitary cells and special techniques are required. This is a section of the adenoma shown in **8.4**, stained by Brooke's stain which colours the cytoplasm of acidophil (eosinophil) cells yellow. With few exceptions the cytoplasm of the cells of the adenoma is stained yellow, confirming that the adenoma is of the acidophil type. Other structures which are well demonstrated are the nuclei and nucleoli, the small thin-walled blood vessels of the stroma and delicate connective tissue fibres (stained blue). Brooke's stain ×860

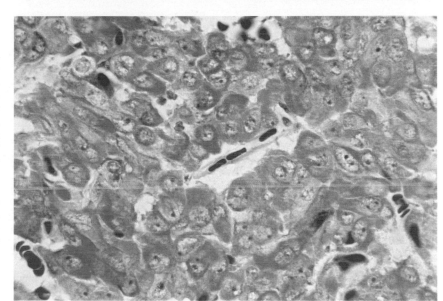

8.6 Acidophil (eosinophil) adenoma: pituitary

The most specific methods for classifying the cells of the pituitary and of pituitary adenomas are immunohistochemical and make use of a specific antibody to a particular hormone. This section of the lesion shown in **8.4** and **8.5** has been subjected to the indirect immunoperoxidase method using a specific antibody to growth hormone. The reaction product is brown, and the cytoplasm of a large majority of the cells in the adenoma gives a strong reaction for growth hormone.
Indirect immunoperoxidase method for growth hormone ×360

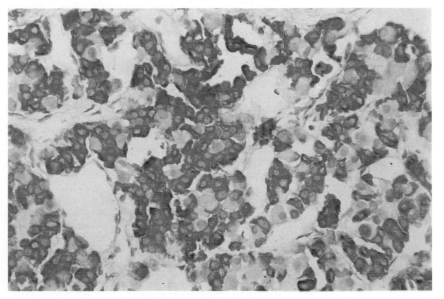

8.7 Secondary carcinoma: pituitary

Secondary deposits of tumour are not infrequently found in the pituitary, most of them from primary carcinoma of bronchus or breast. In this case the primary neoplasm was an oat-cell carcinoma of bronchus. This shows the anterior pituitary. The pituitary cells are necrotic and eosinophilic, with no nuclear staining, and infiltrating between them are compact cells with deeply basophilic ovoid nuclei and no discernible cytoplasm.   HE ×270

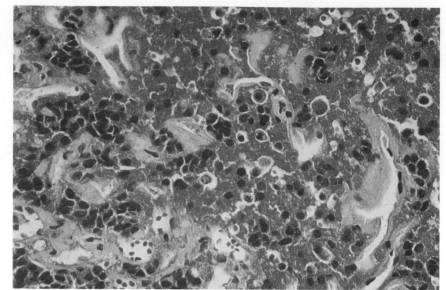

8.8 Craniopharyngioma: pituitary

Craniopharyngiomas arise from epithelial 'rests' derived from the pars tuberalis (Rathke's pouch) which also gives rise to the hypophysis. They tend to be located above the sella turcica (suprasellar cysts). This specimen is from a man of 87 who developed hypopituitarism shortly before his death. His pituitary had been largely replaced by a multicystic tumour, with only a thin rim of pituitary remaining. The remaining pituitary (left) is severely compressed, consisting of cords of flattened atrophic cells. The edge of the tumour is sharply demarcated with a cleft between it and the pituitary. The cells of the tumour have prominent basophilic nuclei and a moderate amount of cytoplasm. The 'spaces' containing amorphous material and fibrils and lined by cuboidal tumour cells are stromal.   HE ×150

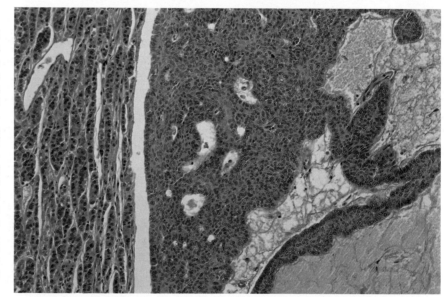

8.9 Craniopharyngioma: pituitary

In this part of the tumour the cells form long trabeculae which enclose 'spaces' containing eosinophilic fibres and small numbers of cells. These 'spaces' which are lined by cuboidal cells with basophilic nuclei are in fact edematous stroma, and in some craniopharyngiomas they are replaced by a stellate reticulum similar to that found in adamantinomas. A few thin-walled blood vessels are also present in the stroma.
HE ×150

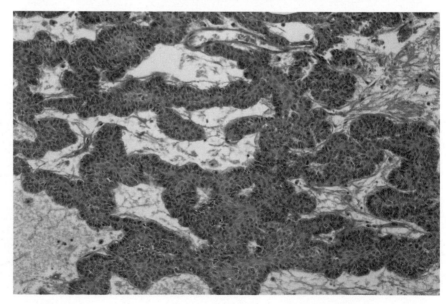

8.10 Thyroglossal duct cyst

The isthmus of the thyroid gland is derived from a tube of epithelium that grows down from the base of the tongue (the foramen cecum). Above the hyoid bone the tube is lined by squamous epithelium and below the bone by ciliated columnar epithelium. If the tube fails to involute, cysts may form, in the midline of the neck. Above the hyoid bone the cyst is a 'lingual dermoid' and below it a thyroglossal cyst. Thyroid tissue may be found in the wall of a thyroglossal cyst. The lumen of the cyst is on the left. The cyst is lined by stratifed squamous epithelium. The wall consists of vascular fibrous tissue and inactive-looking thyroid tissue (right). If large amounts of thyroid tissue are present in the wall, the cyst becomes a 'lingual thyroid'.   HE ×135

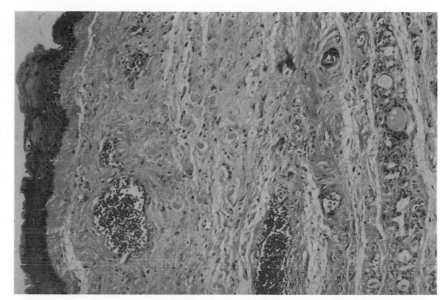

8.11 Thyroglossal duct cyst

The lumen of the cyst is at the top. The lining epithelium is essentially of respiratory type. It consists of ciliated columnar epithelial cells which appear pseudostratified and mucin-secreting goblet cells are also present. The connective tissue in the wall is vascular and lightly infiltrated with plasma cells and lymphocytes.   HE ×235

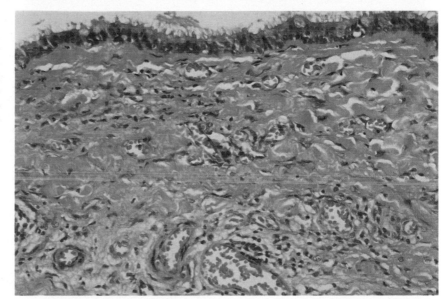

8.12 Nodular goitre: thyroid

In iodine-deficient parts of the world, endemic goitre is common. The enlargement of the thyroid is initially parenchymatous, i.e. it is caused by hyperplasia of the thyroid tissue. The number of follicles is increased and the epithelial cells are enlarged and columnar. Later focal changes occur, with some parts of the glands becoming hyperactive in their uptake of iodine and other parts becoming atrophic. In this way the gland becomes nodular, and its histological structure varies considerably from one part to another. This is a colloid-rich area. The epithelial cells lining the follicles are flat and most of the follicles are distended with colloid. Hemorrhage has occurred into a very large follicle (top). Sometimes the whole of a goitrous thyroid has a structure similar to that shown here: a colloid goitre.   HE ×60

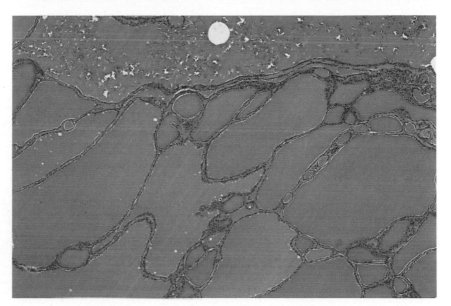

8.13 Graves' disease: thyroid

In Graves' disease a thyroid-stimulating antibody (an auto-antibody) reacts with receptors for thyroid-stimulating hormone (TSH) on the surface of thyroid epithelium. The thyroid secretes excess of hormone with widespread effects on many tissues. In the untreated case the thyroid is diffusely hyperplastic and contains little colloid. Because of the lack of colloid, the cut surface is a pale fleshy, greyish-pink instead of the glistening reddish-brown colour of the normal gland. Most of the follicles are small but others are fairly large. Several contain colloid, probably following pre-operative treatment of the patient with iodine. In the larger follicles the epithelium tends to form papilliform processes. The colloid is vacuolated at the edges, a sign of active resorption. There is a lymphoid follicle in the stroma (centre right). HE ×95

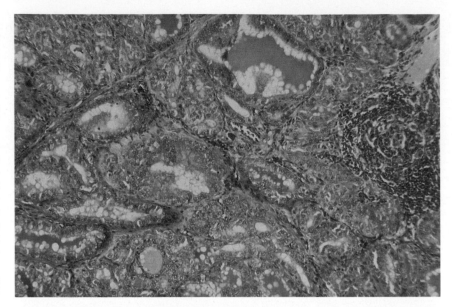

8.14 Graves' disease: thyroid

The lining epithelium is intensely hyperplastic: it is composed of tall columnar cells which have prominent vesicular nuclei and show a marked tendency to form papilliferous projections into the lumen. There is some colloid (left) but it is relatively pale-staining and the edge is excessively scalloped. In the rest of the lumen only strands of pale-staining secretion are evident. There is a prominent fibrous stroma. The stroma is very vascular but the vascularity is not obvious since the small vessels tend to collapse during processing of the tissue. Some smaller follicles which were very numerous throughout the gland are visible at the top. HE ×360

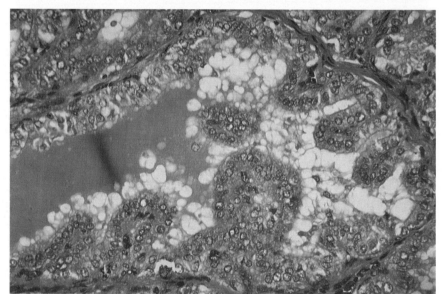

8.15 Graves' disease: thyroid

This case had been treated with iodine prior to surgical resection of the gland. Involution is almost complete, and apart from the relatively small follicles close to the fibrous capsule of the gland (right) the thyroid now consists of large follicles lined by flattened epithelial cells and full of well-stained colloid. Some peripheral vacuolation of the colloid is however still visible, showing that resorption of colloid continues. HE ×135

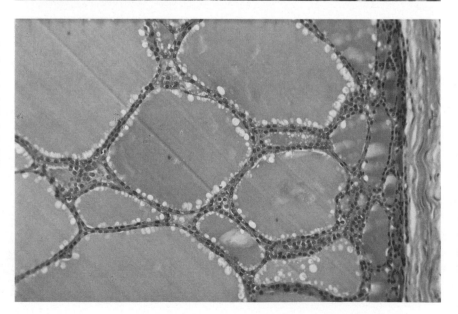

8.16 Riedel's thyroiditis: thyroid

Riedel's thyroiditis is a rare lesion which usually affects only part of the gland. The adjacent muscles are often involved and the condition may be mistaken for an invasive neoplasm, particularly since the affected part of the gland feels very hard. This lesion was typically limited to one pole of the thyroid which was adherent to the surrounding tissues in the neck. No thyroid epithelial cells can be identified. Apart from several small follicles lined by atrophic epithelial cells (arrows) the tissue consists largely of collagenous fibrous tissue in which there are many plasma cells and lymphocytes. Hypothyroidism is not usually present in Riedel's thyroiditis, presumably because it is a focal lesion. The sclerosing 'inflammatory' process in Riedel's thyroiditis has been compared with that in retroperitoneal fibrosis. HE ×235

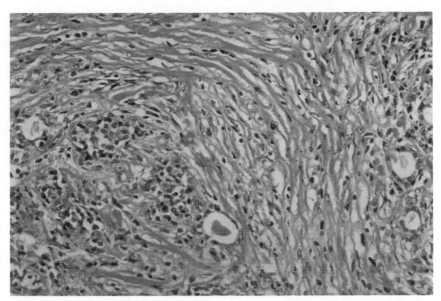

8.17 Giant-cell (de Quervain's) thyroiditis: thyroid

Giant-cell thyroiditis is an inflammatory condition which is possibly caused by the mumps virus. The gland becomes painful and tender and fever is often present. In this case the patient was a woman of 38. Part of the isthmus of the gland was removed surgically. Much of the gland's architecture has been destroyed and replaced by small irregular follicles and clusters of epithelial cells, lying in abundant fibrous stroma. Several larger follicles remain, one filled with colloid (centre left). The tissue is diffusely infiltrated with lymphocytes and a giant-cell granuloma (arrow) is present. HE ×150

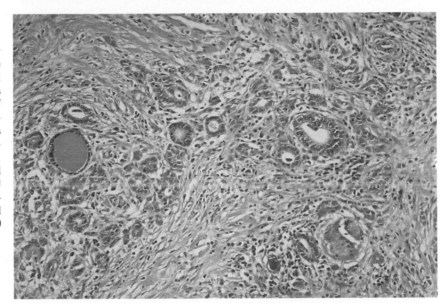

8.18 Giant-cell (de Quervain's) thyroiditis: thyroid

At higher magnification the destruction of the follicles is more obvious. The follicles are small, irregular in shape and greatly reduced in number. They are lined by cuboidal or flattened epithelium, but most show disruption of the epithelial lining. This is particularly obvious in the large follicles (arrow). The stroma consists of loose connective tissue and is abundant. Lymphocytes are present throughout the tissue but are more numerous at the points of disruption of the follicles. HE ×235

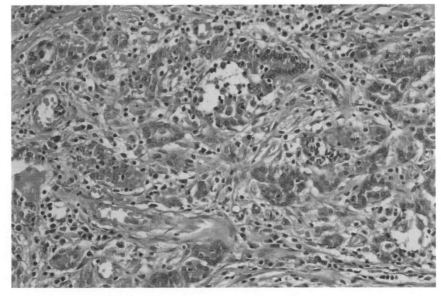

8.19 Hashimoto's thyroiditis: thyroid

Hashimoto's thyroiditis is an autoimmune disease and antibodies are generally present in the serum against thyroglobulin and a component of the endoplasmic reticulum ('microsomes') of thyroid epithelial cells. The thyroid is enlarged and destruction of epithelium generally leads to hypothyroidism. Sometimes goitre does not develop but destruction and shrinkage of the thyroid progresses until the gland is largely functionless (primary myxedema). Many follicles have disappeared. Those remaining are irregular in size and shape, some consisting only of clusters of epithelial cells. Others are lined by cells with pleomorphic nuclei and eosinophilic cytoplasm. Very little colloid is present. The tissue is infiltrated with lymphocytes and plasma cells, and a lymphoid follicle with a germinal centre is present (centre). HE ×200

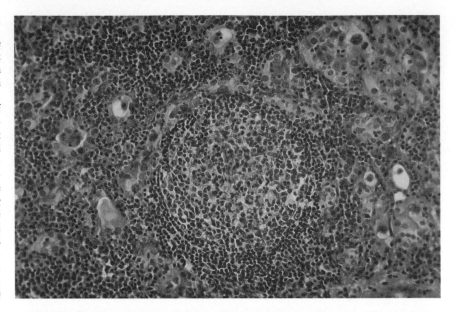

8.20 Hashimoto's thyroiditis: thyroid

The thyroid in this case was markedly enlarged. The remaining follicles (right) are lined by cuboidal epithelium and contain colloid. The other tissue is heavily infiltrated by small lymphocytes and histiocytes. A similar dense cellular infiltrate was present throughout the rest of the thyroid. Its monomorphic nature suggested the development of lymphocytic lymphoma in the gland. There was however no involvement of the cervical lymph nodes. HE ×360

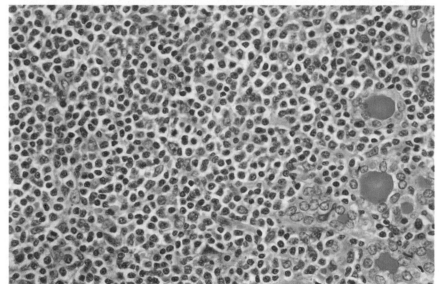

8.21 Adenoma: thyroid

Adenomas of thyroid are common lesions but it is difficult and occasionally impossible to distinguish between a true neoplasm and a localized hyperplastic nodule in e.g. a nodular goitre. It may also be difficult to distinguish a benign lesion from carcinoma. In this gland from a woman of 35 there were multiple well-circumscribed nodular lesions. This one is a so-called fetal adenoma, consisting of closely-packed small follicles lined by cuboidal epithelium within a fibrous capsule (right). Only a few of the follicles contain colloid and the smallest follicles are 'solid'. In some the epithelium is vacuolated. There is no nuclear pleomorphism and no mitoses are present. HE ×235

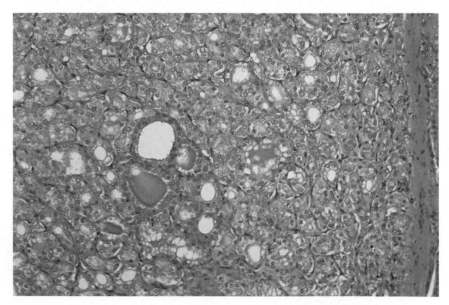

8.22 Papillary adenocarcinoma: thyroid

Papillary adenocarcinoma is the commonest type of carcinoma of thyroid. It may occur in young people who have a history of X-irradiation to the neck. It varies considerably in size and is not encapsulated. Despite the relatively benign appearance of the neoplastic epithelium, the lesion metastasizes readily to the adjacent lymph nodes. Distant metastases are infrequent. This tumour is from a man of 70. It is a well-differentiated papilliferous lesion, lying within a cystic space. The fibrous capsule (right) was incomplete. The papillae have a core of vascular connective tissue and are covered with a single layer of cuboidal epithelium. Eosinophilic secretion is present within the cyst.

HE ×30

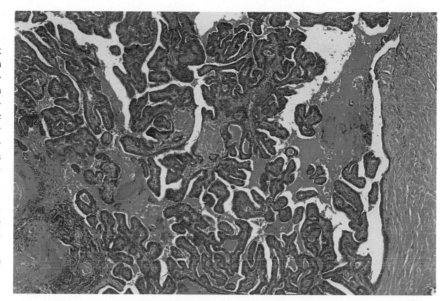

8.23 Papillary adenocarcinoma: thyroid

A number of nodules were present in the left lobe of the thyroid of a man of 47. This nodule has a pseudocapsule of fibrous tissue (right) which was incomplete. It consists of closely-packed papillae covered with cuboidal or low columnar epithelial cells. Follicles lined with similar cells are present in the core of the papillae and were numerous in other parts of the tumour. Only one follicle contains colloid. The nuclei of the epithelial cells are oval or round and pale-staining. Occasional mitotic figures were detectable elsewhere in the lesion, particularly in the follicular areas. HE ×150

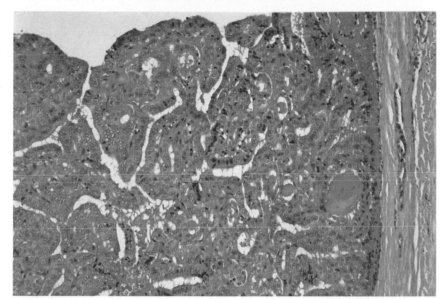

8.24 Papillary adenocarcinoma: thyroid

This is a follicular part of the lesion shown in **8.23**. The epithelial cells are cuboidal. Each cell has a large central nucleus in some of which a nucleolus is visible. As in the papillary areas of the tumour the nuclei are of uniform size and characteristically pale-staining and empty-looking. There are several mitotic figures (arrow). Several nuclei are pyknotic but most of the dark-staining nuclei are lymphocytic. There is eosinophilic amorphous material in the stroma but no colloid within the follicles. HE ×360

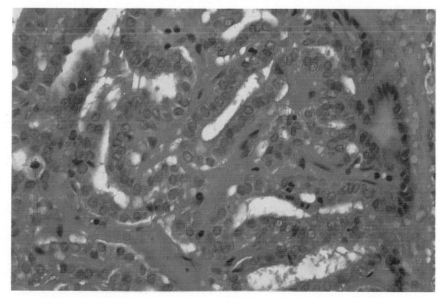

8.25 Follicular carcinoma: thyroid

Histologically follicular adenocarcinoma of the thyroid may be very well-differentiated and difficult to distinguish from 'adenomas', and the undifferentiated forms can easily be confused with lymphomas. This was a large tumour in an 11-year-old girl who had had X-irradiation to her neck in infancy. It is well-differentiated: the epithelial cells are cuboidal with relatively large nuclei, and they form colloid-filled follicles of fairly uniform shape. Although the nuclei of the epithelial cells show no pleomorphism, scattered mitoses were present throughout the tumour (arrow). The tumour was also locally invasive and recurred several times after surgery.   HE ×360

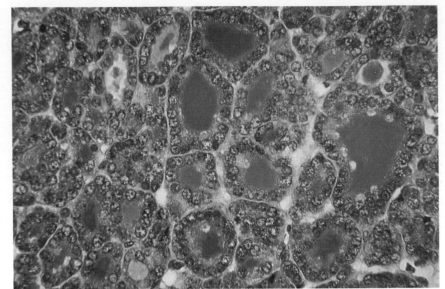

8.26 Follicular carcinoma: thyroid

A woman of 35 had a subtotal thyroidectomy for carcinoma of the thyroid. Macroscopically there were many fleshy nodules throughout the gland, as well as areas of hemorrhage and necrosis. This is part of a nodule. It consists of cells with large vesicular nuclei each containing a prominent central nucleolus. Well-formed follicles are present (thin arrow) filled with pale-staining colloid. There are also very small follicles (thick arrow) containing no colloid and also solid sheets of epithelial cells (lower right). A delicate fibrous stroma is present (double arrow). Many of the epithelial cell nuclei are pyknotic. There is one mitosis in this field but others were present in considerable numbers elsewhere. The tumour was enclosed in a fibrous 'pseudocapsule' but was invading a vein alongside it.
HE ×360

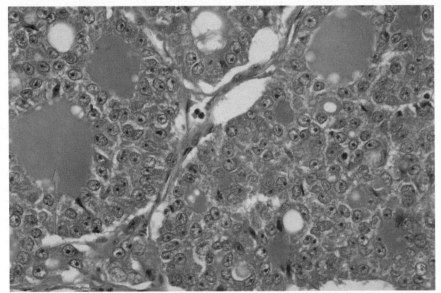

8.27 Hürthle cell carcinoma: thyroid

An encapsulated nodule (4 × 3 × 3cm) in the right lobe of the thyroid of a woman of 65 was removed at operation. The cut surface showed that it was lobulated and foci of necrosis were present. Histologically it is a solid poorly-differentiated carcinoma of Hürthle-cell type, composed of large polyhedral cells which have very pleomorphic nuclei and abundant densely eosinophilic cytoplasm. There is much mitotic activity and some of the mitoses are abnormal (arrow). There is some necrosis (centre left). The stroma is vascular. This is a highly malignant type of neoplasm, with a marked tendency to metastasize to other tissues, particularly bone and lung.   HE ×360

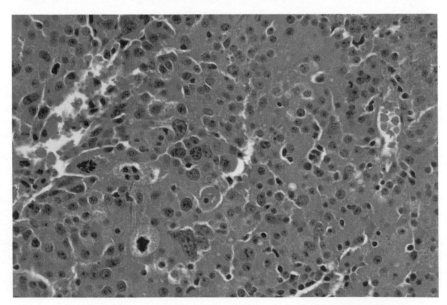

8.28 Medullary carcinoma: thyroid

Medullary carcinoma originates in the para-follicular cells which usually secrete calcitonin (C cells). It forms a solid greyish mass which appears well-demarcated. Histologically it consists of sheets and clumps of compact cells with small, round or oval nuclei and scant cytoplasm. There is no pleomorphism of nuclei and no mitotic figures are present. There is an abundant eosinophilic amorphous stroma (right) which gives the staining reactions of amyloid and is probably derived from hormone secreted by the tumour cells. Many small vessels are also present. Medullary carcinoma grows slowly but is locally invasive. It does not produce a characteristic clinical syndrome, and in some families the tumour occurs in association with pheochromocytoma and parathyroid adenomas.

HE ×235

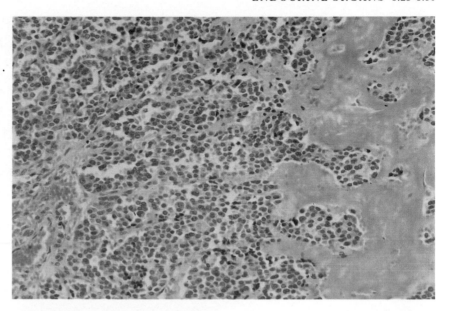

8.29 Adenoma: parathyroid

Parathyroid adenomas are yellowish brown (tan-coloured) nodules. They give rise to primary hyperparathyroidism. Care has to be taken to distinguish adenomas from hyperplasia. The presence of normal parathyroid at the periphery of an adenoma is a helpful feature. The normal parathyroid tissue (right) consists of chief cells. These are compact, with round basophilic nuclei and a moderate amount of cytoplasm. The adenoma (left), which has a well defined periphery and is separated from the normal tissue by a capsule of delicate connective tissue, consists of large cells with pleomorphic nuclei and strongly eosinophilic cytoplasm. These are oxyphil cells. Oxyphil cells occur in adenomas and not in hyperplastic parathyroids. Nuclear pleomorphism may be very marked but is of little significance.

HE ×150

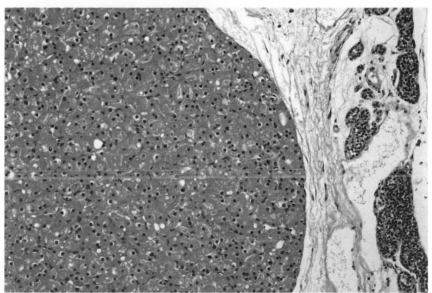

8.30 Adenoma: parathyroid

This was an ovoid mass (2.5 × 2cm) in the right lower parathyroid of a woman of 75. The cut surface was tan-coloured. The predominant cell is the oxyphil cell. Oxyphil cells have abundant deeply eosinophilic cytoplasm and form sheets and cords. The cell boundaries are ill-defined. The nuclei are basophilic and show a striking degree of pleomorphism, ranging from small round nuclei to extremely large hyperchromatic forms. There is however no mitotic activity in this field and little elsewhere in the tumour; and nuclear pleomorphism in these lesions does not denote malignancy. The stroma consists of delicate connective tissue and thin-walled blood vessels (arrow). Adenomas composed entirely of oxyphil cells (oxyphil-cell adenomas) do not as a rule secrete a significant excess of parathyroid hormone.

HE ×235

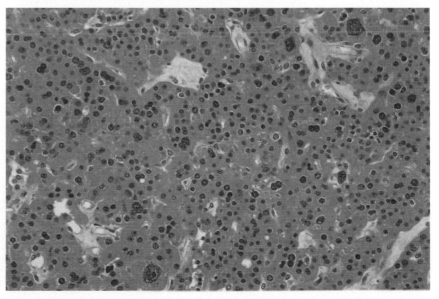

8.31 Adenoma: parathyroid

This was an ovoid nodule (1 × 1.5cm) in the right upper parathyroid of a woman of 46. The cut surface was dark brown. Histologically it consisted mainly of chief cells with occasional groups of oxyphil cells and 'water-clear' cells. This shows a chief cell region. The cells form a closely-packed solid mass, with no acinar formation. They have uniformly round nuclei and a moderate amount of cytoplasm. The stroma is inconspicuous, consisting of delicate connective tissue and small blood vessels. There is a mitotic figure (arrow) but the tumour is benign, carcinoma of parathyroid being extremely rare.   HE ×360

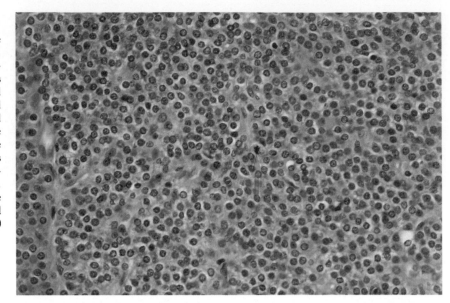

8.32 Adenoma: parathyroid

In this part of the adenoma shown in **8.31** the cells have abundant vacuolated cytoplasm with well-defined cell boundaries. These are 'water-clear' cells and rarely they are the major component of an adenoma. They have uniformly compact round nuclei, similar to those of chief cells. The stroma is inconspicuous, but thin-walled blood vessels (right) are present.   HE ×360

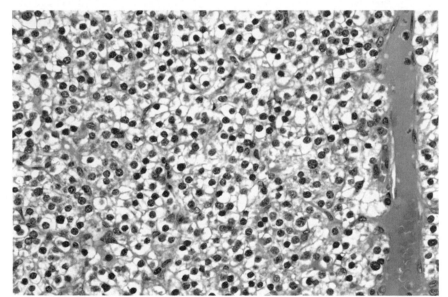

8.33 Adenoma: adrenal cortex

Adenomas of the adrenal cortex form small well-defined nodules, usually yellow from their content of lipid and steroid hormones. A cortical nodule more than 1cm dia is probably an adenoma. The tumour cells secrete hormone(s) and if the amount is sufficient, clinical signs and symptoms are produced. This tumour is from a woman of 53 on whom it had a virilizing effect. It was a firm apparently encapsulated yellowish mass 1.5cm dia. The capsule of the adrenal is on the right, and between it and the adenoma are compressed adrenal cortical cells with vacuolated cytoplasm (arrow). The adenoma (centre and left) is composed of small groups of cells with eosinophilic cytoplasm, separated by hyalinized stroma. Some groups show a pseudoglandular arrangement. The nuclei of the tumour cells are moderately pleomorphic but there are no mitoses.   HE ×150

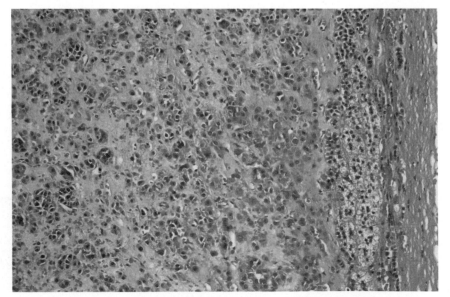

8.34 Carcinoma: adrenal

Carcinoma of the adrenal arises from the cortex and is generally highly malignant. Lymphatic and hematogenous spread readily occurs. It is usually large (more than 20cm dia) at the time of diagnosis. The tumour may secrete hormones, sometimes non-steroidal, but some neoplasms are non-secretory. The cut surface is characteristically yellow but necrosis and hemorrhage are generally prominent. Histologically the degree of cellular pleomorphism varies but is often extreme as in this lesion. The nuclei of the closely-packed tumour cells vary greatly in size and shape, from small round nuclei to very large hyperchromatic forms (centre). Many have prominent nucleoli and eosinophilic inclusions (of cytoplasm) are present in others (arrow). Some cells are multinucleated. The cells have abundant granular cytoplasm.   HE ×235

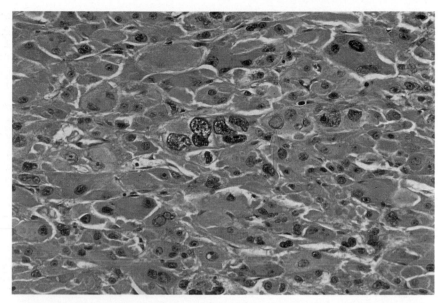

8.35 Pheochromocytoma: adrenal

Pheochromocytoma arises from the chromaffin cells of the adrenal medulla and, uncommonly, from chromaffin cells elsewhere (paraganglioma). It is almost invariably benign but the catecholamines secreted by the tumour cells give rise to a variety of signs and symptoms, notably systemic hypertension. This tumour was a bossellated mass (5 × 4 × 3.5cm), cystic and with a firm capsule, in the right adrenal of a woman of 56. It is composed of closely-packed large polyhedral cells arranged around the stromal blood vessels (arrows). The tumour cells have abundant granular cytoplasm and a large round nucleus containing one or more nucleoli. There is little nuclear pleomorphism although it may be pronounced without denoting malignancy. There are no mitoses.   HE ×360

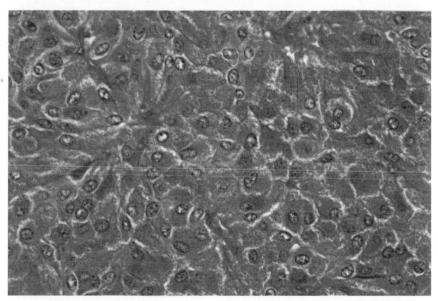

8.36 Neuroblastoma: adrenal

Neuroblastoma is a highly malignant neoplasm of the adrenal medulla and retroperitoneal tissues. It is composed of primitive nerve cells – neuroblasts – and mainly affects infants and children. It tends to form a large soft hemorrhagic mass, often accompanied by widespread secondaries in lymph nodes, liver and bones. It may secrete catecholamines, and their metabolites such as VMA (vanillyl mandelic acid) and HVA (homovanilic acid) may be excreted in the urine. This specimen is a secondary deposit in skeletal muscle. It consists of cells with round or oval deeply basophilic nuclei and very little cytoplasm. The cells, however, form circular rosettes around an eosinophilic mass which consists of very fine filaments originating in the tumour cells. Several small stromal blood vessels are visible (arrow).   HE ×360

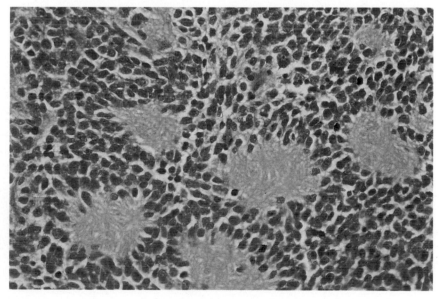

8.37 Diabetes mellitus: islet of Langerhans

In early-onset (insulin-dependent) diabetes the number of beta cells in the islets of Langerhans is reduced. The remaining beta cells are degranulated and sometimes they show hydropic degeneration from accumulation of glycogen. In maturity-onset diabetes these changes are occasionally present and in some individuals extracellular material is deposited. In this case, most of the islet cells have been replaced by an amorphous hyaline material which gave the staining reactions of amyloid. Amyloidosis of the islets is not specific for diabetes however and is found occasionally in older non-diabetic individuals. The exocrine tissue is normal.
HE ×335

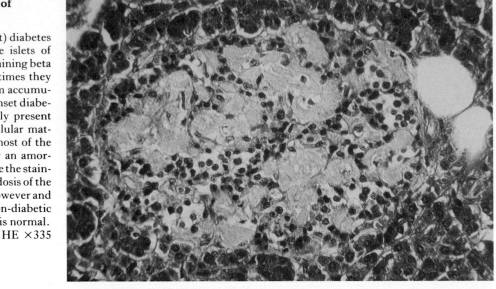

8.38 Islet cell hyperplasia: pancreas

Islet cell hyperplasia sometimes occurs in adults and enough hormone may be secreted to produce clinical symptoms and signs. If the increase is mainly in beta cells, hyperinsulinism and hypoglycemia result. Hyperplasia of islets also occurs in the newborn infants of diabetic mothers, as a response during the pregnancy by the infant's pancreas to its mother's hyperglycemia, and the infant may suffer from severe hypoglycemia. This is the pancreas of an adult subject who presented with the symptoms of hypoglycemia. Two islets are shown, both increased in size, that on the right very markedly. They are surrounded by fibrous tissue. The islets are vascular and the cells show nuclear pleomorphism. Most of the other islets were similarly enlarged. The exocrine tissue of the pancreas (bottom left) is normal.
HE ×150

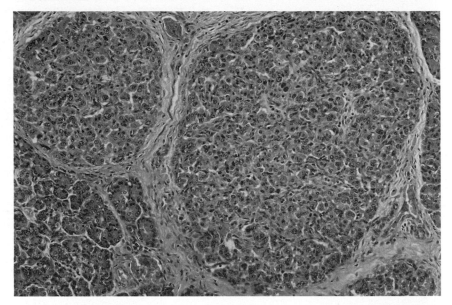

8.39 Islet cell adenomatosis: pancreas

Islet cell tumours can secrete a variety of hormones in addition to insulin and produce a range of clinical syndromes. This patient suffered from hypoglycemic episodes. A soft tumour 2cm dia in the head of pancreas, not encapsulated, was resected. It consists of groups of cells closely associated with a large thin-walled blood vessel. They are surrounded by fat and connective tissue heavily infiltrated by lymphocytes. The tumour cells have ovoid or round nuclei and eosinophilic cytoplasm. There is no fibrous capsule. Multiple other small tumours, separate from the main lesion, were found in the surrounding pancreatic tissue. They varied in size and were poorly defined. It was considered that they represented multiple primary tumours. Special stains gave results consistent with a beta cell origin. HE ×235

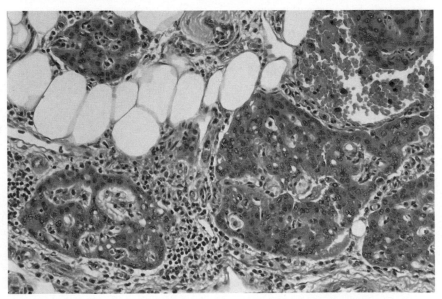

8.40 Malignant islet cell tumour (insulinoma): pancreas

About 10% of islet-cell tumours metastasize and are regarded as carcinomas. This patient suffered from hypoglycemia. An ill-defined mass (approx. 1cm dia) was found within the head of pancreas. The tumour is composed of a uniform population of cells with round nuclei and a moderate amount of cytoplasm. They form cords and sheets, supported by a fibrous tissue stroma. The fibrous tissue is hyalinized and also very vascular. There is some hemorrhage at the edge of this nodule, which is well-demarcated and enclosed by a 'capsule' of fibrous tissue. Other nodules of tumour were not encapsulated in this way however and the tumour was invading adjacent tissues.　　HE ×150

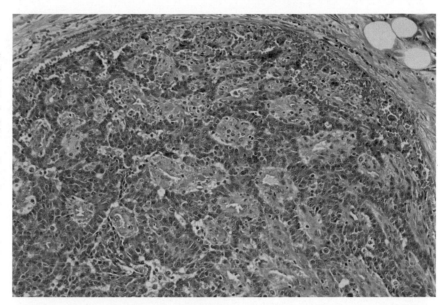

8.41 Malignant islet cell tumour (insulinoma): pancreas

At higher magnification the tumour cells have round or vesicular nuclei. The nuclei are moderately pleomorphic and prominent nucleoli are present in many of them. The cells have a moderate amount of faintly basophilic cytoplasm and lie in cords and sheets intersected by vascular stroma (thin arrow). The fibrous tissue of the stroma is hyalinized. There are no mitoses in this field but there were many throughout the tumour. The cytoplasm of the tumour cells gave a positive argyrophil reaction. It is difficult to predict the behaviour of islet cell tumours from the histological appearances. However in this case a lymph node was almost entirely replaced by tumour and metastases were present in the liver. HE ×360

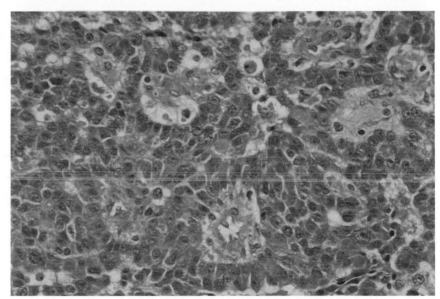

8.42 Malignant islet cell tumour (insulinoma): pancreas

To identify the hormone secreted by the tumour cells paraffin sections were treated by the indirect immunoperoxidase method with antibodies against insulin, glucagon and somatostatin. The sections treated with anti-glucagon and anti-somatostatin gave a negative result. In this section treated with the antibody against alpha-insulin, brown reaction product is present in the basal region of the tumour cells (thin arrow). This confirms that they are beta cells. The abundant stroma is hyaline and amyloid-like and it contains many small blood vessels (thick arrow). The structure of the nuclei of the tumour cells, with their several nucleoli, is also well shown.

Indirect immunoperoxidase method for
alpha-insulin ×470

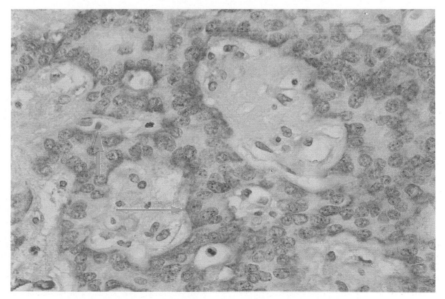

9.1 Subdural hematoma

Many small veins traverse the 'space' between the arachnoid membrane and the dura, and large veins drain into the venous sinuses. Bleeding from these veins occurs not infrequently after a head injury and a large subdural hematoma may form. Although 'organization' of the blood clot takes place, recurrent bleeding causes the mass to increase in size until it produces a life-threatening increase in intracranial pressure. Chronic subdural hematoma tends to affect old people but this lesion was in a woman of 26. This is the 'membrane' that was present in the subdural space. Its upper part (top), which was in close contact with the dura, is vascular connective tissue (arrow). Most of the membrane consists of denser fibrous tissue and its other surface (bottom) was in contact with the arachnoid membrane. HE ×60

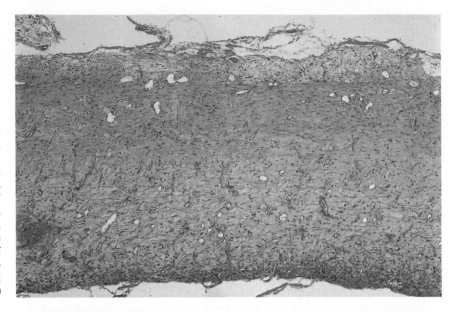

9.2 Subdural hematoma

This part of the 'membrane' present in the subdural space consists largely of blood clot (top). Organization of the extravasated blood is taking place, with invasion of the blood clot by capillaries, fibroblasts and macrophages; and a layer of fibrous tissue and granulation (organization) tissue (bottom) has already formed. Many of the cells in this tissue are macrophages full of hemosiderin (siderophages). HE ×150

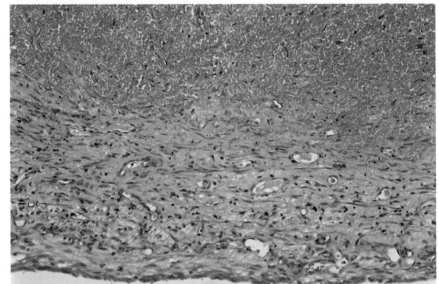

9.3 Subdural hematoma

This shows some of the blood clot (left) that was present beneath the dura and the tissue in contact with it, i.e. the most superficial part of the 'membrane'. The latter is rich in capillary blood vessels, macrophages and elongated fibroblasts. Small numbers of lymphocytes are also present. The tissue is a kind of 'granulation tissue' produced by organization of the blood clot, and the cytoplasm of most of the macrophages (siderophages) is dark brown from the presence of hemosiderin from digested red blood cells. The many small blood vessels in this tissue are themselves liable to bleed in the event of even mild trauma. HE ×360

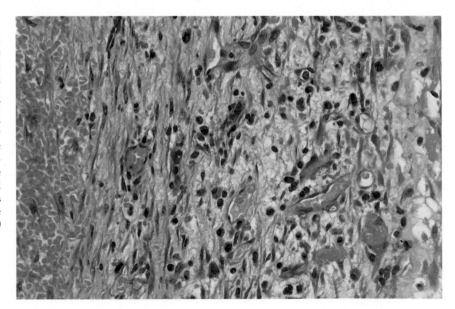

9.4 Infarct: brain

Infarction of the brain is caused by a sudden reduction in the blood flow, for example by thrombotic occlusion of an atheromatous cerebral, internal carotid or vertebral artery. Severe ischemic damage may also occur in the absence of thrombotic occlusion or significant stenosis as a result of a sudden reduction in blood flow, e.g. when severe hypotension develops. Infarcted brain becomes soft (a 'softening'). This infarct of white matter is of about 6 weeks' duration. The tissue has a spongy structure, many of the myelinated fibres having undergone ischemic necrosis and disappeared. The cells with foamy cytoplasm lying in the spaces between the surviving fibres are macrophages ('compound granular corpuscles') which have phagocytosed lipoproteins from the necrotic tissue. A number of astrocytes are also present.　　HE ×335

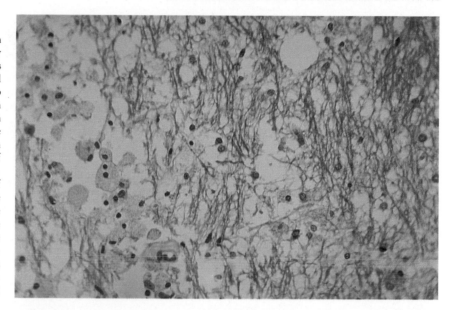

9.5 Cerebral hemorrhage: brain

This is the edge of a hemorrhage into the globus pallidus. The blood clot is on the left. There is patchy necrosis of the brain adjacent to the blood clot (centre and right). Many nerve cells and glial cells have disappeared and the tissue is edematous and vacuolated. The many vacuoles contained watery fluid. Scattered ischemic nerve cells survive as basophilic round bodies (arrow).　　HE ×135

9.6 Fat embolism: brain

When fatty tissue is traumatized, or a large bone such as the femur fractures, fat may enter the lacerated blood vessels and form emboli. The emboli are mostly retained by the lung capillaries but sometimes they pass through. In the brain they produce small infarcts, and if these are in vital centres death may occur. In this case there were multiple small (2–3mm dia) hemorrhagic areas in the brainstem. Each consists of a circular ('ring') hemorrhage around a dilated small blood vessel which is blocked by red cells and polymorph leukocytes and is probably thrombosed. The tissue between the vessel and the extravasated blood is necrotic. There is also a necrotic zone round the other blood vessel (top left). Macrophages have infiltrated the blood clot and necrotic tissue. A frozen section showed fat globules in the lumen of the blood vessels.　　HE ×135

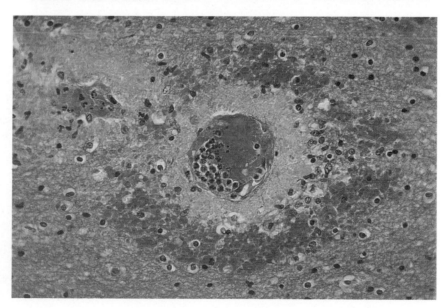

9.7 Acute purulent meningitis: brain

In acute purulent meningitis, pyogenic organisms (*Neisseria meningitidis, Streptococcus pneumoniae*, etc.) spread throughout the subarachnoid space. This may happen very rapidly and the person may die within a few hours. The leptomeninges become acutely inflamed, an inflammatory exudate collects in the subarachnoid space, and greenish pus-like material covers part or all of the surface of the brain. The surface of the cerebral cortex is on the right and the arachnoid membrane on the left. The subarachnoid space is distended with inflammatory exudate rich in fibrin and polymorph leukocytes (left). Polymorphs have infiltrated the arachnoid membrane but there is no increase in the cellularity of the cerebral cortex. The small blood vessels (arrows) on the surface of the brain however are dilated. HE ×120

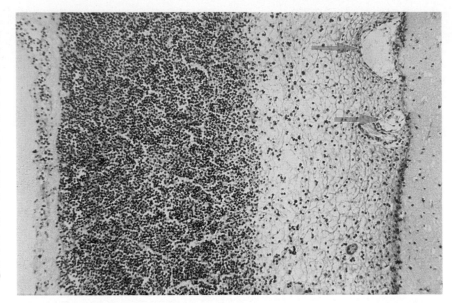

9.8 Abscess: brain

A brain abscess is a localized focus of suppurative encephalitis. Many types of organism can cause abscess formation in the brain and more than one type may be present. The infection reaches the brain by the bloodstream or by direct spread from a suppurative lesion in adjacent tissues such as the middle ear or mastoid. This is the wall of a developing abscess in the white matter. The centre of the lesion is out of the picture on the left. The white matter adjoining it (left) is necrotic and heavily infiltrated with polymorphs. That on the right is edematous and vacuolated, with fewer polymorphs. The small blood vessels are dilated and slight hemorrhage has occurred. If the patient successfully localizes the infection, a pyogenic membrane consisting of connective tissue, capillaries, glial cells and macrophages forms around the pus. HE ×160

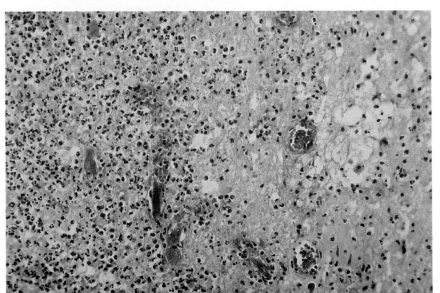

9.9 Tuberculous meningitis: brain

The tubercle bacillus (*Mycobacterium tuberculosis*) spreads to the brain by the bloodstream. In the subarachnoid space a thick greyish-green gelatinous exudate forms in which the characteristic tubercles (1–3mm dia) are detectable. The subarachnoid space (left half of picture) is full of inflammatory exudate which consists of fibrin and cellular debris (staining purplish-red). The small blood vessel (arrow) is necrotic and probably thrombosed. The exudate is adherent to the cortex (right half of picture) and the boundary between the cortex and the exudate is blurred. The cortex is edematous and vacuolated, and infiltrated by macrophages and lymphocytes. This superficial encephalitis has been caused by the inflammatory reaction in the meninges but necrotizing arteritis is also an important factor. HE ×160

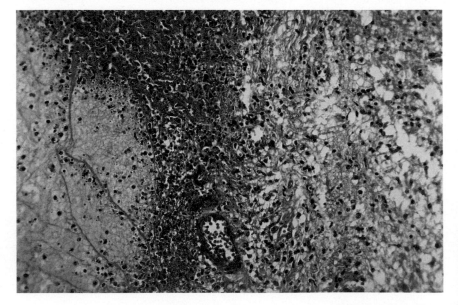

9.10 Aspergillosis: brain

Fungal infections of the brain are secondary to lesions elsewhere, usually in the lungs, and take the form of meningitis or multiple 'abscesses' which may resemble infarcts macroscopically. Frequently the patient's defences have been weakened, e.g. by drugs which depress immunity. Aspergillus is a branching filamentous organism and in the hyphae (3–4μm dia) there are many transverse septa. In this brain purplish-red hyphae are growing around and into a venule (thin arrow). The venule contains fibrin. The microorganism is also alongside and within the smaller blood vessels (thick arrow). There is a moderate infiltrate of polymorph leukocytes in the vicinity of the vessels but little reaction in the adjacent white matter. The patient was of girl of 13 who had received chemotherapy for Hodgkin's disease.   HE ×200

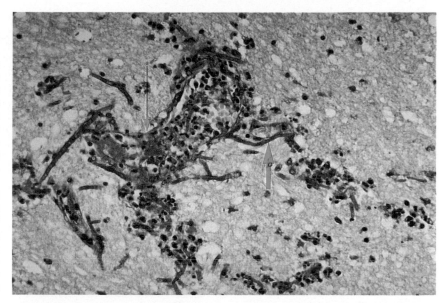

9.11 Cryptococcosis: brain

The yeast *Cryptococcus neoformans* most often produces lesions in the lungs. It may spread to the nervous system, however, and cause subacute meningitis and lesions in the cerebral cortex (meningo-encephalitis). The inflammatory reaction may be mild, despite the presence of gelatinous masses containing large numbers of the organism. As with other fungal infections, depression of the immune system is often as a predisposing factor. This cyst in the superficial cortex is typically flask-shaped and contains large numbers of the microorganism. The microorganism (arrow) stains very lightly and is enclosed in a colourless thick mucoid capsule. The capsular material gives the lesions their mucoid naked-eye appearance. There are small numbers of lymphocytes in the surrounding tissues but no macrophage or glial response.   HE ×360

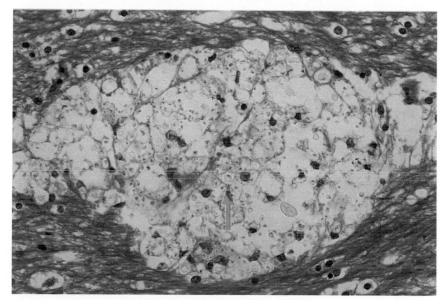

9.12 Malaria: brain

Malaria caused by *Plasmodium falciparum* (subtertian malaria) can affect many tissues and organs, as a result of blockage of small blood vessels by red cells containing the parasite. This is a particularly important feature of cerebral malaria, a serious and often fatal condition. This shows one such small vessel in the brain of a fatal case. It is completely occluded by parasitized red cells, the malaria parasite appearing as small basophilic 'rings' (arrow). There is extensive hemorrhage around the blocked vessel. Macroscopically there were many similar petechial hemorrhages throughout the brain.   HE ×580

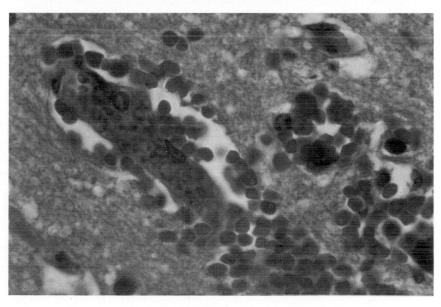

9.13 Tuberous sclerosis: brain

In tuberous sclerosis the child is mentally deficient and there are multiple gliomatous nodules in the cerebral hemispheres. There are also a variety of lesions in other organs, including angiofibromatous papules in the skin of the face, rhabdomyoma of the heart and malformations of the kidneys and liver. This is part of a nodule in the cerebral cortex. The normal cortex has been replaced by a tissue consisting of abundant glial fibres and numerous bizarre giant cells. Some of the giant cells, which are characteristic of tuberous sclerosis, have the features of neurons and others of astrocytes.   HE ×150

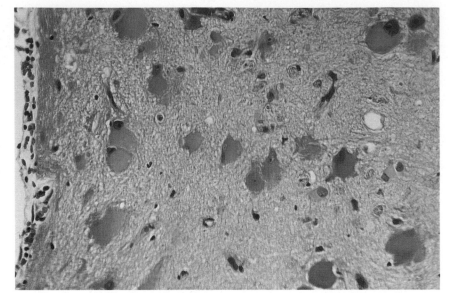

9.14 Gumma: brain

In the tertiary stage of syphilis gummas may develop. They are characterized by extensive necrosis of tissue. Absorption of the necrotic tissue tends to take place relatively slowly and considerable scarring of the involved organ may result. This is a gumma of brain. Part of the necrotic centre of the gumma is visible on the left. The necrotic material is being phagocytosed by macrophages many of which have granular cytoplasm full of lipid-rich ingested material. A layer of fibrous tissue (centre) has formed and peripheral to it (right) is a cellular zone rich in plasma cells and lymphocytes. Small numbers of plasma cells, lymphocytes and macrophages are also present in the fibrous tissue (centre). *See also 1.18, 5.13*

HE ×235

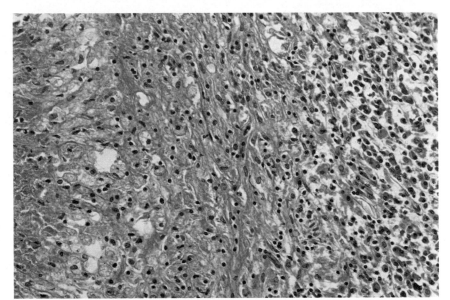

9.15 Tabes dorsalis: spinal cord

Tabes dorsalis is a late manifestation of syphilis. The posterior roots of the spinal nerves, particularly those entering the lumbar enlargement of the cord, and their upward extensions in the posterior columns of the spinal cord slowly and progressively degenerate. This is a section through segment L4 of the cord, stained by the Loyez method for myelin. There is an area of pallor in each of the posterior columns (arrows) in the middle root zone, caused by the loss of the myelinated fibres from those areas. These changes lead to loss of proprioreceptor sense in the muscles and joints of the legs, as well as loss of pain sense. Similar changes are present in the affected roots (not shown here) and they may occur also in sensory cranial nerves including the optic nerves.

Loyez method ×11

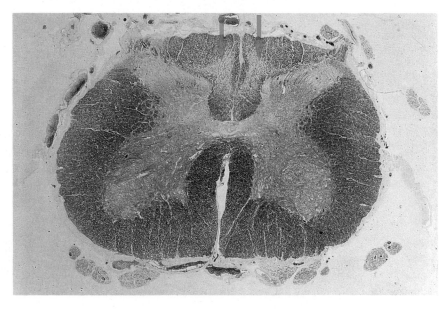

9.16 Alzheimer's disease: brain

Alzheimer's disease is similar to senile dementia but appears about a decade earlier ('pre-senile' dementia). There is widespread atrophy of the cortex of the frontal and temporal lobes, with loss of many neurons and increase in glial tissue. This biopsy specimen from the cortex of a man of 63 has been treated by the periodic acid-silver method. The subarachnoid space and the surface of the cortex are at the top. Within the cortical grey matter there are many dark-staining (argyrophilic) round plaques. Alzheimer plaques consist of masses of small argyrophilic granules and filaments; and amyloid is also often present in the centre of the plaque. Many small blood vessels (stained red) are also shown, as well as larger vessels in the subarachnoid space.

Periodic acid-silver ×90

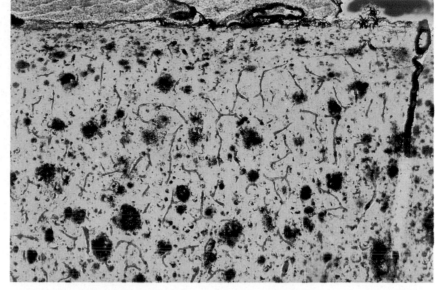

9.17 Poliomyelitis: spinal cord

Most infections with the poliomyelitis virus are clinically 'silent' and only a small minority (about 1%) get paralytic lesions from destruction of the motor neurons in the spinal cord or in the brainstem. This is the ventral horn in a patient who died 6 days after the onset of the illness. The tissue is infiltrated with inflammatory cells, mostly lymphocytes and macrophages. All the neurons are degenerate, with no nucleus and little or no Nissl substance. They are shrunken and lying in vacuoles. The surrounding tissues are edematous and vacuolated. Several necrotic neurons are being phagocytosed by macrophages and polymorphs (neuronophagia) and in places where a neuron has disappeared only swollen macrophages remain (thin arrow). The small blood vessels (thick arrow) are dilated and have a cuff of inflammatory cells. HE ×235

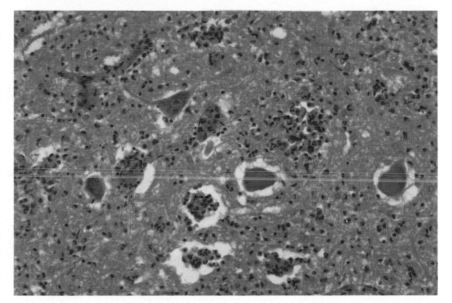

9.18 Poliomyelitis: spinal cord

This is the ventral horn from a person who died 1 week after the onset of the illness. There is very extensive destruction and disruption of tissue, with loss of all neurons except for one degenerate and shrunken neuron (arrow). The tissue is infiltrated with large numbers of macrophages, lymphocytes and plasma cells. The macrophages are swollen, their granular cytoplasm being full of lipid following ingestion of necrotic neurons and tissue. The inflammatory reaction is still acute and the small blood vessels are dilated. *See also 1.16.*

HE ×235

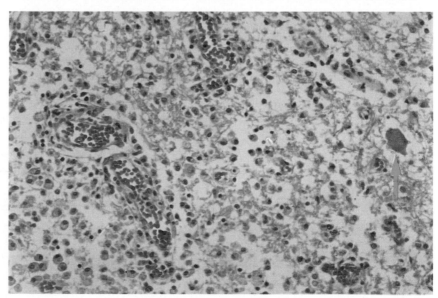

9.19 Multiple sclerosis: spinal cord

In multiple sclerosis, parts of the central nervous system undergo rapid loss of myelin (acute demyelination). Axons are preserved however and may function for a long time. The presence of these plaques of patchy demyelination causes increasing disturbance of both motor and sensory functions. This is a section of the cervical spinal cord stained by the Weigert-Pal method which colours myelin black. Two plaques of demyelination are present: a small round one in the ventrolateral part of the cord (thin arrow) and a much larger one (thick arrow) affecting most of the posterior columns. Note the irregular shape of the large plaque, the lack of conformity with anatomical structure, the complete loss of myelin within the plaque and the sharp line of demarcation between the plaque and the surrounding tissues.   Weigert-Pal ×9

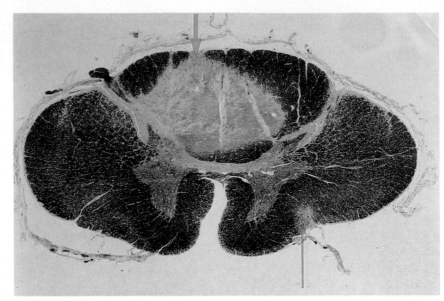

9.20 Multiple sclerosis: brain

The lesions in multiple sclerosis are randomly distributed and develop at irregular intervals. This is the edge of a plaque in the white matter in the brain. The plaque is on the left and the white matter on the right. The plaque is pale-staining and appears virtually structureless, with complete loss of the eosinophilic myelin (the stain does not demonstrate axons). Only a few astrocytes remain. The white matter on the right is more or less normal in appearance, most of the many cells in it being oligodendrocytes.
HE ×150

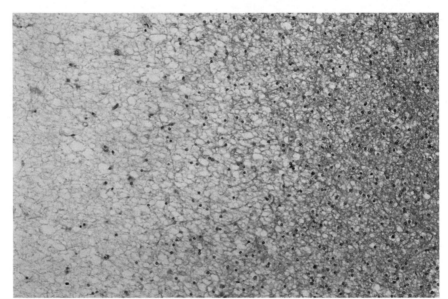

9.21 Multiple sclerosis: brain

This is a frozen section of a recent lesion in the cerebrum, stained with Sudan IV to show fat. The plaque, located in the central white matter, is coloured orange-red, from the presence of much stainable fat. The fat has come from the breakdown of the complex lipids of the myelin of the medullary sheaths of the axons. Above and to the left of the plaque are the subcortical blue-staining 'U' fibres which have been spared by the demyelinating process.   Sudan IV ×11

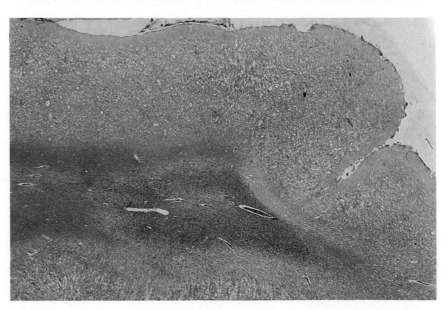

9.22 Multiple sclerosis: brain

Gliosis of the plaques develops slowly but eventually converts the plaques from soft yellowish areas to firm grey lesions, easily visible macroscopically. This is an old plaque in the wall of a lateral ventricle, a common site for lesions. The ependymal lining of the ventricle is on the left. The plaque now consists entirely of astrocytes and glial fibres. HE ×200

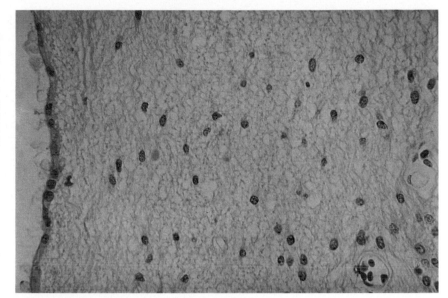

9.23 Motor neuron disease: spinal cord

In motor neuron disease the motor neurons degenerate spontaneously and progressively. When the neurons are in the spinal cord the muscles supplied by them atrophy (progressive muscular atrophy) and when the neurons are in the brainstem the condition is progressive bulbar palsy. When upper motor neurons in the brain are affected, fibres are lost from the corticospinal tracts and there is spastic paralysis (amyotrophic lateral sclerosis). This is the spinal cord from a patient with amyotrophic lateral sclerosis, stained for myelin. There is loss of staining (pallor) in the lateral and anterior columns, affecting both the crossed cerebrospinal tracts (thin arrows) and the direct tracts (thick arrows). Loss of myelinated fibres is more pronounced in the crossed cerebrospinal tracts than in the direct tracts. Loyez method ×9

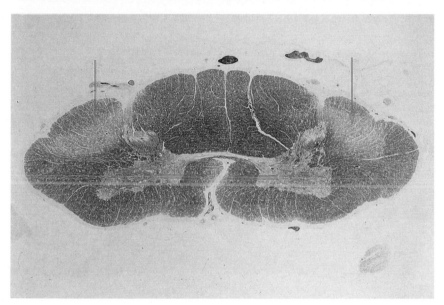

9.24 Motor neuron disease: spinal cord

This is the ventral horn of the spinal cord from a patient who had progressive muscular atrophy. It has been stained with thionin to demonstrate the motor neurons selectively. The number of motor neurons is much smaller than normal. Those that remain are degenerate, with evidence of karyolysis and chromatolysis. Degeneration and loss of the motor neurons produce atrophy and paralysis of the muscles which they supply. Thionin ×80

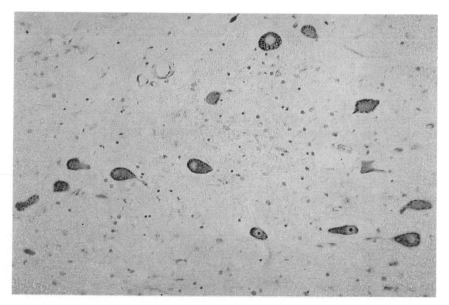

9.25 Subacute combined degeneration: spinal cord

Subacute combined degeneration is caused by vitamin B12 deficiency. Myelinated fibres in the dorsal and lateral columns of the spinal cord degenerate and disappear. The lower thoracic region of the cord is most often involved, but degenerative changes spread upwards and downwards, as high as the cerebral cortex, from the affected part. This section of the spinal cord (at level C2) has been stained by the Weigert-Pal method for myelin. The nerve roots are unaffected but the following tracts are degenerate and unstained: the posterior columns (gracilis and part of cuneate) (thin arrows); the anterior (direct) cerebrospinal (thick arrows); and, on the lateral aspects of the cord, the anterior and posterior spinocerebellar and lateral (crossed) cerebrospinal tracts (double arrows). Weigert-Pal ×9

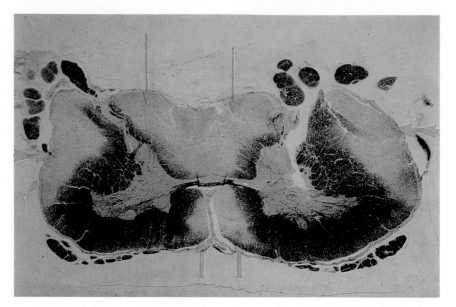

9.26 Polyneuropathy: sural nerve

A man of 43 who had drunk 25 pints of beer a day for many years presented with the signs and symptoms of polyneuropathy. A small portion of sural nerve was removed for diagnostic purposes. In this section myelin is coloured blue. There is marked demyelination of all the nerve fibres. Special stains showed however that the axons were intact. The pattern of breakdown of myelin is typical of the segmental degeneration of Gombault in which the degeneration primarily affects segments of myelin lying between two nodes of Ranvier. If the etiological agent is removed, remyelination takes place with recovery of function. Gombault's segmental degeneration is found in other conditions including the polyneuropathy of malnutrition and acute infectious polyneuritis (Guillain-Barré syndrome).

Solochrome cyanin ×335

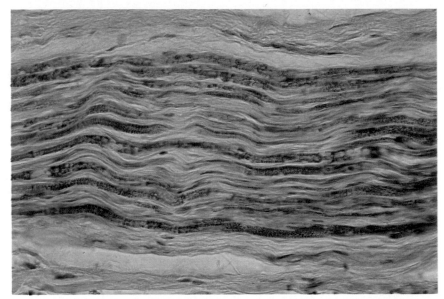

9.27 Ganglioneuroma

This tumour arises in the adrenal medulla or in the sympathetic nerve chain. It is usually benign but when neuroblasts are present (ganglioneuroblastoma) the prognosis may be less good. Conversely a neuroblastoma may undergo differentiation into a ganglioneuroma, with a corresponding improvement in prognosis. This is a typical ganglioneuroma consisting of Schwann cells with slender elongated nuclei and a number of large round ganglion cells (neurons) (arrows). The eosinophilic fibrillary tissue has been formed by the Schwann cells and like a Schwannoma it is rich in reticulin fibres but contains relatively little collagen.

HE ×235

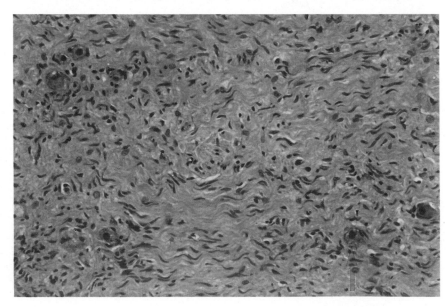

9.28 Meningioma

Meningiomas arise from the 'arachnoidal fibroblasts' of the arachnoidal granulations, which return the cerebrospinal fluid to the bloodstream. They are firm lobulated tumours attached to the dura, usually in the vicinity of the venous sinuses. They may grow to a large size and compress the brain but they do not invade it and are usually resectable. This is a cellular transitional meningioma. The cells are polygonal with round nuclei of fairly uniform structure. There are many pyknotic nuclei but no mitoses. The cells have a marked tendency to form whorls of various sizes. The eosinophilic cells at the centre of the whorls (arrows) are undergoing hyalinization and calcification (some of the whorls look like blood vessels). Calcification which is just beginning in some of the whorls converts them into psammoma bodies. HE ×235

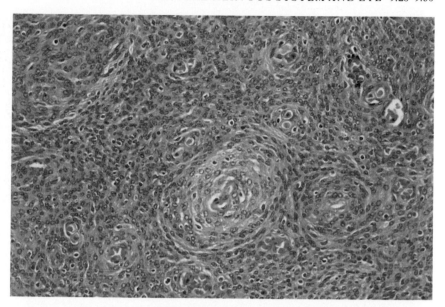

9.29 Meningioma

A meningioma may invade the overlying skull bone and cause it to thicken. This lesion, which arose in a woman of 59, invaded the dura which reacted and became much thicker. It is a transitional type of meningioma. The tumour cells have uniform ovoid vesicular nuclei and abundant eosinophilic cytoplasm which is vacuolated in some cells. There is no pleomorphism of nuclei and no mitoses are present. The cells form whorls (top left) with a small central cavity. The cells at the centre of the whorls eventually become hyalinized and may calcify to form 'psammoma' bodies. Sometimes there is a blood vessel at the centre of a cellular whorl. A meningioma with many psammoma bodies has a gritty texture when it is cut. HE ×360

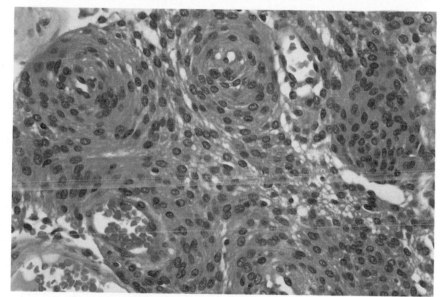

9.30 Oligodendrocytoma: brain

The oligodendrocytoma is derived from oligodendrocytes, the small glial cells with few cell processes which play an important role in the maintenance of the myelin sheaths of the axons. It is a slow-growing, soft fleshy tumour most often located in the white matter of the cerebral hemispheres. The prognosis is generally better than with astrocytomas but a minority are aggressive in their behaviour. The histological structure is characteristic. It is highly cellular, each cell having a small round nucleus and clear cytoplasm bounded by a well-defined cell membrane—an appearance termed 'boxing' of the nucleus. There are numerous small thin-walled blood vessels (arrow). Calcification (not evident here) is a frequent occurrence in oligodendrocytomas and is demonstrable in almost half of them radiologically. HE ×360

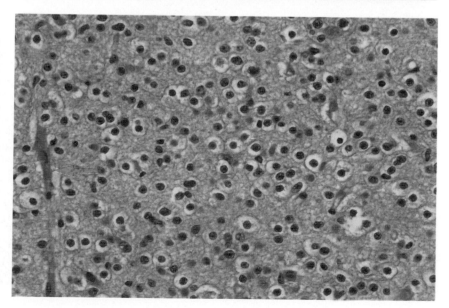

9.31 Astrocytoma: brain

Astrocytes are the predominant glial-forming cell and astrocytomas are the most common form of primary neoplasm, in the central nervous system. They vary considerably in malignancy, from Grade I, the best differentiated, to Grade IV, the most malignant. It is often difficult however to allocate a tumour precisely to one grade, because of variations in its structure from one part to another. Astrocytomas are not encapsulated and resection of even a Grade I neoplasm may be impossible if it is in a vital part of the brain. This is a fibrillary astrocytoma of Grade I or possibly Grade II: the astrocytes are mature and their fibrillary processes are well developed. The dense red bodies (arrows) are Rosenthal fibres. There are also collections of fluid (microcysts) in the interstices of the fibrillary processes of the astrocytes. HE ×200

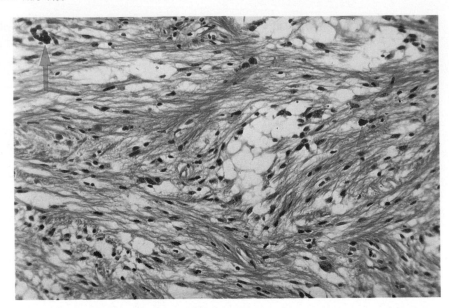

9.32 Astrocytoma: brain

This tumour was in the right frontal lobe of a man of 52. It is a fairly well-differentiated astrocytoma (Grade II). The nuclei are fairly large and well-stained and show some pleomorphism. There are no mitoses however. The cells have abundant eosinophilic cytoplasm and tend to be round, with a well-defined cytoplasmic boundary. This form of astrocyte is a gemistocyte and when it is the predominant form in an astrocytoma, the tumour is termed a gemistocytic astrocytoma. The many vacuoles between the cells were filled with fluid. Sometimes fluid collects on a large scale in an astrocytoma, with the formation of a cyst into which neoplastic tissue projects. HE ×360

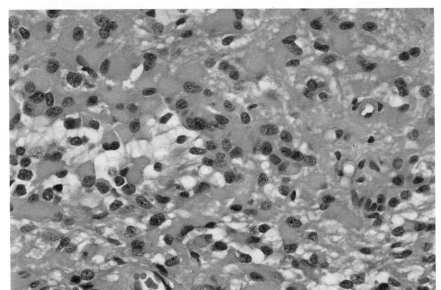

9.33 Astrocytoma: brain

A poorly differentiated astrocytoma (Grade III) was removed surgically from the left fronto-parietal region of a man of 26 but recurred 4 years later. This is the recurrent lesion. The cells are pleomorphic, with wide variations in the size and shape of their nuclei which are also deeply basophilic. Some are multinucleated (arrow). There are no mitoses in this field but elsewhere moderate numbers of mitotic figures could be detected. The neoplastic astrocytes have abundant pale-staining cytoplasm but special stains showed that the cytoplasmic processes were shorter and thicker than normal. Numerous small blood vessels are present, some lined by plump endothelial cells. No necrosis is evident but necrotic areas were present in other parts of the neoplasm. HE ×415

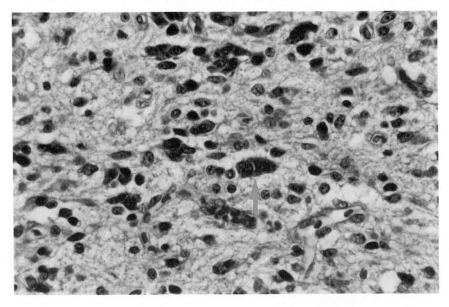

9.34 Glioblastoma multiforme: brain

The term glioblastoma multiforme is sometimes still applied to highly malignant astrocytomas of Grades III and IV. This lesion, a Grade IV astrocytoma, was a round (4cm dia) cream-coloured hemorrhagic mass in the left parietal lobe. The cells vary greatly in size and shape but most are elongated with long fibrillary processes. Their nuclei are pleomorphic and hyperchromatic with prominent nucleoli. Many are multinucleated. There are also numerous mitoses (arrows), some of abnormal form. Part of a large area of necrosis is just visible at the bottom, containing much nuclear debris. Other extensive areas of necrosis were present throughout the tumour.   HE ×360

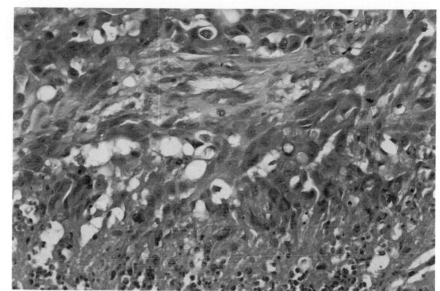

9.35 Polar spongioblastoma: brain

Polar spongioblastoma is a rare highly malignant tumour of astrocytes or their precursors (spongioblasts) which occurs in young individuals usually in the region of the 3rd and 4th ventricles of the brain. This lesion however was in the right frontal lobe of a boy of 16. The tumour cells (e.g. right of centre) are primitive, with little cytoplasm and round basophilic nuclei containing one or more nucleoli. Many nuclei are pyknotic and there are occasional mitoses. The tumour is very vascular and the small blood vessels (thin arrows) are lined with large endothelial cells with abundant cytoplasm. Cellular 'buds' resembling miniature glomeruli (thick arrows) project from the surface of the vessels. Proliferation of the endothelial cells of small stromal blood vessels is a feature also of less well-differentiated astrocytomas (Grades III and IV).   HE ×360

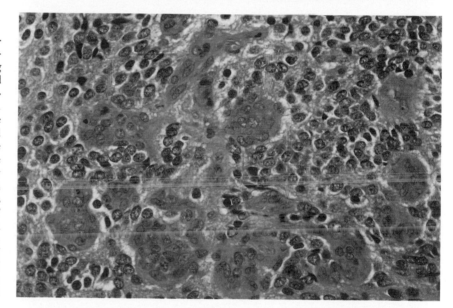

9.36 Medulloblastoma: brain

Medulloblastoma is a highly malignant tumour composed of primitive nerve cells, occurring mostly in children. It is usually located in the cerebellum, frequently invading the fourth ventricle. It spreads readily by the cerebrospinal fluid to the surface of the brain and spinal cord, and to the other ventricles. This tumour is from a woman of 24. The malignant cells are very uniform, with round or ovoid basophilic nuclei and inconspicuous cytoplasm. Many nuclei are pyknotic. There are many small spaces and vacuoles between the cells but small rosettes of the type found in some medulloblastomas are not present. The stroma consists largely of thin-walled blood vessels (centre).   HE ×360

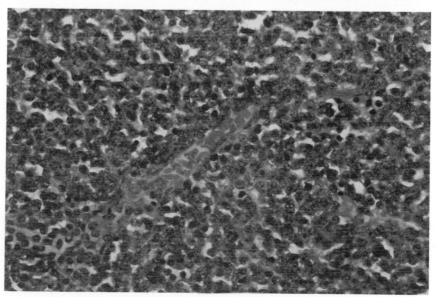

9.37 Ependymoma: brain

Ependymomas tend to be located in the ventricles of the brain, and are therefore liable to block the flow of cerebrospinal fluid. Many are in the vicinity of the 4th ventricle and difficult to resect. As with astrocytomas the degree of malignancy varies widely. This is a cellular type of ependymoma composed of closely-packed polygonal cells with large, round or ovoid basophilic nuclei and abundant eosinophilic cytoplasm. In each nucleus there is a prominent nucleolus which tends to be located centrally. Some nuclei have more than one nucleolus. There are no mitotic figures in this field. The cell boundaries are eosinophilic and distinct. Tumour cells are orientated around and attached by their elongated bases to two blood vessels to form pseudorosettes (arrows). HE ×360

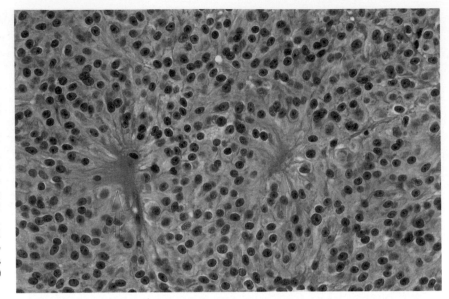

9.38 Ependymoma: brain

This tumour was a circumscribed ovoid mass (2.8 × 1.5 × 1.5cm) attached to the cauda equina. The cells are polygonal with eosinophilic cytoplasm. The nuclei are ovoid and vesicular and contain several nucleoli. There is little pleomorphism and no mitoses are present. The tumour cells are oriented around two blood vessels with thick hyalinized walls and attached to them by their filamentous vacuolated bases. These are pseudorosettes. Elsewhere in the tumour there was extensive myxomatous change in the stroma. HE ×360

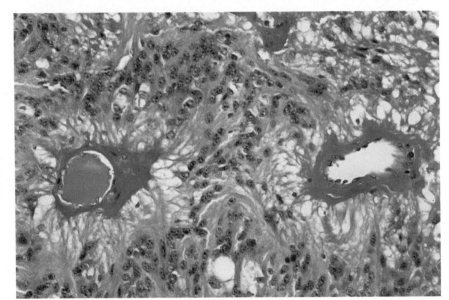

9.39 Myxopapillary ependymoma: sacrum

Myxopapillary ependymoma most often arises from the filum terminale of the spinal cord. Macroscopically it has a gelatinous appearance and it slowly envelops the cord and nerve roots. Histologically this part of the tumour has a papillary structure, the papillae consisting of a delicate core of vascular connective tissue (arrows) and on their external surface a single layer of cuboidal epithelial cells. The spaces between the papillae are occupied by colourless secretion. This type of structure resembles normal choroid plexus but the epithelial cells are taller and more prominent than in the normal tissue. In most of the tumour the connective tissue of the papillae had undergone myxomatous change. HE ×150

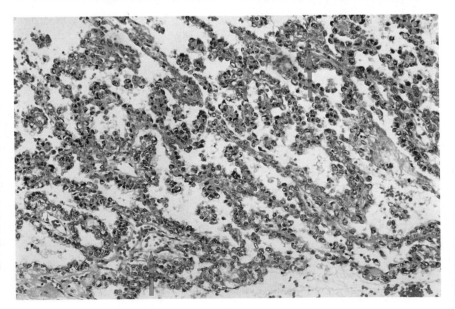

9.40 Capillary telangiectasis: brain

Vascular malformations (hamartomas) occur not infrequently in the central nervous system. They vary considerably in their composition and complexity of structure, ranging from small lesions consisting of capillaries to large thick-walled vessels. They are mostly located on the surface of the brain and they tend therefore to bleed into the subarachnoid space. They may also give rise to a large intracerebral hemorrhage. This was a solitary lesion. It consists of abnormal capillaries, some of large calibre. The vessels of the lesion are separated by neural tissue and not fibromuscular tissue as in an ordinary capillary or cavernous hemangioma, and complete resection of lesions of this type may be difficult or impossible. There is however no evidence of previous hemorrhage in this lesion.   HE ×80

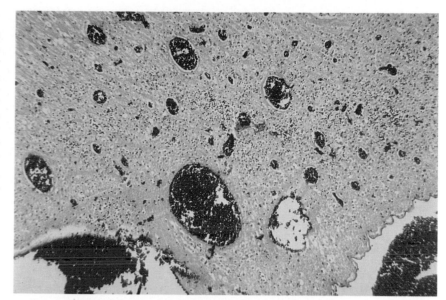

9.41 Hemangioblastoma: cerebellum

The cerebellum is the commonest site for hemangioblastoma although it may be found occasionally in the cerebral hemispheres. It is a true neoplasm of vascular origin and not a hamartomatous malformation. It may occur by itself or as part of Lindau's disease. It is a malignant tumour but when located in the cerebellum it is often resectable. However this lesion in the cerebellum of a 79-year-old woman, typically forming a cystic hemorrhagic mass, was not completely resectable. The tumour consists of large numbers of thin-walled dilated capillary-type blood vessels and large round cells with basophilic pleomorphic nuclei and grey or weakly eosinophilic granular cytoplasm. The walls of the two larger blood vessels (arrows) are infiltrated with hyaline material.   HE ×150

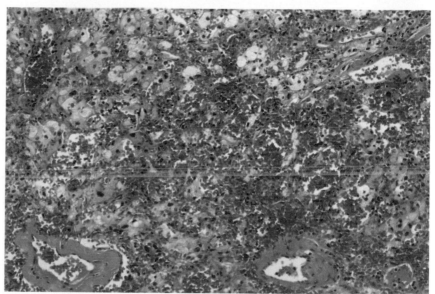

9.42 Hemangioblastoma: cerebellum

At higher magnification the pleomorphism of the nuclei of the closely-packed large cells is pronounced. There are no mitoses however. The cytoplasm of the cells generally has a finely-granular pale 'frosted-glass' appearance but in some it is eosinophilic. The cells are macrophages and their cytoplasm is granular or foamy from its content of lipid. The presence of these cells is characteristic of this type of tumour. A considerable number of small lymphocytes is present (left). Only a few of the many capillary-type blood vessels that made up the bulk of the tumour are shown in this field.   HE ×235

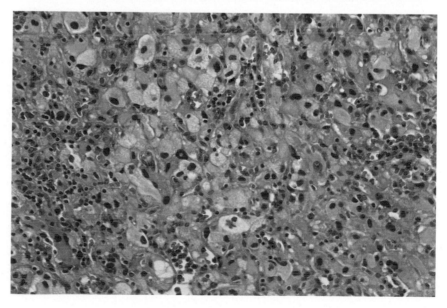

9.43 Schwannoma: cranial nerve

Schwann cells form the myelin sheaths, the neurilemma, around the axons in peripheral nerves. A Schwannoma (neurilemmoma) is a benign tumour, forming a well-circumscribed, round or lobulated mass which is often soft and even cystic. Its location may however make resection very difficult. This lesion (an 'acoustic neuroma') arose from the 8th nerve (vestibular portion) where it entered the internal auditory meatus (the 'cerebello-pontine angle'). Histologically it contains two types of tissue; compact groups of spindle-shaped cells with eosinophilic cytoplasm and nuclei which tend to arrange themselves in parallel rows (palisading) – Antoni type A tissue; and two, loose reticular tissue - Antoni type B tissue. The spaces in this tissue contained watery fluid and they are large and cystic in some lesions. Very little collagen is present.   HE ×150

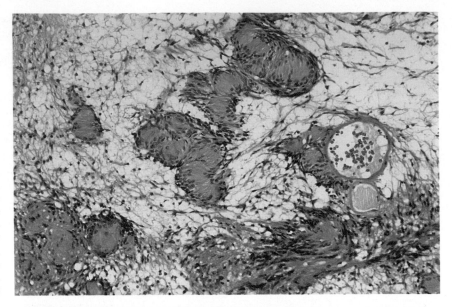

9.44 Schwannoma: spinal nerve

This tumour is from the spinal cord of a woman of 67 with neurofibromatosis (von Recklinghausen's disease). It consists of loose vacuolated Antoni type B tissue (thin arrow) and cellular Antoni type A tissue (thick arrows). In the latter the elongated tumour cells form long cords and compact ovoid bodies (Verocay bodies) (double arrows) in which the eosinophilic cytoplasm is at the centre and the palisaded nuclei are at the periphery. The hyalinization of the walls of the blood vessels present here is commonly seen in Schwannomas. An infiltrate of small lymphocytes is present in the Antoni type B tissue.   HE ×235

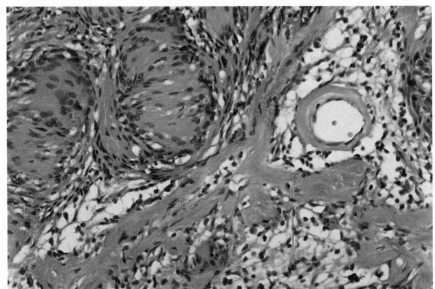

9.45 Malignant Schwannoma: buttock

Although the great majority of Schwannomas are benign, occasional lesions occur which are more cellular, with evidence of mitotic activity. Such a lesion may recur after surgical removal, as happened in this case. It was in the buttock. This shows the recurrent tumour invading the fatty tissue of the buttock. It is frankly sarcomatous; the elongated cells are haphazardly arranged and have relatively large pleomorphic nuclei with prominent nuceli. There is considerable mitotic activity (arrows).   HE ×360

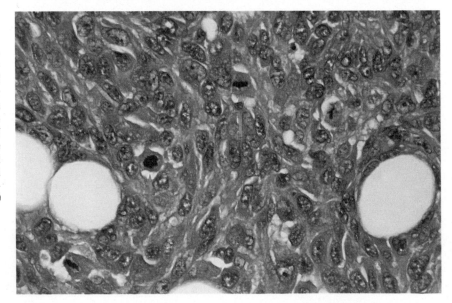

9.46 Secondary carcinoma: brain

The central nervous system is a common site of metastasis of malignant tumours. Generally there are multiple secondaries, characteristically round and apparently well-demarcated from the surrounding white matter. Cords of tumour cells are invading the white matter of the brain. They are papillary structures, with a central core of vascular connective tissue and an outer layer of large cuboidal or columnar cells. The nuclei of the tumour cells are round and fairly uniform in shape but mitoses are present. Some of the cells have large droplets of mucin in their cytoplasm. There is only slight lymphocytic infiltration of the white matter around the tumour. Secondary tumours tend to have a histological structure similar to that of the primary and the primary in this case was a papillary adenocarcinoma of breast. HE ×150

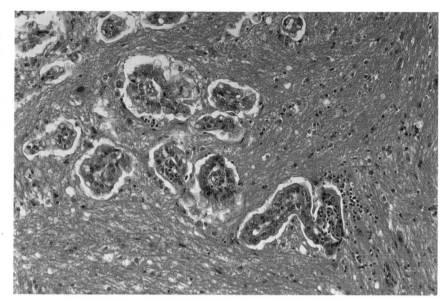

9.47 Secondary carcinoma: brain

Sometimes the metastatic tumour is confined to the leptomeninges, the malignant cells growing in the subarachnoid space. The clinical signs and symptoms can be very suggestive of bacterial or fungal meningitis, and since the tumour cells metabolize the sugar in the cerebrospinal fluid the fall in the sugar level in the CSF may increase the diagnostic difficulties, as happened in this case. The cerebral cortex covered with the pia mater is at the bottom, and the arachnoid membrane at the top. In the subarachnoid space tumour cells are growing freely. They are very pleomorphic. A few are shrunken and necrotic but most are large or very large compared, e.g. with the small lymphocytes at the top. The primary was a carcinoma of bronchus. HE ×160

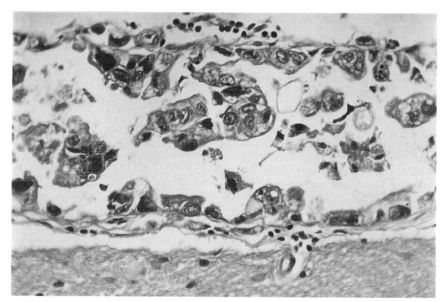

9.48 Retinoblastoma: eye

Retinoblastoma is a not uncommon tumour of early childhood. Some of the cases are familial. It is malignant and very invasive locally. It does not metastasize but in about half of the cases tumours are present in both eyes. Macroscopically it is solid fleshy mass within the retina, with areas of necrosis and calcification frequently present. Histologically it is similar to neuroblastoma and medulloblastoma, consisting of small cells of uniform size and shape, with deeply basophilic round or slightly ovoid nuclei. Occasional mitoses are present. The cells tend to arrange themselves around eosinophilic fibrillary material, to form 'rosettes'. The fibrils are nerve fibrils originating in the tumour cells. Melanin spilled from the disrupted retina is also present (lower left). HE ×360

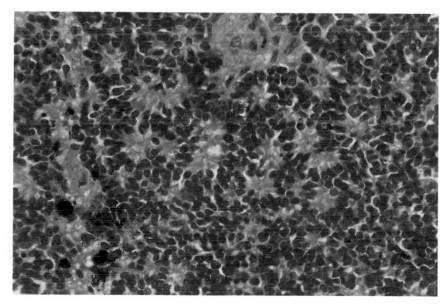

10.7 Infarct: kidney

The branches of the renal artery are end-arteries and when one is suddenly blocked, the anastomoses do not provide enough blood to maintain the viability of the tissue and it undergoes infarction. This is the edge of a recent infarct. The cells lining the tubules (left) are necrotic: there is no nuclear staining and the cytoplasm is deeply eosinophilic. The tissues in the right half of the field are probably viable, though the left half of the glomerulus appears necrotic. The capillaries of the boundary zone (centre) are greatly distended and the stroma is edematous, with separation of the tubules. If the patient had survived, macrophages, polymorph leukocytes and fibroblasts would have appeared within a few days in the boundary zone and organization of the dead tissue would have led to its removal or encapsulation in fibrous tissue. HE ×135

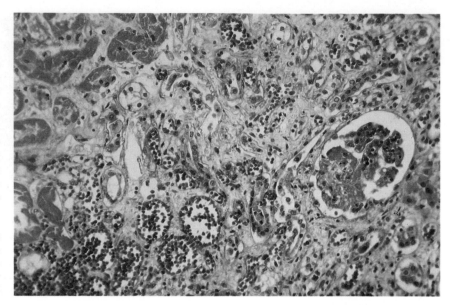

10.8 Acute tubular necrosis: kidney

Necrosis of the epithelium of the renal tubules may occur on a large scale. The consequence is sudden onset of anuria or severe oliguria and acute renal failure. The lesion is associated with shock following trauma and incompatible blood transfusion. Some chemicals also are toxic to tubular epithelium. The epithelium has considerable powers of regeneration and renal function can be restored, being heralded by profuse diuresis. This patient died from renal failure seven days after an operation for relief of constrictive pericarditis. Most of the epithelium lining the collecting tubules has died and sloughed into the lumen. The surviving cells have made considerable attempts at repair and already the tubules are lined by flat elongated cells. Recovery of renal function did not occur in time to save the patient's life. HE ×200

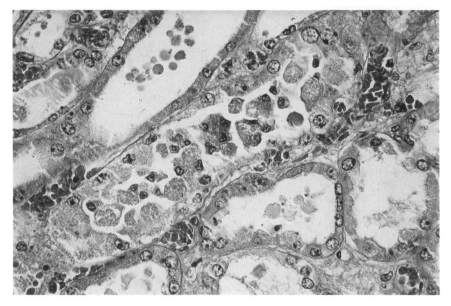

10.9 Potassium deficiency: kidney

In a variety of conditions including prolonged diarrhea, Cushing's syndrome, untreated diabetic coma etc. the cells of the body lose potassium and the plasma potassium level falls. When this happens to a marked extent, the renal tubular epithelium may become very swollen and hydropic, as shown here. The lesion is reversible and the cells return to normal when potassium levels are restored. A similar change can be induced experimentally by injecting concentrated sucrose solution into animals. The glomerulus (right) is unaffected. HE ×150

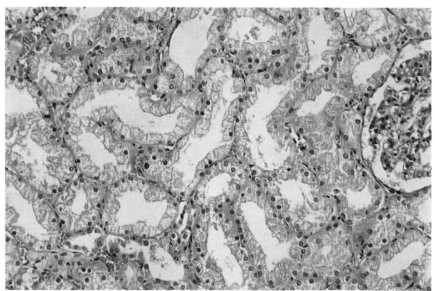

10.10 Fat embolism: kidney

When fatty tissues such as the marrow of a large bone are severely traumatized, particles of fat may enter the venous circulation. Most are filtered off by the capillaries of the lungs but some get through the lung and form small emboli in the systemic circulation. This is the kidney of a young woman who died from multiple injuries including fracture of a femur. The glomerular capillaries are distended with fat globules and a small amount of fat is present in the subcapsular space (right). The gaps in some droplets were caused by fat dissolving in the stain (Sudan IV). Fat embolism of the kidneys and other organs happens not infrequently after severe injury without altering the clinical outcome, unless it produces lesions in a vital part of the central nervous system (**9.6**). Sudan IV ×335

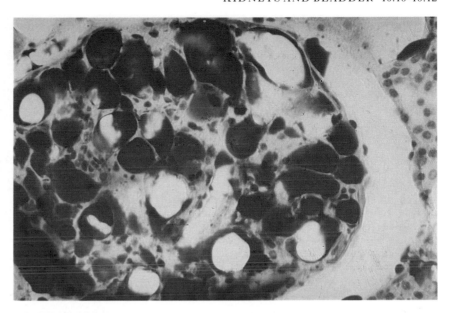

10.11 Diabetic glomerulosclerosis: kidney

Lesions develop in the glomeruli of a significant proportion (up to 50%) of diabetic individuals in the form of deposits of hyaline material in the mesangium of the lobules of the glomerulus. It may be deposited diffusely and more or less evenly throughout the glomerulus or unevenly as one or more nodules. The two types of lesion are often present together as in this case. There is diffuse infiltration of the glomerular tuft with eosinophilic material and also heavy focal deposition. The diffuse infiltrate appears to be in the basement membranes of the capillaries and the capillary bed has been obliterated in places. In time this can lead to more or less complete hyalinization of many glomeruli. The afferent arteriole (arrow) shows hyaline change. HE ×335

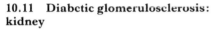

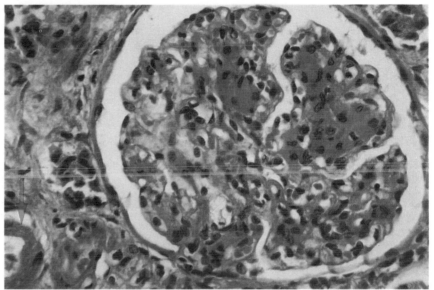

10.12 Diabetic glomerulosclerosis: kidney

This is the classical Kimmelstiel–Wilson lesion. The hyaline material has been deposited in round, practically acellular nodules (thin arrows) in the glomerular tuft. The cuff of red cells around the larger of the two main nodules is a greatly dilated capillary blood vessel. There is also a considerable deposit of hyaline material in Bowman's capsule and in the subintima of the arteriole (thick arrow). Lesions in the arterial system, including atheroma, are common in diabetes mellitus. HE ×235

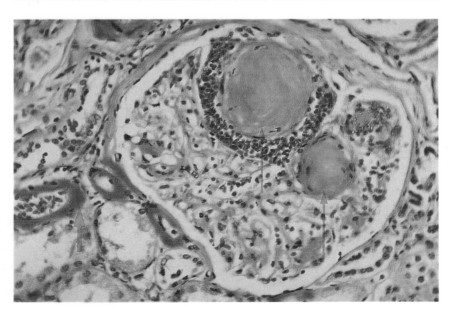

10.19 Minimal change glomerulonephritis: kidney

Minimal change glomerulonephritis occurs most often in young children (1–4 years). The child usually presents with a nephrotic syndrome. Most children recover completely but a minority eventually develop chronic renal failure. This glomerulus has been stained by the periodic acid-Schiff method. There is no cellular proliferation in the glomerulus and no thickening of the basement membrane, the amount of purplish-red at the centres of the glomerular lobules being within the normal range of variation. The walls of the capillaries, therefore, show no thickening. Electron microscopy showed that the foot processes of the epithelial cells of the glomerulus were fused, bringing the basement membrane into contact with epithelial cell cytoplasm.

PAS ×375

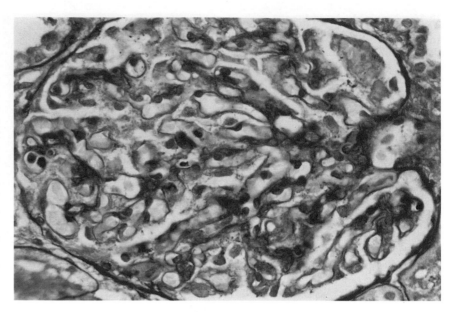

10.20 Membranous nephropathy: kidney

As in minimal change glomerulonephritis, patients with membranous nephropathy generally present with the nephrotic syndrome. Adults are affected more than children and the term prognosis is much less favourable than in minimal change glomerulonephritis. Steroid treatment is not effective but remissions are common and it may be many years before chronic renal failure develops. The characteristic lesion is a diffuse thickening of the basement membrane, as shown in this glomerulus stained by the periodic acid-Schiff method. The change is diffuse, affecting all capillaries in the glomerulus equally, and all the glomeruli in both kidneys showed the same changes. There is no proliferation of endothelial or mesangial cells and no infiltration of the mesangium by leukocytes. PAS ×375

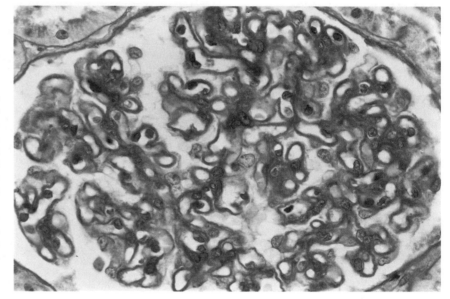

10.21 Membranous nephropathy: kidney

It is occasionally difficult to distinguish between membranous and membrano-proliferative glomerulonephritis and the use of thin (1μm) sections stained by the periodic acid-methenamine silver method is helpful. In this preparation several capillaries show changes typical of membranous glomerulonephritis: instead of being a smooth black line the glomerular basement membrane now consists of an inner continuous black line with stumpy black 'bristles' projecting from its outer surface. Granular epimembranous deposits of IgG were detected by immunohistochemistry in many glomerular capillary loops but no IgA or IgM.

Periodic acid methenamine silver ×1200

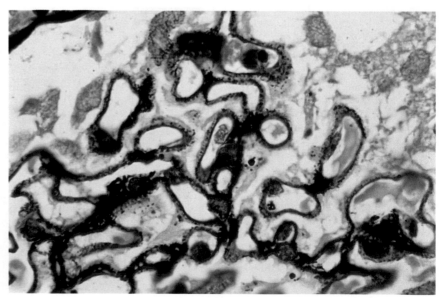

10.22 Membranoproliferative glomerulonephritis: kidney

Membranoproliferative glomerulonephritis may present with malaise, pyrexia, hematuria and edema or more commonly with the nephrotic syndrome. Sometimes the first sign is symptomless proteinuria. The disease tends to follow a slowly progressive course over several years with development of the nephrotic syndrome and eventual renal failure. A small minority nevertheless do recover. There is a diffuse proliferative change in the glomerulus, with increase in the numbers and size of the endothelial and mesangial cells. The mesangial changes are particularly marked and have produced diffuse eosinophilic sclerosis of the glomerulus. The lobular structure of the glomerulus is more obvious than normal, this being a feature ('lobular glomerulonephritis'). HE ×335

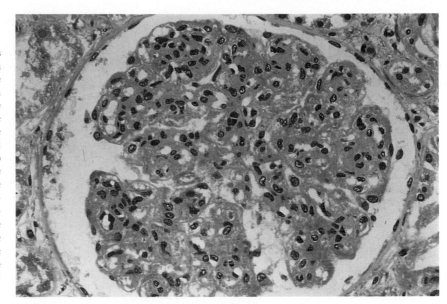

10.23 Membranoproliferative glomerulonephritis: kidney

As with membranous glomerulonephritis the use of a thin (1.0 μm) section and the periodic acid-methenamine silver method makes it possible to see the structural changes in the glomeruli in membranoproliferative glomerulonephritis more clearly. In this glomerulus there is an increase in the endothelial cells as evidenced by the number of nuclei. There is also glomerulosclerosis in the form of fine black-staining fibres (thin arrow); and in some areas the glomerular basement membrane is a double line (thick arrows) as a result of new fibres being laid down within the existing basement membrane.

Periodic acid–methenamine silver ×450

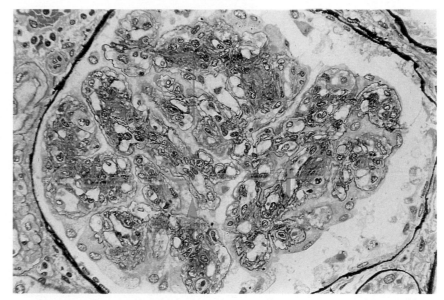

10.24 Goodpasture's syndrome: kidney

Goodpasture's syndrome affects mainly the lungs and kidneys. There is acute focal glomerulonephritis, with hematuria and proteinuria, which tends to develop into rapidly progressive glomerulonephritis. Uniquely, an autoantibody is present in the serum which reacts specifically with the basement membrane of the glomerular capillaries. This shows a severely affected glomerulus. It is shrunken and atrophic and surrounded by, and closely adherent to, a large epithelial crescent (arrow) which fills the subcapsular space. A considerable number of pyknotic and probably necrotic nuclei are also present in the glomerulus. There is eosinophilic proteinaceous material in the adjacent tubules. The tubules are lined by atrophic epithelium, probably as a result of secondary ischemic damage. HE ×310

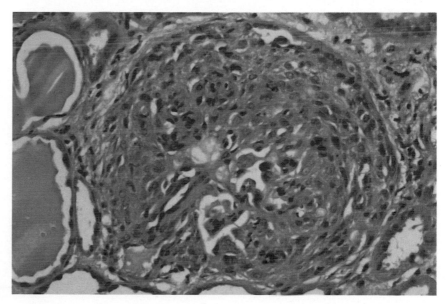

10.25 Proliferative glomerulonephritis: kidney

This form of nephritis develops 1–4 weeks after an infection with beta-hemolytic streptococci (Group A), often in the throat. There is malaise, fever and edema. Hematuria is also a feature. Children and young adults generally recover completely in a week or two, but a minority of adults develop rapidly progressive glomerulonephritis or the condition persists in a clinically silent form. The glomerular tuft shows marked increase in cellularity, as a result of proliferation of endothelial cells and infiltration with polymorph leukocytes. The increased number of swollen endothelial cells has blocked the lumen of many capillaries, and the swollen tuft has almost obliterated Bowman's space. The interstitial tissues (left) are edematous, causing separation of the tubules.   HE ×335

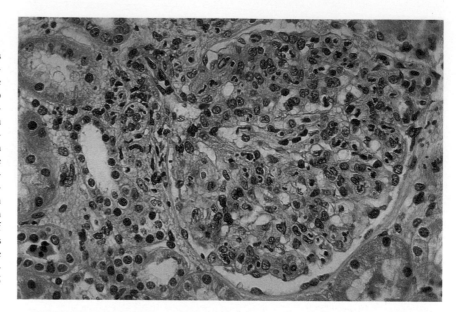

10.26 Proliferative glomerulonephritis: kidney

This patient had an attack of acute diffuse proliferative glomerulonephritis which did not clear up. The parietal epithelial cells of Bowman's capsule have proliferated to form a large cellular crescent which fills Bowman's space (extracapillary glomerulitis). Crescent formation is associated with previous severe damage to the glomerulus and the glomerulus is shrunken. Pressure from the crescent may also have played a part in causing the glomerular atrophy. The tubules (left) are widely separated from one another by edematous interstitial tissue. The normal epithelium of the tubules is replaced by a low simple epithelium, possibly as a result of regeneration following previous tubular damage.

HE ×270

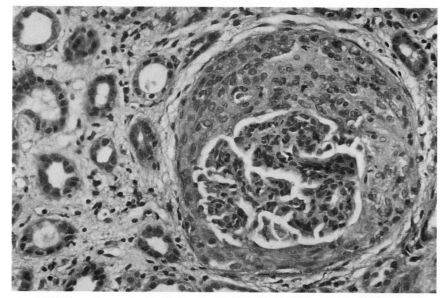

10.27 Proliferative glomerulonephritis: kidney

A boy of 16 had an attack of acute diffuse proliferative glomerulonephritis (post-streptococcal). This biopsy specimen was taken 12 weeks later because of persistent proteinuria. There is fairly pronounced proliferation of mesangial cells throughout both glomerular tufts without much increase of mesangial matrix. The endothelial and epithelial cells and the capillary basement membranes appear normal. Increased numbers of cells in the mesangium may persist for many weeks after an attack of acute proliferative glomerulonephritis and they may provide retrospective confirmation of the diagnosis. The lobular pattern of the glomeruli and the mesangial proliferation in this case are therefore not unexpected.   HE ×360

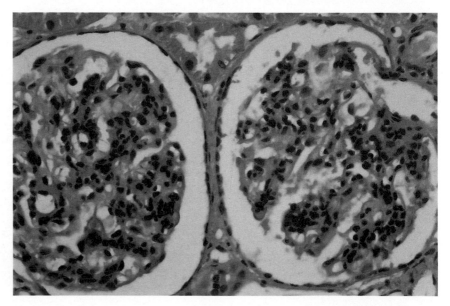

10.28 Proliferative glomerulonephritis: kidney

This is the same case as shown in **10.27**. The changes in the glomerulus are seen more clearly in this thin (1 μm) section stained by the periodic acid–methenamine silver method than in ordinary HE sections. The lobules of the glomerulus are clearly demonstrated. There is no abnormality of the endothelial or epithelial cells or of the peripheral capillary basement membranes; and Bowman's capsule is not thickened. In all the lobules however the mesangium is more prominent than normal, from increase of nuclei. At higher magnification a doubled basement membrane was detectable in several capillary loops. The changes in the basement membrane did not allow a firm diagnosis of early membranocapillary glomerulonephritis.

Periodic acid–methenamine silver ×440

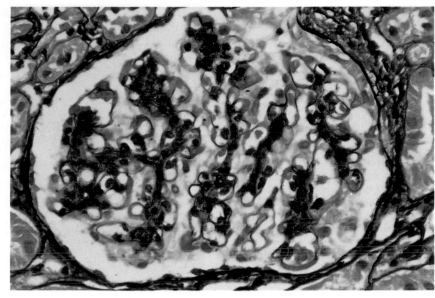

10.29 Lobular membranoproliferative glomerulonephritis: kidney

Lobular glomerulonephritis is a form of membranoproliferative (or mesangioproliferative) glomerulonephritis which is characterized by accentuated lobulation of the glomerular tuft with solidification of the centrolobular or mesangial region. The lobules assume a club shape, and all glomeruli are affected. The changes are demonstrated particularly effectively in thin (1 μm) resin-embedded sections stained by the periodic acid-methenamine silver method, and in this example, the increased size and argyrophilia of the mesangium are very obvious.

Periodic acid-methenamine silver ×440

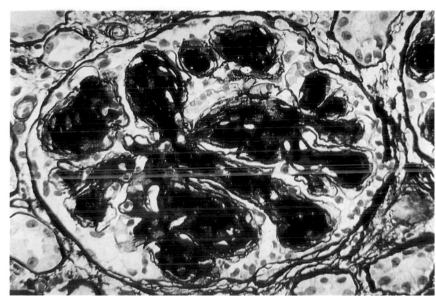

10.30 Chronic glomerulonephritis: kidney

Chronic glomerulonephritis is the end-stage of various forms of glomerulonephritis. The patient is uremic, hypertensive and in chronic renal failure. The glomerular tuft is sclerosed and many of its capillaries are obliterated. There are extensive adhesions between the tuft and Bowman's capsule, and Bowman's space is reduced to several narrow slits lined by proliferated enlarged epithelial cells. There is marked periglomerular and interstitial fibrosis. The tubules on the left are so damaged that it is not possible to identify them with certainty, but they are probably grossly altered proximal convoluted tubules. Changes similar to these were present throughout both kidneys, involving all the glomeruli.   HE ×335

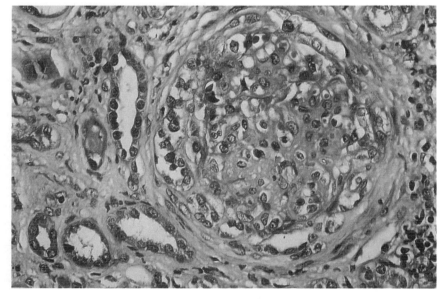

10.37 Rejection: renal allograft

A transplanted kidney induces an immunological reaction in the host. Cellular rejection develops during the first few weeks after transplantation and is generally controllable by immunosuppressive drug therapy. This kidney transplanted 7 months previously into a 13-year-old girl shows a severe cellular rejection reaction. The glomerulus (centre) is shrunken and the endothelial cells are swollen. There is severe tubular damage, with replacement of the epithelium by flattened cells. Many of the epithelial cells are vacuolated and some appear necrotic. The interstitial tissues are edematous with wide separation of the tubules. In the interstitial tissues there are small hemorrhages and an infiltrate of chronic inflammatory cells which tends to concentrate around the tubules (arrow).

HE ×150

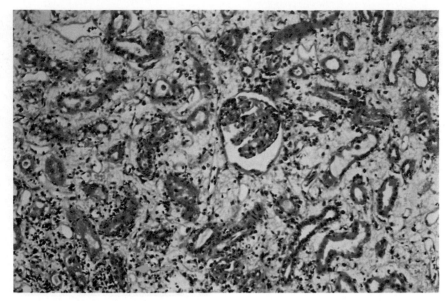

10.38 Acute vascular and cellular rejection: renal allograft

This renal allograft of 14 days' duration in a man of 34 was removed from the recipient because of a severe vascular and cellular rejection reaction. It was swollen (265g) and its surface was mottled and hemorrhagic. This shows the collecting tubules in longitudinal section. There is extensive necrosis of their epithelium (thin arrow) and several tubules contain eosinophilic fluid or blood (thick arrow). There is also extensive hemorrhage into the interstitial tissues (double arrow). The arteries also showed severe rejection damage (**10.39**) and some arterioles were necrotic. The glomeruli were hypercellular. HE ×235

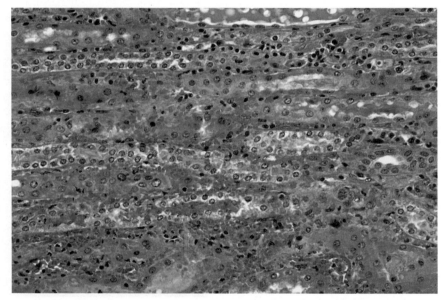

10.39 Acute vascular and cellular rejection: renal allograft

This is the same kidney as that shown in **10.38**. It is a small muscular artery in cross-section. The media appears normal apart from the presence of a few lymphocytes. The intima however is greatly thickened by proliferated and swollen endothelial cells and an infiltrate of mononuclear inflammatory cells. The lumen is so small that the vessel is probably non-functional. Its adventitia and the interstitial tissues of the kidney are packed with an infiltrate of small lymphocytes, plasma cells and histiocytes. HE ×360.

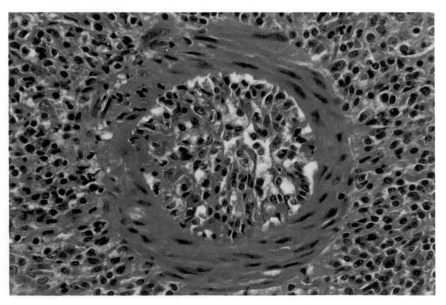

10.40 Severe chronic vascular rejection: renal allograft

In chronic rejection of a renal allograft, vascular lesions predominate and may cause renal failure through ischemia. Vascular rejection is less amenable to therapy than cellular rejection. This is an allograft of 19 months' duration in a man of 53. The most marked change is the loss of tubules; and many of those remaining (e.g. bottom left) are small and lined by very atrophic epithelium. The interstitial tissues are severely fibrosed and heavily infiltrated by chronic inflammatory cells. One glomerulus is atrophic. There is some epithelial and mesangial proliferation in the others but they are fairly well preserved.

HE ×150

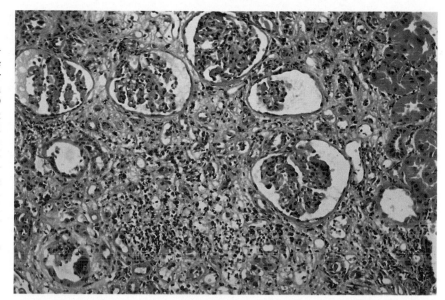

10.41 Severe chronic vascular rejection: renal allograft

This is the same kidney as that shown in **10.40**. The two glomeruli are fairly well preserved but there is both mesangial and extra-capillary cell proliferation. The basement membrane shows ischemic corrugation. There is severe loss of tubules and those surviving are very atrophic (arrows). An infiltrate of inflammatory cells is present in the interstitial tissues. HE ×360

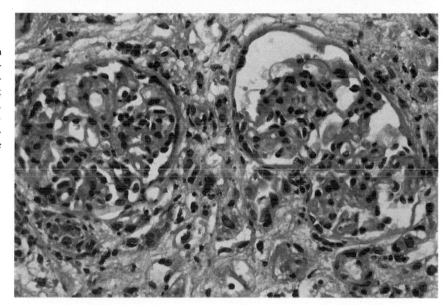

10.42 Chronic vascular rejection: renal allograft

This is a biopsy specimen from a renal allograft of approx. 3 months' duration in a boy of 15. The intima of the small arteries is a very thick layer of edematous poorly vascular fibrous tissue and the lumen is correspondingly reduced in size. In the glomerulus (lower right) there is endothelial swelling and collapse of capillary loops, with wrinkling of the basement membranes. The cellular lining of Bowman's capsule is incomplete in places. There is loss of many tubules and those remaining are atrophic (arrow). The interstitial tissues are fibrous but the cellular infiltrate is slight. These changes in the arteries are evidence of active vascular rejection of the allograft and the damage to the other structures has probably been caused by chronic ischemia following the vascular changes. HE ×235

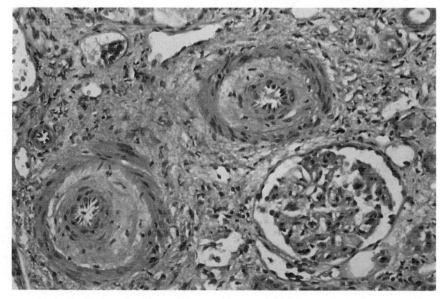

10.43 Adenoma: kidney

Adenomas of the kidney are generally located in the cortex. They are benign although a small minority behave as malignant lesions. They are yellowish nodules which appear encapsulated. The tumour (left) has a branching papillary structure, consisting of delicate cords of connective tissue covered with a single layer of cuboidal epithelial cells. The epithelial cells have small round nuclei which show no pleomorphism or mitotic activity. There are many pyknotic nuclei, some belonging to necrotic desquamated epithelial cells. The cytoplasm of the epithelial cells is weakly eosinophilic and in many cells it is vacuolated (the vacuoles contain lipid). There is no capsule and the tumour is bounded (right) by normal tubules lined by large epithelial cells. The wall of the arteriole is hyalinized (arrow). HE ×150

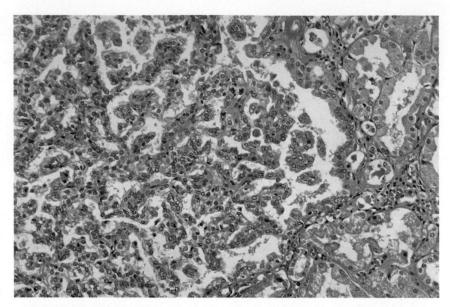

10.44 Adenoma: kidney

This shows the boundary between the tumour and the normal kidney (left) at higher magnification. There is no fibrous capsule between the tumour and the kidney. The epithelial cells covering the finger-like processes are well-differentiated, with round or ovoid vesicular nuclei which show no pleomorphism or mitotic activity. Their cytoplasm is granular and vacuolated. It is sometimes difficult histologically to distinguish between an adenoma and a small primary renal adenocarcinoma but lesions less than 2.5cm can generally be regarded as adenomas, since they almost never metastasize. The wall of the arteriole is hyalinized. HE ×335

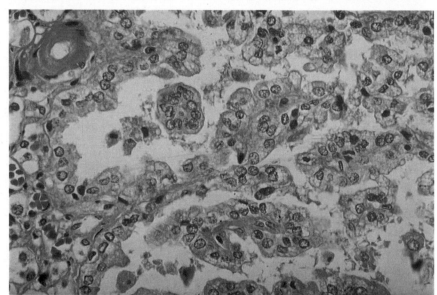

10.45 Wilms' tumour: kidney

Wilms' tumour (embryoma or nephroblastoma) is the commonest malignant tumour of infants and young children. It may form a large fleshy mass. It invades blood vessels and tends to metastasize to the lungs and elsewhere. It arises from embryonic nephrogenic tissue, though neural elements may be present. This tumour (3.5 × 3.5 × 3.0cm) in a girl of 3, was yellow and seemed to be encapsulated. It contains the two elements usually present: rudimentary glomerular ('glomeruloid') structures (arrows), and a loose immature spindle-cell stroma (e.g. lower right). The epithelial cells of the glomeruloid structures have ovoid deeply basophilic nuclei, and several mitotic figures are present. Tubules and muscular elements are often also present. The fibrous capsule was incomplete.

HE ×360

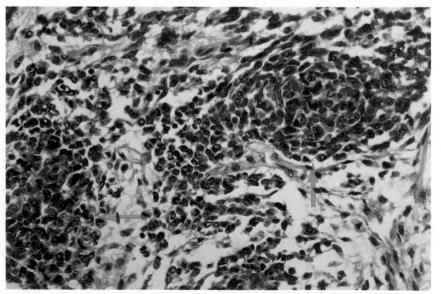

10.46 Carcinoma: kidney

Renal carcinoma originates from renal tubular epithelium and shows a marked tendency to invade the renal vein Macroscopically it is often yellow, with extensive areas of necrosis and hemorrhage. Hematuria is frequently the presenting sign. This tumour consists of large pale cells with abundant cytoplasm, foamy from the presence of lipid and glycogen. The nuclei are vesicular and fairly pleomorphic. They have a prominent nucleolus but there are no mitoses (they were present elsewhere in the tumour). The cytoplasmic lipid which gives the tumour its yellow colour may cause the cytoplasm to appear 'clear': clear-cell carcinoma. The scanty stroma consists largely of thin-walled blood vessels which tend to rupture and bleed (bottom left).　　　　HE ×135

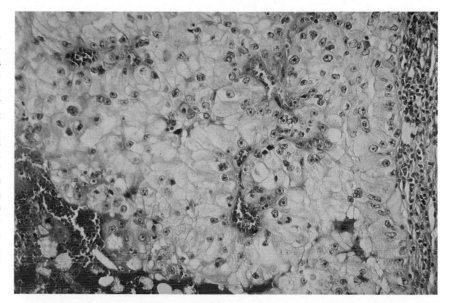

10.47 Carcinoma: kidney

The cells of renal carcinoma vary considerably in their morphology from 'clear' cells to 'solid' cells with eosinophilic cytoplasm, and not infrequently the various types are found within one tumour. The cells of this tumour are round and have eosinophilic cytoplasm (thin arrow). They are forming acini which contain pale granular secretion (thick arrow). Their nuclei are round and vesicular and fairly uniform in structure. There is no mitotic activity. The small deeply-staining nuclei are either lymphocytes or pyknotic nuclei of tumour cells. The stroma consists of thin-walled blood vessels. Cells with granular eosinophilic cytoplasm are termed oncocytes and sometimes renal carcinomas of the type shown here are called oncocytomas.　　　　HE ×360

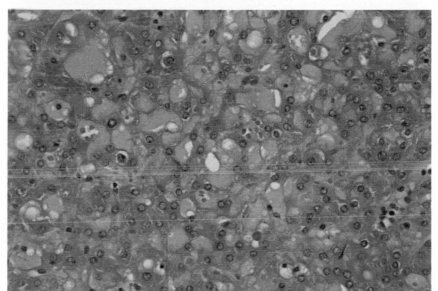

10.48 Transitional cell papilloma: renal pelvis

The tumour consisted of long delicate papilliform processes and this is the tip of one of them. It has a core of vascular connective tissue covered with a layer of fairly well-differentiated transitional epithelium. The cells show some loss of polarity and their nuclei are more pleomorphic than normal. The epithelial basement however is intact. A mitotic figure (arrow) is present. As in the bladder, papillomatous tumours of transitional epithelium, however well-differentiated, are liable to recur and should be regarded not as benign papillomas but as well-differentiated papillary carcinomas (Grade 1). The thin-walled blood vessels in the connective tissue core of the papillae are liable to bleed and hematuria is often the presenting sign.　　　　HE ×335

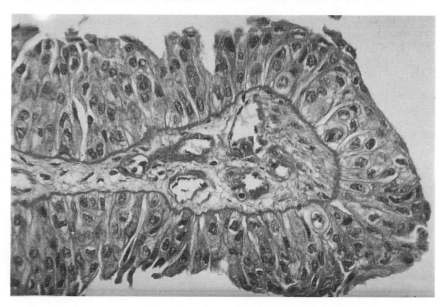

10.49 Chronic pyelitis: renal pelvis

In pyelitis the renal pelvis and calyces are inflamed. Infection spreads to them more readily from the bladder via the ureters when there is an obstruction to urinary outflow. Further spread to the renal parenchyma leads to pyelonephritis. The organism is commonly *Escherichia coli* but a mixed infection not infrequently establishes itself. The inflammatory reaction may be acute with excretion of large numbers of polymorph leukocytes in the urine. In this case the lesion was chronic. The transitional epithelium of the renal pelvis is on the left. It is intact and appears normal, apart from the presence within it of a few lymphocytes. The tissues beneath the epithelium are edematous and hyperemic and heavily infiltrated by chronic inflammatory cells, mostly plasma cells. HE ×135

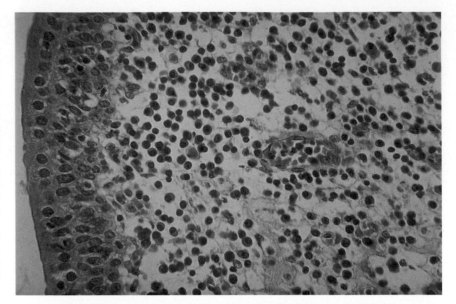

10.50 Idiopathic retroperitoneal fibrosis

Retroperitoneal fibrosis is a chronic inflammatory lesion of the para-aortic region which leads to the formation of dense fibrous tissue around the ureters. They are drawn medially and narrowed; and complete obliteration of the lumen may occur. There is an association with idiopathic fibrosing lesions in other sites (pseudotumour of the orbit, Riedel's thyroiditis, etc.) and an autoimmune basis for the disorder is suspected. In this example the ureter is on the left. The para-aortic tissues (on the right) are heavily fibrosed and the dense fibrous tissue has spread into the muscular wall (thin arrow) of the ureter and into the submucosa. As a result there is marked narrowing of the lumen. There are two granulomas (thick arrows) in the muscle coat of the ureter.

HE ×13

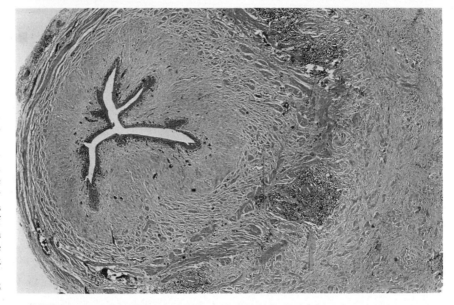

10.51 Chronic cystitis: bladder

Women are more liable to cystitis than men, although prostatic obstruction is a major predisposing factor in older men. A woman of 60 had repeated attacks of cystitis and polypoid bladder mucosa was removed for diagnostic purposes. The transitional epithelium (left) is hyperplastic and thicker than normal. Slight nuclear pleomorphism is evident. There are no mitoses, however, and the polarity of the cells is preserved. The epithelium is infiltrated with polymorph leukocytes (bottom left) and occasional lymphocytes. The underlying lamina propria (centre and right) is edematous and hyperemic and infiltrated by chronic inflammatory cells, mostly plasma cells. The appearances are those of chronic cystitis and the epithelial changes were presumed to be secondary to the persistent inflammatory reaction. HE ×235

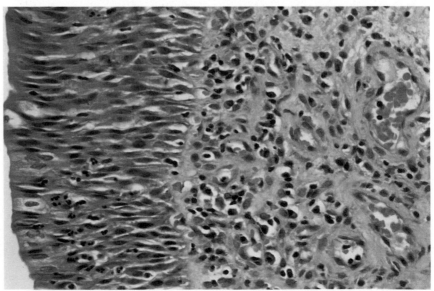

10.52 Pseudomembranous trigonitis: bladder

Sometimes in chronic cystitis the trigone of the bladder is particularly affected and the transitional epithelium there undergoes squamous metaplasia. Macroscopically it loses the normal translucency of transitional epithelium and appears whitish and opaque. Histologically it consists of a thick layer of well-differentiated squamous epithelium with elongated interpapillary processes. Many of the epithelial cells are vacuolated, and can be shown histochemically to contain glycogen. The resemblance therefore to vaginal epithelium is close and the condition is sometimes called 'vaginal metaplasia'. The underlying submucosa is hyperemic but the number of chronic inflammatory cells is small. HE ×150

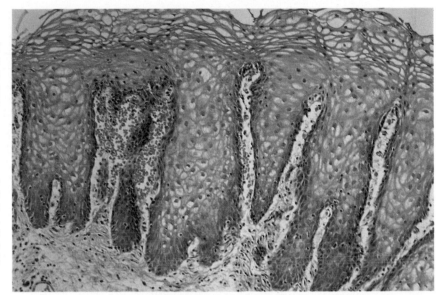

10.53 Ureteritis cystica: ureter

In a chronically inflamed bladder (cystitis) or ureter (ureteritis) the transitional epithelium may proliferate downwards into the lamina propria to form the cell-nests of von Brunn. Fluid collects in the centres of these cell-nests and with increase in the fluid they become cystic. This shows cell-nests of various sizes. The lumen of the bladder is on the left. Two large cell nests, filled with eosinophilic fluid and lined by flattened transitional epithelium, project into the lumen of the bladder. Smaller cell nests (arrows) also containing eosinophilic fluid are being formed by epithelial downgrowths. HE ×120

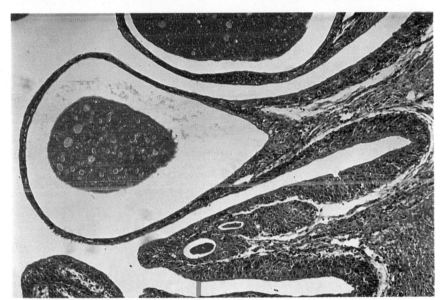

10.54 Schistosomiasis (bilharziasis): bladder

The parasitic worm *Schistosoma haematobium* tends to lodge in the veins of the bladder where the adult female deposits its ova in the adjacent venules. Embryos escape from the eggs into the bladder and leave the body in the urine, to start the next phase of the worm's life-cycle in snails. A gravid adult female worm (lower right) lies in the submucosal connective tissues of the bladder wall. There is an intense cellular infiltrate (of eosinophil leukocytes) around the worm and throughout the edematous submucosa. The transitional epithelium (top left) is hyperplastic, with downgrowths of finger-like processes into the lamina propria. This hyperplasia sometimes proceeds to malignancy, and carcinoma of bladder is common in those countries in which schistosomiasis is endemic. *See also 1.39, 5.18.* HE ×55

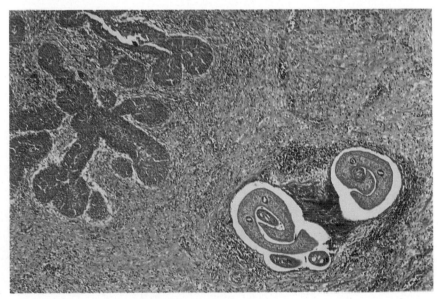

10.55 Transitional carcinoma: ureter

Transitional cell carcinomas vary widely in malignancy from Grade I the best-differentiated 'papilloma' to Grade IV the most malignant. However even a well-differentiated 'papilloma' of transitional epithelioma may recur locally and should be regarded as a low-grade carcinoma. Bladder carcinomas tend to arise in the trigone region and usually present with hematuria. This neoplasm was an irregular friable mass ($6 \times 2.5 \times 1.5$cm) blocking the distal end of the left ureter in a man of 75 and the kidney was hydronephrotic. It is a papilliform lesion consisting of short broad papillae. This shows parts of several papillae. They have a fibrovascular core (arrow) covered with transitional epithelium several times thicker than normal epithelium. The polarity of the epithelial cells is reasonably well preserved.   HE ×150

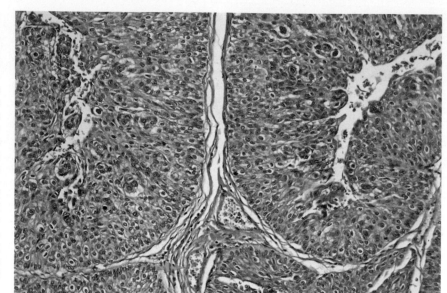

10.56 Transitional carcinoma: ureter

This shows the fibrous core (thin arrow) of a papilla and the basal layers of the covering epithelium. The epithelial cells show some loss of polarity. Their nuclei are large and vesicular and in each there is a prominent nucleolus. They are slightly pleomorphic and one mitotic figure is present (thick arrow). This tumour was classified as a fairly well-differentiated transitional cell carcinoma of ureter (Grade II). There is a close association between carcinoma of the renal tract and exposure to carcinogenic substances in the environment, presumably because they or their derivatives are excreted in the urine, and workers exposed to aniline dyes, benzidine, etc. are particularly at risk.   HE ×360

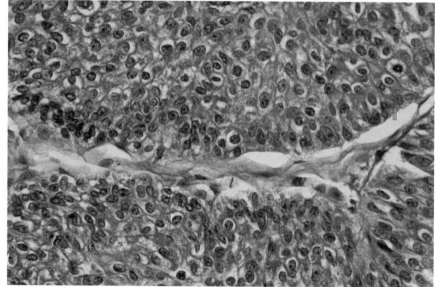

10.57 Squamous cell carcinoma: bladder

Squamous cell carcinoma of bladder arises from transitional epithelium that has undergone metaplasia to squamous epithelium as a result of chronic 'irritation', e.g. from calculi present in the bladder. This lesion in a man of 87 contained areas of both transitional cell carcinoma and squamous cell carcinoma. This area consists of cords of fairly well-differentiated squamous cells which form keratin (thin arrow). The more 'basal' cells towards the surfaces of the cords are pleomorphic and mitotic activity is evident (thick arrow). The stroma is fibrous and infiltrated with chronic inflammatory cells. The squamous cell carcinoma component but not the transitional cell part was invading the muscle coat of the bladder and was also present in perineural lymphatics, indicating its more aggressive nature.

HE ×235

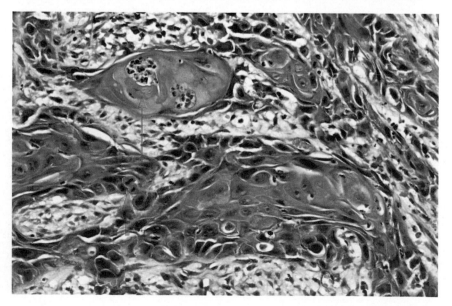

10.58 Inverted papilloma: bladder

Inverted papilloma of the bladder is sometimes confused with transitional cell carcinoma or papilloma but it is a separate entity. It is a benign lesion and its structure is not papilliferous. Instead it is generally a smooth-surfaced mass, formed by the neoplastic epithelium growing beneath the epithelial lining of the bladder. This lesion was in a man of 69. It consists of cords of basophilic epithelial cells lying in a pale-staining loose connective tissue. The epithelial cells are well-differentiated and there is no nuclear pleomorphism. The cells on the outsides of the cords are columnar and similar to the basal layer of transitional epithelium. Small cystic spaces containing secretion are present within the epithelium (arrow). The blood vessels in the connective tissue are dilated and there are small numbers of lymphocytes. HE ×150

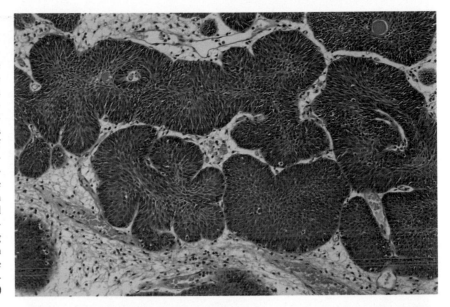

10.59 Inverted papilloma: bladder

At higher magnification the tumour epithelium is seen to be well-differentiated, resembling a double layer of normal transitional epithelium. The epithelial cells at the periphery of the cords of epithelial cells, i.e. the 'basal' layers, are columnar and palisaded, and similar to the basal layer in normal transitional epithelium. They rest on a well-formed basement membrane. The stroma is very loose connective tissue and occasional lymphocytes are present in it along with a thin-walled blood vessel. There is a small cystic space containing eosinophilic secretion (arrow). Inverted papilloma is generally located in the region of the trigone or bladder neck and it presents with either hematuria or obstruction of the urethra. There is a close association with chronic cystitis. HE ×360

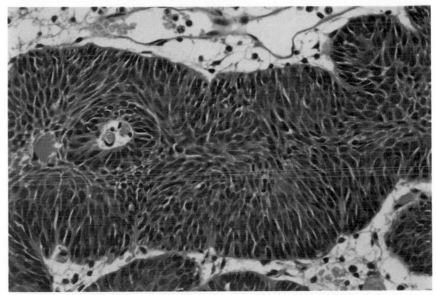

10.60 Pheochromocytoma: bladder

Pheochromocytoma occasionally arises from sympathetic paraganglia in sites other than the adrenal. This tumour arose in the bladder, a rare occurrence. The tumour cells form cords separated by connective tissue stroma. Their nuclei are large and round and the nucleoli small. Nuclear pleomorphism is moderate. The brown colour of the cytoplasm was produced by the reaction of the catecholamines in it with the dichromate in the tissue fixative (Orth's solution) – the chromaffin reaction. The stroma is vascular but the blood vessels are inconspicuous. Some families tend to develop pheochromocytomas, sometimes in association with medullary carcinoma of thyroid, parathyroid adenoma, etc. (multiple endocrine adenoma syndrome).
HE ×360

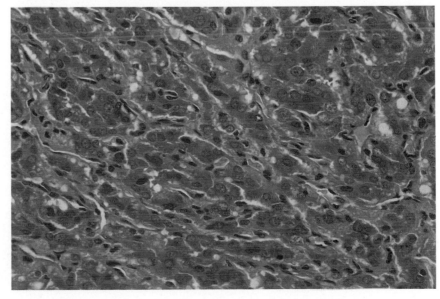

11.1 Gynecomastia: breast

The male breast often enlarges moderately at puberty but enlargement in later life should be regarded as pathological. A man of 65 developed a 'lump' in his right breast and a portion of tissue (6 × 4 × 1.5cm) was removed surgically. This shows several small ducts. There is marked epithelial hyperplasia, with several layers of cells lining each duct. The cells show considerable nuclear pleomorphism but no mitotic figures are present. The stromal connective tissue around the duct is myxoid and similar to that in a fibroadenoma. Similar appearances were present elsewhere in the resected tissue. These changes are typical of simple gynecomastia. Sometimes the epithelial hyperplasia is more marked and liable to be mistaken for carcinoma. HE ×360

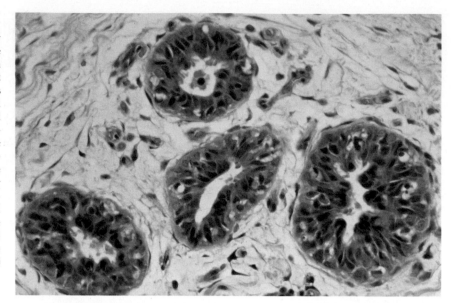

11.2 Queyrat's erythroplasia: penis

Queyrat's erythroplasia takes the form of a bright red plaque on the penis. It is in an area of squamous cell (epidermoid) carcinoma-in-situ similar to Bowen's disease, and in a small minority of cases the lesion becomes invasive after a long period. In this example the squamous epithelium of the penis is greatly thickened, with very marked acanthosis; and the surface (left) is hyperkeratotic. The rete ridges are greatly elongated. There is no evidence of invasion. The epithelial cells show some loss of stratification but nuclear pleomorphism is slight. Large pigment-laden cells are present in the basal layers of the epithelium, and in the underlying connective tissues there is a dense infiltrate of chronic inflammatory cells. HE ×85

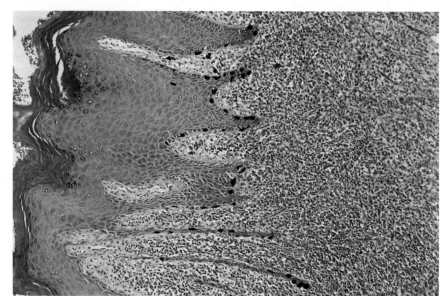

11.3 Queyrat's erythroplasia: penis

This is the tip of one of the very elongated rete ridges. The acanthotic epithelial cells show some loss of polarity and the nuclei are moderately pleomorphic. There is no mitotic activity. The large pigment-laden cells are in the basal layer of the epithelium or immediately beneath it. The infiltrate of chronic inflammatory cells in the underlying connective tissues consists almost entirely of plasma cells (arrow). HE ×335

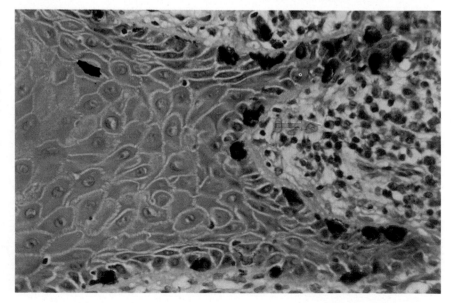

11.4 Primary syphilitic chancre: penis

The primary chancre of syphilis takes the form of an indurated papule usually on the penis, vulva or cervix. It may be 2 or 3 cm dia but is often less. The surface is ulcerated. If untreated the ulcer heals and the induration disappears in 6–8 weeks leaving only a very small scar. A man of 19 had an ulcer on his foreskin which was biopsied for diagnostic purposes. The surface of the chancre (left) is ulcerated and in place of the squamous epithelium the surface is covered with a fibrin network containing a mixture of polymorphs, lymphocytes and macrophages. Under it is a layer of granulation tissue and beneath that there is connective tissue containing numerous thick-walled small blood vessels (arrows) and an intense infiltrate of chronic inflammatory cells, many of them plasma cells.　　　　HE ×60

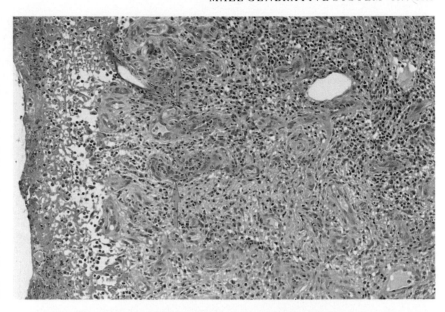

11.5 Primary syphilitic chancre: penis

The small blood vessels in the chancre are thick-walled and lined by large endothelial cells (arrows). Their lumen is inconspicuous. The inflammatory cell infiltrate is very polymorphic, consisting of plasma cells, lymphocytes, macrophages and polymorph leukocytes; and elongated fibroblast-like cells are present around the blood vessels. A silver stain demonstrated many spirochetes in the tissue.　　HE ×360

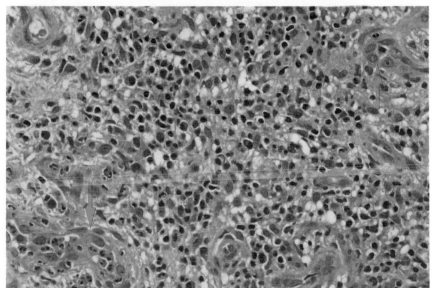

11.6 Primary syphilitic chancre: penis

Deeper in the chancre the chronic inflammatory cells tended to be located in the adventitia of the blood vessels, 'cuffing' them. In some vessels, however, like this venule the intimal changes are marked. There is very pronounced endophlebitis with a high concentration of plasma cells and lymphocytes in the thickened intima; and a similar, but much less intense, infiltrate in the other coats of the vessel. The lumen of the vessel is much reduced. When many vessels are affected in this way, and especially when thrombosis occurs to occlude them completely, the tendency to necrosis and ulceration is increased.　　HE ×135

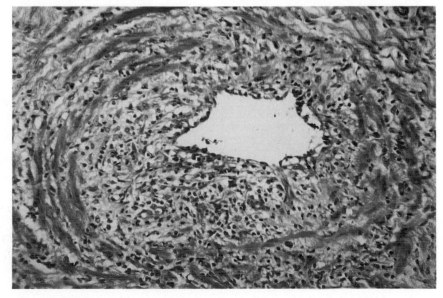

11.7 Lichen sclerosis et atrophicus: penis

Lichen sclerosis et atrophicus is a disorder of collagen but is probably not associated with autoimmune phenomena. In addition to affecting the skin, it may involve the vulva or penis. This lesion was on the foreskin of a man of 23. The surface of the epithelium (left) shows dense lamellar hyperkeratosis and lymphocytes are present within it. Beneath the epithelium there is a broad band of edematous hyalinized and poorly cellular collagen. There are vacuoles close to the epithelial basement membrane but the basal cells of the epithelium are intact (liquefaction degeneration of this layer is often present). The deeper tissues (right) are infiltrated with chronic inflammatory cells and the blood vessels are dilated (see also 12.33).

HE ×150

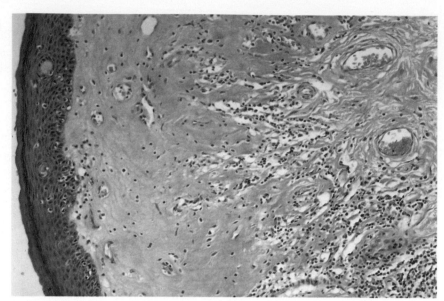

11.8 Benign nodular hyperplasia: prostate

Hyperplasia of the prostate is common in older men. The prostate may weigh several hundred grams. It consists of yellowish nodules of hyperplastic glandular tissue. The condition tends to affect the central parts of the prostate and cause obstruction to the outflow of urine. The risk of infection of the renal tract is greatly increased. Bilateral hydronephrosis and renal failure are possible complications. The specimen is from a man of 71. This is one of the nodules. It consists of acini and abundant fibromuscular stroma. The acini are tortuous and papilliform growth of epithelium into the lumen has occurred in many of them. A few acini contain pale-staining secretion. There is an infiltrate of chronic inflammatory cells around many of the acini.   HE ×60

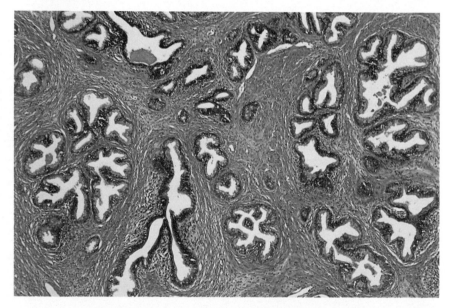

11.9 Benign nodular hyperplasia: prostate

At high magnification the cells lining the glandular acini are tall, well-differentiated columnar cells with vacuolated eosinophilic cytoplasm and uniform round basal nuclei. There is no mitotic activity. The cells are similar to the normal prostatic epithelium but show little evidence of secretory activity. They show, however, a marked tendency to intra-acinar proliferation. A laminated corpus amylaceum and fragments of others are present in one acinus. The stroma between the acini is abundant, consisting of eosinophilic smooth muscle and pale-staining fibrous connective tissue.

HE ×150

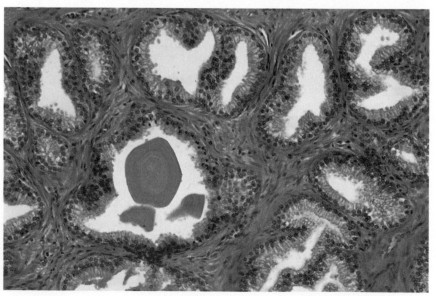

11.10 Infarct and squamous metaplasia: prostate

Enlarged hyperplastic prostates are liable to undergo infarction from vascular occlusion which may be associated with catheterization, infection, etc. This is the edge of an area of infarction. The blood vessels are greatly dilated (arrow) and there is hemorrhage into the stroma. The acini are lined by a thick layer of epithelial cells resembling squamous epithelium. In several acini the epithelial cells have proliferated to fill the lumen (right). The two larger acini contain corpora amylacea and necrotic cell debris. Squamous metaplasia occurs not uncommonly around infarcts of prostate and care must be taken not to confuse the change with carcinoma. Although the metaplastic epithelium resembles squamous epithelium, intercellular bridges are not usually present and keratin is not formed.   HE ×120

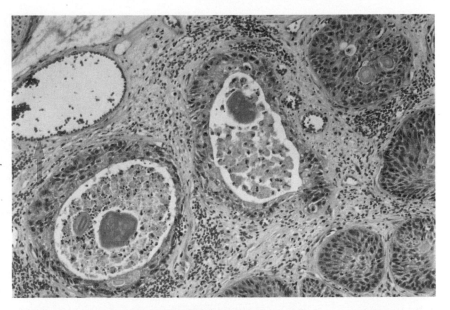

11.11 Squamous metaplasia and carcinoma: prostate

Adenocarcinoma of prostate is often treated with estrogens. The therapy tends however to cause squamous metaplasia in the normal glandular epithelium. A man of 72 developed an adenocarcinoma of prostate for which he was treated with stilboestrol. A year later fragments of the prostate were removed *per urethram*. This shows several acini. They are enlarged and filled with stratified squamous epithelium in which small foci of keratin formation are present (thin arrows). The nuclei of the basal layer of cells are basophilic and moderately pleomorphic. There is however no evidence of malignancy. In places the normal columnar epithelium has survived (thick arrows). The stroma is abundant, with probable increase in fibrous tissue.   HE ×150

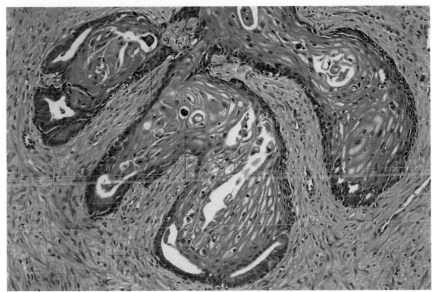

11.12 Carcinoma: prostate

After stilbestrol therapy the adenocarcinomatous cells often show marked regressive changes and may appear as collections of small pyknotic nuclei without detectable cytoplasm. In this case however in addition to the squamous metaplasia in the non-neoplastic acini, there were also small cords of malignant cells which still look fully viable. One of these is shown here. The tumour cells have round or ovoid nuclei, each with a fairly prominent nucleolus. They have abundant cytoplasm and form irregular acini. The stroma consists of eosinophilic smooth muscle fibres and pale-staining fibrous tissue. *See also 1.47, 1.48, 4.48.*   HE ×360

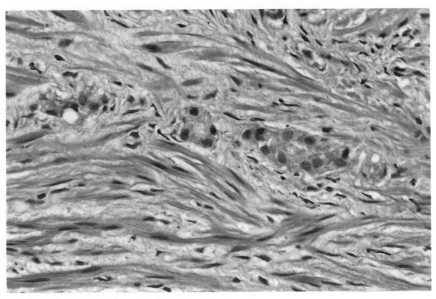

11.13 Carcinoma: prostate

In contrast to benign hyperplasia carcinoma of the prostate tends to arise in the peripheral parts of the organ and often presents with symptoms and signs caused by local growth and invasion of adjacent tissues. Tissues removed via the urethra are not ideal for early diagnosis and cytology is of little value. Transrectal biopsy specimens are often used with success. The tumour is an adenocarcinoma, consisting of closely-packed irregular acini lined by a single layer of cuboidal cells. The cells are fairly uniform, with a large, round nucleus containing a strikingly prominent nucleolus. Some of the acini are very small and there are groups of malignant cells which are not parts of acini. Mitotic figures are present. The stroma consists of fibrous tissue and smooth muscle. HE ×360

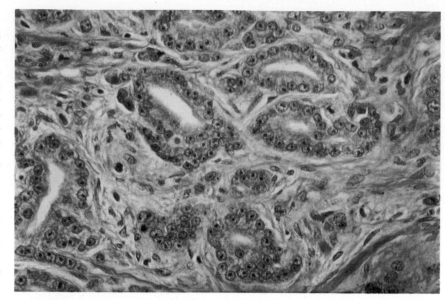

11.14 Carcinoma: prostate

Carcinoma of prostate may be well-differentiated and resemble normal prostate fairly closely. This lesion from a 68-year-old man however does not form acini. Solid cords of large cells (arrow) with abundant pale foamy cytoplasm and small nuclei are invading the eosinophilic muscular tissues. The nuclei of the tumour cells are round and regular in size and shape; and in only a few of them is a nucleolus visible. There are no mitotic figures. Carcinoma of prostate tends to spread to regional lymph nodes and the bone marrow, and multiple metastases in the skeleton may be the first clinical sign. HE ×360

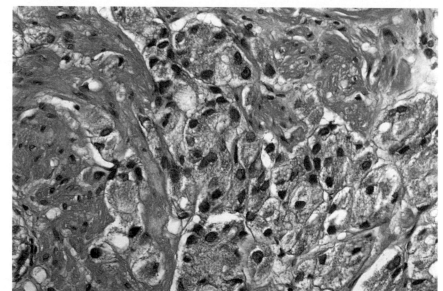

11.15 Carcinoma: prostate

A man of 80 had a prostatectomy for benign hyperplasia. Histology confirmed the diagnosis but also revealed a small carcinomatous focus consisting of several cords and clusters of malignant cells (arrows) in the fibromuscular stroma adjacent to the hyperplastic nodules. There is only rudimentary acinus formation by the tumour cells, which have relatively large nuclei each containing one or more prominent nucleoli. There are no mitotic figures. Their cytoplasm is scanty. Careful examination of tissues removed for benign hyperplasia of the prostate often demonstrates small foci of carcinoma, the incidence increasing with the age of the patient. These lesions must be regarded as latent microcarcinomas of no clinical significance for the individual concerned and treatment for carcinoma is not indicated. HE ×360

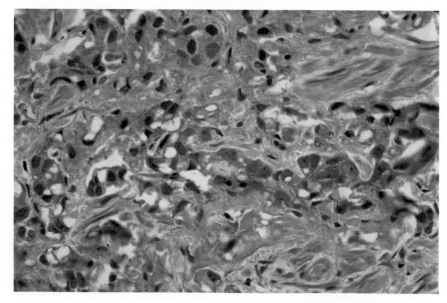

11.16 Undescended testis (cryptorchidism)

The testes are normally in the scrotum by the age of 4 and if a testis is still in the inguinal region or abdomen at 6 years of age treatment should be given. Otherwise irreversible atrophy will take place in the testis; and the risk of a malignant tumour developing in later life is greatly increased (30–50 times). This is an undescended testis that was removed surgically. It was smaller than normal (2.5 × 2 × 0.5cm). Histologically there is loss of many seminiferous tubules. The remaining tubules are atrophic and their basement membranes are greatly thickened. The smaller tubules (arrow) are hyalinized and the larger tubules contain only Sertoli cells with vacuolated cytoplasm. The interstitial (Leydig) cells are hyperplastic, filling the spaces between the tubules. HE ×150

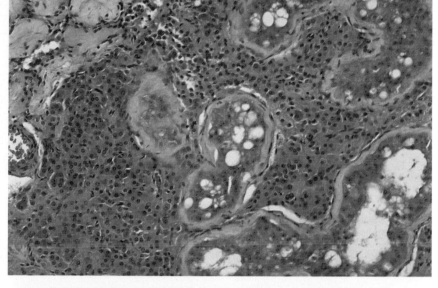

11.17 Undescended testis (cryptorchidism)

At higher magnification the tubular atrophy is striking. The basement membrane is very thick and hyaline and the cells in the lumen are Sertoli cells, with round or ovoid vesicular nuclei, containing a prominent nucleolus, and heavily vacuolated cytoplasm. There is no evidence of spermatogenesis. The tissues around the tubules are packed with interstitial (Leydig) cells with their characteristically deeply eosinophilic cytoplasm and round nuclei. The hyperplasia of the Leydig cells appears greater because of the loss of so many tubules from the testis. HE ×360

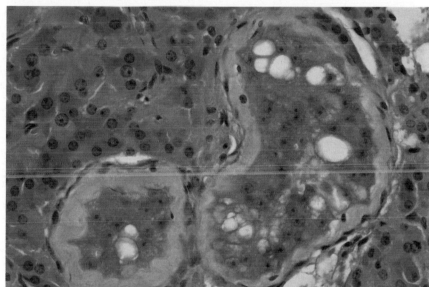

11.18 Torsion: testis

Occasionally a testis undergoes spontaneous torsion. The veins in the spermatic cord are occluded, usually at the inguinal ring. The testis becomes extremely congested, swollen and hard and if the condition is not treated in time, infarction follows. This is the testis of a boy of 15. It measured 4.4 × 4 × 3cm (with epididymis) and was hemorrhagic on sectioning. The fibrous tunica albuginea is just visible on the right. There is extensive hemorrhage beneath it. Although some pyknotic nuclei are present near the tunica the whole of the testicular tissue is necrotic. The tubules contain only eosinophilic cell debris, all the cells having lost their nuclear staining. HE ×150

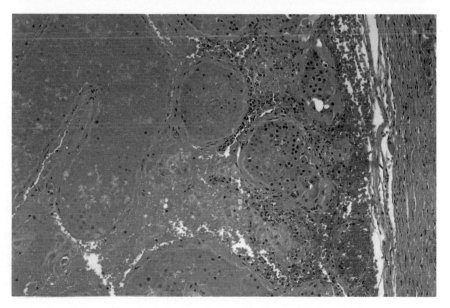

11.19 Klinefelter's syndrome: testis

A person with Klinefelter's syndrome is an infertile male with gynecomastia and small testes and is often mentally retarded. The disorder is chromosomal in origin, the majority of cases having an extra X chromosome in their cells (XXY). Much of the tissue in the atrophic testes in Klinefelter's syndrome consists of interstitial (Leydig) cells and this shows a group of large interstitial cells with their granular eosinophilic cytoplasm and round vesicular nuclei. In some of the nuclei the extra X chromosome is visible as a sex chromatin body (arrow). The sex chromatin body differs from a nucleolus in being smaller and attached to the nuclear membrane. It is also basophilic whereas nucleoli are eosinophilic. The small dark nuclei at the periphery of the group of interstitial cells are lymphocytic. HE ×940

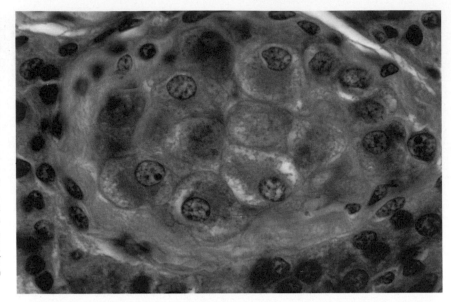

11.20 Granulomatous orchitis: testis

Granulomatous orchitis tends to occur in middle-aged men. It usually presents as a painful swollen testis. The swelling gradually subsides to leave a testis that is firmer than normal and less sensitive to pressure. Histologically there is an intense granulomatous reaction in the testis which is centred on the seminiferous tubules. This gives it a follicular distribution which is liable to be mistaken for tuberculosis. This lesion was in the left testis of a man of 42. Three seminiferous tubules and the peripheral parts of two others are shown. The seminiferous tissue has been destroyed and each tubule is full of eosinophilic epithelioid cells and giant multinucleated cells of the Langhans type (arrows). The interstitial tissue is heavily infiltrated with chronic inflammatory cells which are mainly plasma cells.
HE ×150

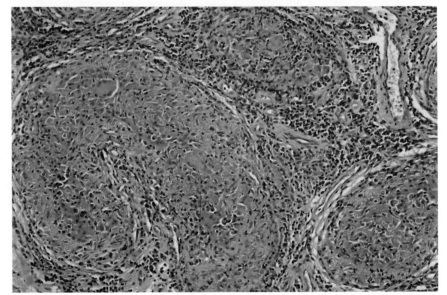

11.21 Granulomatous orchitis: testis

This shows a seminiferous tubule in cross-section. The cells within the lumen and the basement membrane have been completely destroyed and replaced by an ovoid collection of pink-staining epithelioid cells. No Langhans-type giant cell is present in this tubule but they are frequently present. Chronic inflammatory cells, nearly all of which are plasma cells, are present around the periphery and the resemblance to a tuberculous follicle is close. However there is no necrosis and tubercle bacilli are never demonstrable. The etiology of granulatomous orchitis is not known. HE ×360

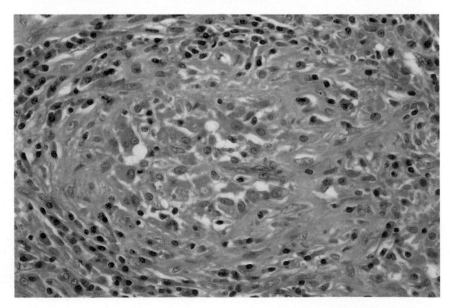

11.22 Seminoma: testis

Seminoma of testis arises from the germinal (seminiferous) epithelium of the mature testis. Macroscopically it usually has a uniform pale creamy colour with occasional areas of necrosis. The tumour tends to occur in early middle age (around 40 years of age) and with early diagnosis and adequate treatment including radiotherapy the prognosis is very good. The histological structure is characteristic. The tumour cells are uniformly large and polyhedral and closely packed. They have clear or foamy cytoplasm (containing glycogen) and the nucleoli are prominent in the round or ovoid nuclei. The delicate fibrous stroma (arrow) which runs between the groups of tumour cells is heavily infiltrated with lymphocytes. HE ×335

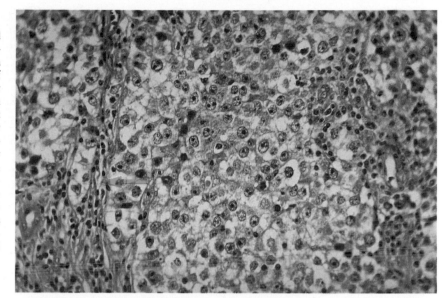

11.23 Seminoma: testis

In this seminoma the tumour cells lie singly or in small clusters and the cytoplasm of each cell instead of having the characteristic foamy appearance is vacuolated (the glycogen in the cytoplasm probably caused this). The inflammatory cell infiltrate in the stroma is intense, consisting of clusters of epithelioid histiocytes, lymphocytes and plasma cells. Langhans-type giant cells were present elsewhere in the tumour. HE ×150

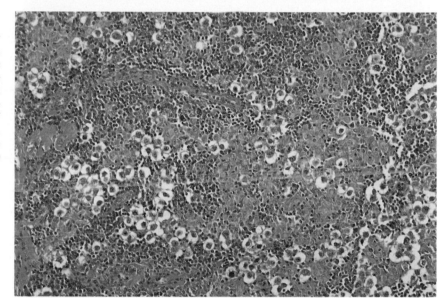

11.24 Seminoma: testis

At higher magnification the tumour cells (arrows) have large nuclei and vacuolated 'clear' cytoplasm, with only wisps of cytoplasm adherent to the nucleus. The epithelioid histiocytes have eosinophilic slightly granular cytoplasm and pale-staining ovoid or elongated vesicular nuclei. They are closely packed and the cell boundaries are indistinct. The cells with small deeply-staining nuclei are lymphocytes and plasma cells. The presence of a large population of lymphocytes in seminomas suggests a strong defensive reaction on the part of the host to the tumour, and this view is strengthened by the presence in many seminomas of the epithelioid histiocytes, a reaction similar to that seen in tuberculous or sarcoid follicles. The combination of lymphocytes and epithelioid follicles is suggestive of a delayed hypersensitivity type of reaction. HE ×360

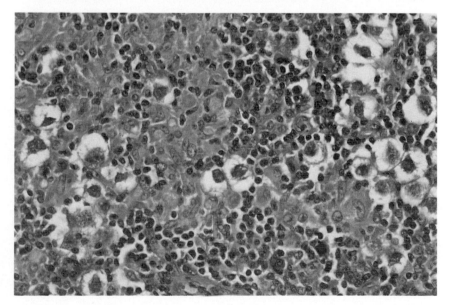

11.25 Teratoma: testis

Teratomas are true 'mixed' tumours, containing all types of tissue in varying proportions and degrees of differentiation. Occasionally the tissues adopt an organoid arrangement. Teratoma of testis, a rare tumour, is invariably malignant, unlike its ovarian counterpart which in a high proportion of cases is a benign 'dermoid'. This is a 'differentiated' (or 'mature') teratoma consisting of mature-looking tissues. The large cyst, just visible on the left, is lined by flattened epithelium. The smaller cysts and ducts are lined by tall columnar epithelial cells. The stroma is a mixture of dense fibrous tissue (left) and highly cellular connective tissue, e.g. around the small ducts, for example. The mixture of cystic glands and fibrous tissue gives a teratoma a characteristic naked-eye appearance (fibrocystic disease). HE ×135

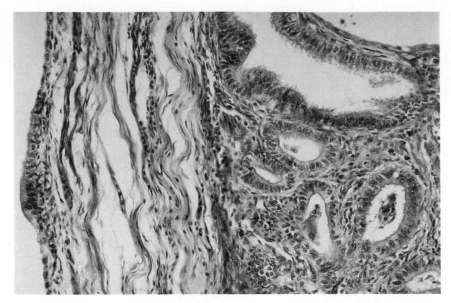

11.26 Teratoma: testis

This, too, is a differentiated (mature) teratoma. In differentiated teratomas squamous epithelium is often abundant and this shows a fibrous-walled cyst full of keratinous squames derived from the lining of stratified squamous epithelium (right). Part of the cyst (arrow) is lined by tall columnar epithelial cells. The surrounding fibrous tissue is infiltrated by lymphocytes. Differentiated teratomas of the testis are rare and even though they appear to be composed of mature-looking tissues, it should not be assumed that they will behave like a benign neoplasm, since malignant areas are readily overlooked during histological examination; and a proportion of differentiated (mature) teratomas of testis do metastasize. HE ×150

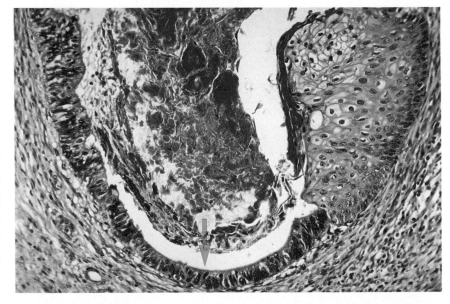

11.27 Yolk sac tumour: testis

The yolk sac (endodermal sinus) tumour is a malignant teratoma in which embryonal yolk sac tissue predominates. This lesion presented as a lump in the left testis of a man of 20. The testis was enlarged (8 × 4 × 3cm) and was resected. The testicular tissue had been almost entirely replaced by highly vascular tumour in which there was extensive necrosis. It consists of a very edematous mesenchymal stroma with numerous clear (microcystic) spaces and complex tubular structures composed of cuboidal cells with vacuolated cytoplasm and large nuclei in each of which there is a prominent nucleolus. The cells tend to form papillary projections into the lumen of the tubules, producing structures which vaguely resemble glomeruli. These are Schiller-Duval bodies and are a diagnostic feature of yolk sac tumours. HE ×150

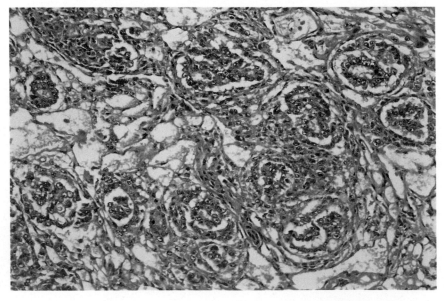

11.28 Choriocarcinoma: testis

Choriocarcinoma of testis is a highly malignant tumour which is assumed to develop in
a teratoma. A man of 29 presented with a
swollen left testis which was resected surgically. It measured 9 × 6 × 5cm and a
tumour was present, the cut surface of
which was light brown and faintly lobulated. Histologically the tumour contains
cytotrophoblast (thin arrow) consisting of
cells with clear or vacuolated cytoplasm and
well-defined cell boundaries; and syncytiotrophoblast (thick arrow) consisting of
sheets of eosinophilic cytoplasm in which
there are very large nuclei with very prominent nucleoli. Individual multinucleated
giant cells (syncytiotrophoblastic) are also
present. There is also abundant eosinophilic
fluid containing fairly numerous polymorph
leukocytes (e.g. bottom left).   HE ×150

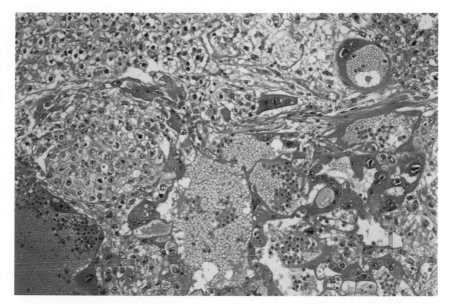

11.29 Lymphoplasmacytoid lymphoma: testis

Lymphoma of testis is rare but in older men
it is the commonest type of tumour. This
lesion was in a man of 64. The testis was
greatly enlarged (7 × 5 × 4cm) weighing
approx. 110 g. Its cut surface was firm and
creamy-white, with flecks of hemorrhage.
Atrophic thick-walled seminiferous tubules
containing only Sertoli cells and cell debris
are surrounded by a highly cellular tissue
consisting of small cells with darkly-staining
nuclei, and small numbers of large pale-
staining histiocytes. At higher magnification the small cells were neoplastic plasma
cells. Numerous mitoses were present. The
whole testis had the same structure and a
diagnosis of lymphoplasmacytoid lymphoma was made. The prognosis of testicular lymphoma is poor since lesions are usually present elsewhere.   HE ×60

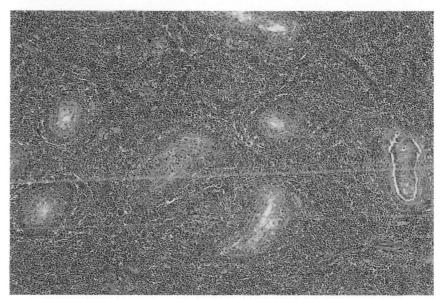

11.30 Mullerian duct cyst: appendix of testis

A man of 47 developed a swelling in his
scrotum and a cyst filled with clear fluid and
attached to the testis was removed surgically. This shows part of the cyst wall. It
consists of vascular connective tissue lined
by a single layer of closely-packed columnar
cells, some of which are ciliated. The lesion
was diagnosed as a Mullerian duct cyst of the
appendix of the testis. The appendix of the
testis is a remnant of the paramesonephric
duct (Mullerian duct) and is located
between the top of the testis and the head of
the epididymis. It corresponds to the
fimbriated end of the Fallopian tube. It is
sometimes pedunculated and may undergo
torsion.   HE ×360

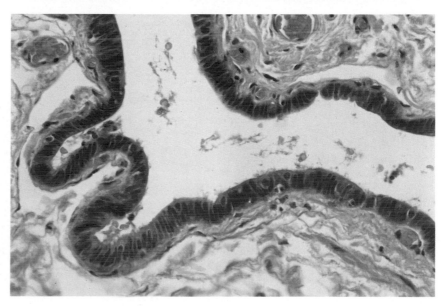

11.31 Adenomatoid tumour: epididymis

The adenomatoid tumour of epididymis is a benign encapsulated tumour of mesothelial origin. It usually occurs in men in their twenties and thirties as a relatively small greyish-white mass. This lesion, however, was a firm smooth-surfaced mass (3cm dia) in a 42-year-old man. The cut surface was cream-coloured. It consists of a complex arrangement of thin-walled cystic spaces and clefts, lined for the most part by flattened cells, lying in abundant connective tissue stroma. Some of the spaces contain small groups of cells.

HE ×150

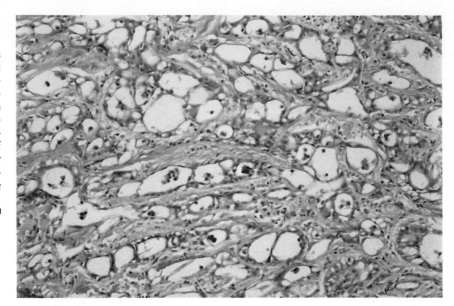

11.32 Adenomatoid tumour: epididymis

At higher magnification the cells lining the spaces and clefts are seen to be very flat and attenuated. The spaces appear empty but in appropriately fixed tissue mucinous secretion can be demonstrated within them. Clusters of cells with round nuclei and individual cells are present in some of the spaces (top left). These are probably tumour cells and the elongated cells (arrow) lying between the spaces are probably also neoplastic. The stroma is fibrous but in other parts of the tumour smooth muscle was present. The histogenesis of adenomatoid tumour is uncertain but a mesothelial origin seems likely.   HE ×360

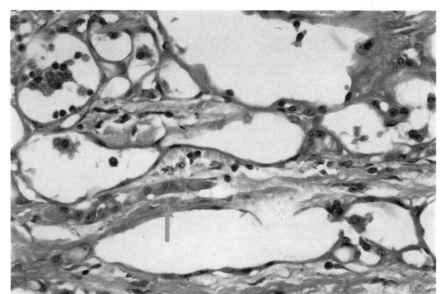

11.33 Paratesticular leiomyosarcoma: testis

Paratesticular malignant tumours of non-germinal cell origin in children are generally rhabdomyosarcomas, but in adults other types of tumour are more likely. This tumour (7 × 5 × 5cm) was in the spermatic cord of a man of 52. It appeared to arise from the upper part of the epididymis. The cut surface was lobulated and greyish-white, with areas of necrosis and hemorrhage. Histologically it is a spindle-cell sarcoma, consisting of interlacing bundles of smooth muscle cells with elongated nuclei with round ('blunted') ends. The nuclei are hyperchromatic and moderately pleomorphic and there are numerous mitotic figures (arrows). The cells have abundant cytoplasm in which special stains demonstrated myofibrils. Even when apparently well-differentiated, leiomyosarcomas tend to recur. *See also 1.53, 4.20.*   HE ×360

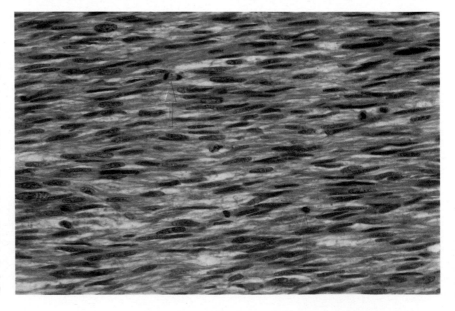

11.34 Embryonal rhabdomyosarcoma: spermatic cord

Embryonal rhabdomyosarcoma is a highly malignant tumour of the first two decades of life but most occur before the age of 5. This tumour however was in the spermatic cord of a man of 21. It was a large lobulated homogeneous mass 7cm dia which was compressing the testis. On section it was of firm consistency and cream-coloured but one area was gelatinous and spongy. The cells show considerable pleomorphism of both nucleus and cytoplasm. Most cells are round with basophilic nuclei (bottom). Others are plump and fusiform, with large nuclei, and many are strap-like (centre) with abundant eosinophilic cytoplasm. The cytoplasm of many cells is highly vacuolated. There is abundant loose stroma between the tumour cells.   HE ×150

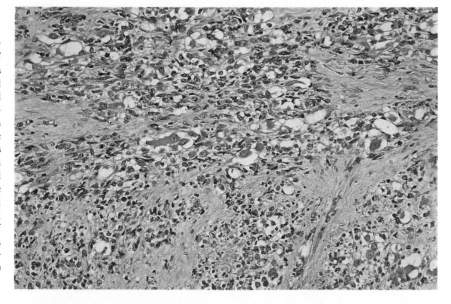

11.35 Embryonal rhabdomyosarcoma: spermatic cord

At higher magnification the extreme pleomorphism of the tumour cells, both cytoplasmic and nuclear, is evident. The nuclei are large and although the majority are round or ovoid, others are elongated. Nucleoli are present and some are very prominent (thin arrow). There are many large vacuoles which are apparently intra-cytoplasmic. Several tumour cells are very elongated and strap-like with abundant cytoplasm which is strongly eosinophilic (thick arrow). The appearance suggests that they are neoplastic muscle cells. The stroma (top) is fibrillary in structure.   HE ×360

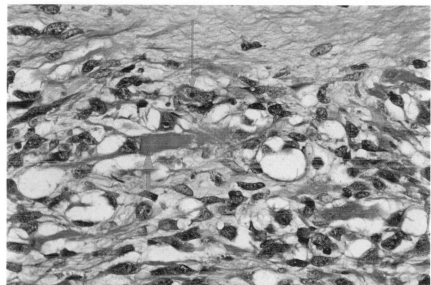

11.36 Embryonal rhabdomyosarcoma: spermatic cord

A section of the tumour shown in **11.34** and **11.35** was stained with toluidine blue which demonstrates the cross-striations in muscle fibres much more effectively than sections stained by hematoxylin and eosin. The cytoplasm of several of the elongated 'strap-like' cells show cross-striations in their cytoplasm identical to that of striated muscle fibres (arrow), confirming that the tumour is a rhabdomyosarcoma. The stain also emphasizes the basophilia and pleomorphism of the nuclei of the tumour cells.

Toluidine blue ×800

12.1   Fat necrosis: breast

Fat necrosis occurs not infrequently in the breast on a scale sufficient to produce a firm but not stony-hard mass which may simulate carcinoma. The lump is often attached to the overlying skin but not to the deeper tissues and it does not cause retraction of the nipple. It occurs more often in pendulous breasts and there is some association with trauma to the breast. Fat released from necrotic fat cells provokes a vigorous reaction which, however, soon becomes more chronic and granulomatous. This lesion (1.5cm dia) was in the axillary tail of the breast of a woman of 76. It is a granulomatous mass, consisting of histiocytes, lymphocytes and plasma cells. Several multinucleated giant cells are also present, as well as areas of necrosis. The vacuoles and small cysts contained extracellular neutral fat released from necrotic fat cells.   HE ×150

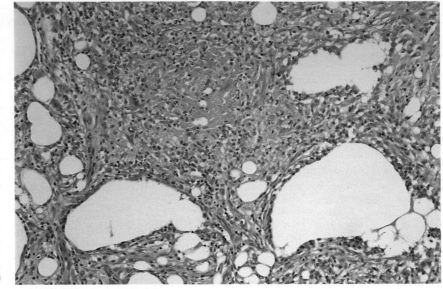

12.2   Fat necrosis: breast

This shows several droplets of extracellular fat (the fat dissolves out during processing of the tissues). The cells in contact with the fat are macrophages (thin arrows), and many macrophages with foamy (lipid-laden) cytoplasm (thick arrows) are present in the surrounding tissues. Considerable numbers of plasma cells (double arrows) are also present, as well as occasional lymphocytes. Fibroblasts are starting to lay down connective tissue fibres *See also 1.20.*

HE ×360

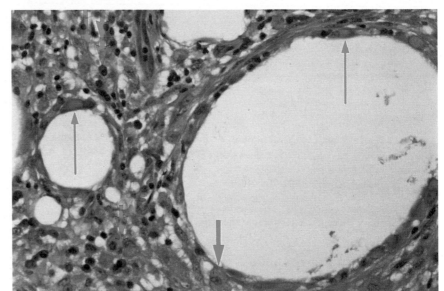

12.3   Fibroadenoma: breast

Fibroadenomas are very common benign tumours of the breast and tend to be found in younger women (20–35 years). They increase in size in pregnancy but often regress with age and may calcify. They very rarely become malignant. A fibroadenoma forms a firm well-defined mass, up to about 3cm dia. Fibroadenomas are 'mixed' tumours, in having both epithelial and connective tissue components. Some compressed normal breast is visible on the right, and between it and the fibroadenoma (left) there is a well-formed capsule of fibrous tissue. The fibroadenoma has a pericanalicular pattern, i.e. it consists of ducts surrounded by myxomatous connective tissue. The lumen of some of the ducts contains eosinophilic secretion. The ducts of the tumour are larger and more basophilic than those in the normal breast tissue.   HE ×95

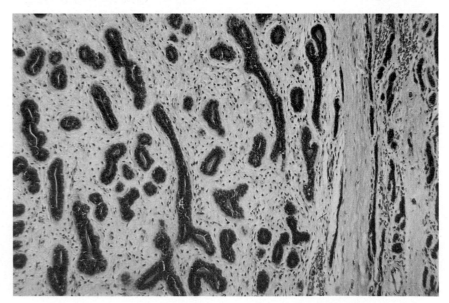

12.4 Fibroadenoma: breast

Sometimes in a fibroadenoma the connective tissue element proliferates more actively than the epithelial component. In effect it 'grows into' the ducts which then appear to be stretched out over the lobules of connective tissue, giving an 'intracanalicular' pattern. It is clearly seen in this fibroadenoma. The connective tissue component (thin arrow) is myxomatous and pale-staining and it forms lobular masses intersected by blue-staining ducts (thick arrow). Some of the ducts are filled with proliferated epithelial cells. The nodules of intracanalicular fibroadenomatous tissue are surrounded by bands of denser fibrous tissue (double arrow). Intracanalicular and pericanalicular patterns are not uncommonly present in the same fibroadenoma. The different patterns have no clinical significance.

HE ×38

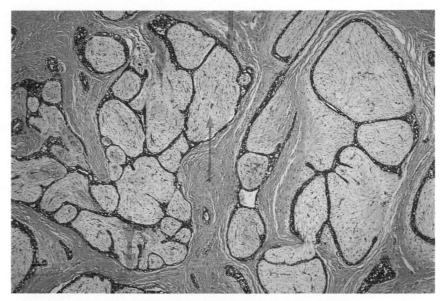

12.5 Cystosarcoma phyllodes: breast

Cystosarcoma phyllodes is an imprecise term based on a particular ('leaf-like') appearance, naked-eye. Typically the tumour occurs in middle-aged women and is often large. Histologically it consists of connective tissue which is sarcomatous and of ductal structures which are benign. This tumour was a well-circumscribed nodular white mass 5cm dia in a woman of 61. Characteristically the predominant element is highly cellular connective tissue composed largely of interlacing bundles of fibroblast-like cells (thin arrows) among which there are numerous mitoses (thick arrows). The connective tissue is invading adjacent fatty tissue (right). The epithelial element of the neoplasm consists of small ducts lined by cuboidal epithelium. The epithelium of the ducts is well-differentiated and shows no mitotic activity. HE ×235

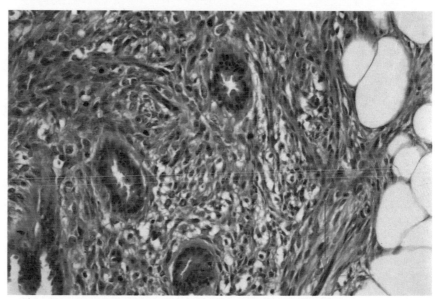

12.6 Intraduct papilloma: breast

Papillomas of the breast vary considerably in their structure, from obviously benign lesions to malignant growths. They may be multiple or single ('isolated' or 'solitary') lesions. The latter are usually small (not more than 1cm dia) and generally present with bleeding from the nipple, as happened with this benign lesion in a dilated duct in the breast of a woman of 61. The wall of the duct (right) consists of dense fibrous tissue, lined by a single layer of flattened epithelium. The papilloma consists of small irregular glands in a dense fibrous stroma. Some of the glands are lined by cuboidal cells and others by atrophic epithelium. There is no nuclear pleomorphism or mitotic activity in the epithelial cells. The surface of the papilloma is smooth and covered with a single layer of very attenuated epithelial cells which is deficient in places. HE ×150

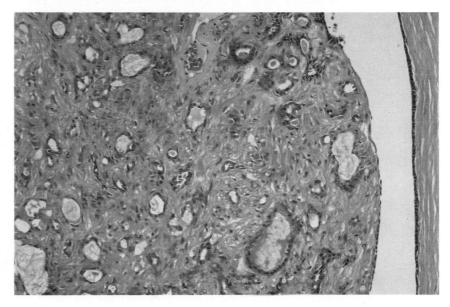

12.7 Intraduct papilloma: breast

This papilloma is from a woman of 61. The duct is cystic and its wall (right) consists of dense hyalinized fibrous tissue lined by flattened epithelial cells. The delicate papilliferous processes of the papilloma have a core of vascular connective tissue covered with epithelium. Most of the surface is covered by tall columnar cells accompanied by a layer of myoepithelial cells. The presence of a double layer of cells indicates that the lesion is benign. Over the surface however and at other areas of probable compression the columnar epithelium is replaced by a single layer of flat cells. The thin-walled blood vessels in the stroma of the papilliform processes tend to bleed, and bleeding from the nipple is a common presenting sign.

HE ×150

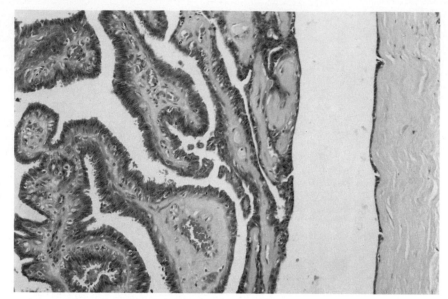

12.8 Generalized cystic mastopathy: breast

In generalized cystic mastopathy the epithelial hyperplasia takes a variety of forms, one of which is papilliform ingrowth of epithelium into the ducts, producing duct papillomatosis. The patient in this case was a woman of 21 who already had florid fibroepithelial hyperplasia of the breast, including duct papillomatosis. This duct is filled with papillary ingrowths which form a kind of cribriform structure, consisting of a fibrous stroma and glandular spaces lined by epithelium. The epithelium is mostly double-layered but in some parts it consists of a single layer of cuboidal or flattened cells. At a higher magnification however no mitoses or nuclear pleomorphism were seen. There are small, round calcified bodies and cellular debris in the lumen.   HE ×150

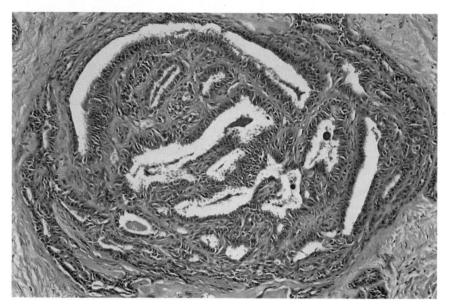

12.9 Generalized cystic mastopathy: breast

This is another example of duct papillomatosis in the same breast as that shown in **12.8**. The papilliform structures are more cellular and the epithelial hyperplasia (arrows) is more pronounced. The hyperplastic epithelium is supported by an eosinophilic stroma. The cells show some variation in size and shape but there is no mitotic activity and the lesion is benign.

HE ×150

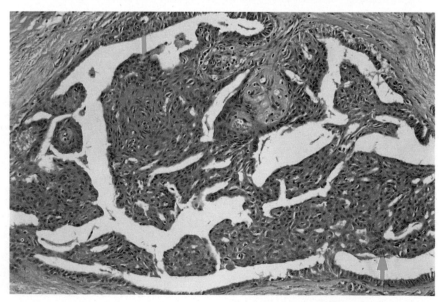

12.10 Generalized cystic mastopathy: breast

Another feature of generalized cystic mastopathy is dilatation of ducts (duct ectasia). Dilatation may be great, with production of multiple cysts, sometimes of considerable size. The dilated ducts and cysts are filled with clear fluid or more turbid material. A piece of tissue measuring $4 \times 2.5 \times 2$cm was removed for diagnostic purposes from the breast of a woman of 46. On section it was fibro-fatty in nature, with 'flecks' of white cheesy material. This shows two ducts, both greatly dilated and cystic, and filled with amorphous material. Each duct is lined with a single layer of flat or cuboidal basophilic cells, with a few small papilliform epithelial ingrowths into one (arrow). Both ducts are surrounded by dense fibrous tissue in which there are collections of small lymphocytes.

HE ×60

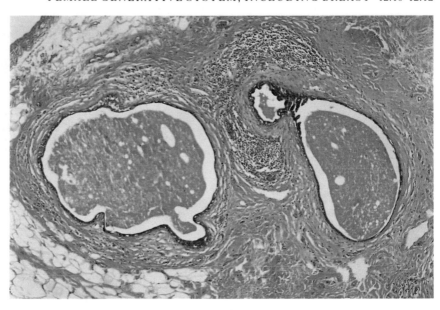

12.11 Generalized cystic mastopathy: breast

This is the same breast as that shown in **12.10**. A duct full of amorphous granular debris has ruptured at one point (thin arrow), with spillage of the contents of the duct into the surrounding tissues where there are many macrophages including a multinucleated giant cell (lower left). In the wall of the duct there are broad bands of hyaline material (thick arrow). This material reacts positively with special stains for elastic tissue and the change in the duct wall is termed 'elastosis'. Peripheral to the elastotic material the breast tissue is fibrous and infiltrated with lymphocytes.

HE ×150

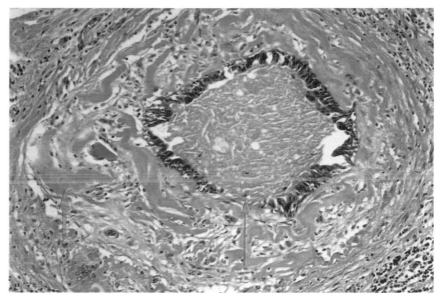

12.12 Generalized cystic mastopathy: breast

This is the lesion shown in **12.11** at higher magnification. The material extravasated into the surrounding tissues has provoked a vigorous cellular response, consisting mainly of macrophages. Some of the macrophages are binucleated and multinucleated (foreign-body) giant cells (arrow). The macrophages have abundant cytoplasm containing ingested material. Lymphocytes and plasma cells are also present and fibroblasts have formed connective tissue. The vacuole (bottom left) contained lipoid debris which dissolved during processing of the tissue.   HE ×235

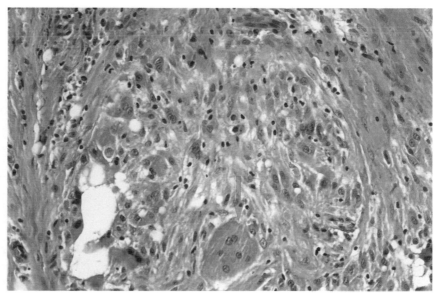

12.19   Intraduct carcinoma: breast

This is part of the large duct shown in **12.18**. The cells in the lumen have ovoid pleomorphic vesicular nuclei in which there is a central nucleolus. The eosinophilic cytoplasm is abundant but the cell margins are indistinct, the cells forming sheets and cords. There is considerable mitotic activity (arrow). The cells enclose numerous gland-like spaces which contain eosinophilic secretion. The periphery of the duct is also ill-defined with no distinct myoepithelial layer or basement membrane. There is no evidence of elastosis in the fibrous tissue around the duct (top).   HE ×360

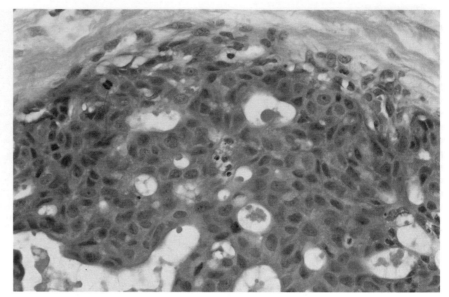

12.20   Intraduct carcinoma: breast

In intraduct carcinoma the proliferating malignant cells may eventually convert many ducts and ductules into tubes full of cancer cells; and when the distended ducts are cut, malignant cells can be squeezed out like toothpaste. The expressed material has been compared to a 'comedo' ('blackhead') and the condition is sometimes called comedo carcinoma. This example is from a woman of 42 whose left breast was removed because of the presence of several hard irregular nodules thought to be malignant. All the ducts in this part of the breast are filled with tumour cells. The malignant cells have large pleomorphic nuclei many of which are hyperchromatic. Eosinophilic necrosis of the tumour cells is present in most of the ducts. The stroma around the ducts consists of myxoid connective tissue.   HE ×95

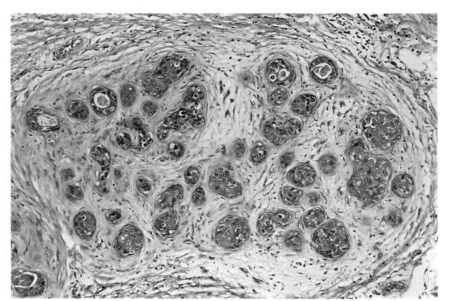

12.21   Intraduct carcinoma: breast

This shows two of the ducts in **12.20** at higher magnification. The nuclei of the tumour cells are very pleomorphic and many contain one or more prominent nucleoli. Several are hyperchromatic. The cytoplasm of the tumour cells is vacuolated and cell boundaries are indistinct. In one duct there is extensive central necrosis (arrow) and in the other duct several tumour cells are necrotic. The walls of the ducts are intact. The stroma around the ducts is loose and myxoid. No tumour was found in the axillary lymph nodes.   HE ×415

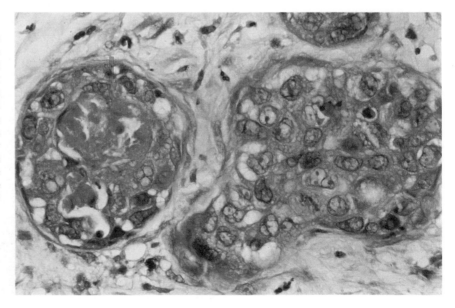

12.22 Carcinoma: breast

The large majority of carcinomas of the breast are invasive ductal carcinomas, i.e. they have their origin in the epithelium lining the ducts of the breast. The tumour forms a firm, sometimes hard, ill-defined lump in the breast. This tumour was in the breast of a woman of 36. The tumour cells form irregular duct-like spaces. Cords of cells are also present (thin arrow). The malignant cells have large pleomorphic nuclei, some containing one or more large nucleoli, and abundant eosinophilic cytoplasm. There is considerable mitotic activity (thick arrows) and some nuclei are pyknotic. The stroma is fibrous and infiltrated with lymphocytes and plasma cells. The connective tissue around the two normal ducts (top right) is hyalinized. In undifferentiated ductal carcinomas there are no duct-like structures.   HE ×150

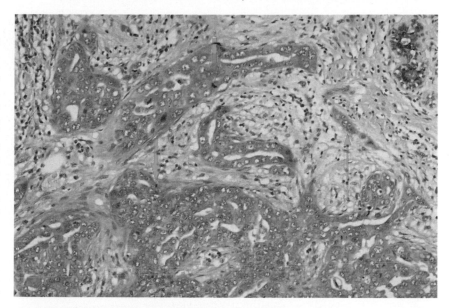

12.23 Carcinoma: breast

In invasive carcinomas (ductal or lobular) of breast, thick bands of eosinophilic hyaline tissue frequently form around the non-malignant ducts of the breast. This tissue stains with elastic tissue stains and the condition is called elastosis of the ducts. It is not specific for carcinomatous lesions however but may be seen in association with benign lesions. This shows a normal duct from the breast in which there was a mucin-secreting carcinoma. The duct is lined with cuboidal epithelium (thin arrow) and surrounded by collagen (stained purplish-red) and a band of closely-packed wavy black elastic fibres. Carcinoma cells (thick arrows) are present in the adjoining collagenous tissue.

Elastic–van Gieson ×360

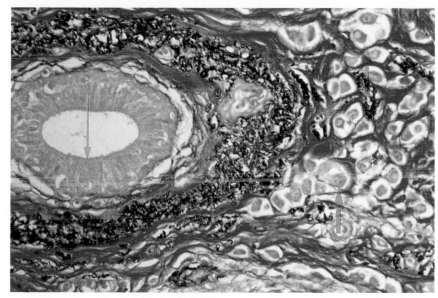

12.24 Carcinoma: breast

The malignant cells in breast carcinomas not infrequently contain droplets of mucin in their cytoplasm. The term mucoid carcinoma is however reserved for those carcinomas of breast in which the malignant cells produce mucin in such large quantities that it makes up a significant part of the bulk of the neoplasm. Mucoid carcinoma is an uncommon type of breast carcinoma. It is reputedly less liable to metastasize by the lymphatics than other types of ductal carcinoma but this is unproven. In this example compact clusters of malignant cells with basophilic pleomorphic nuclei and eosinophilic cytoplasm float in large quantities of mucin (almost colourless) secreted by tumour cells. A stromal blood vessel (arrow) is also present.   HE ×135

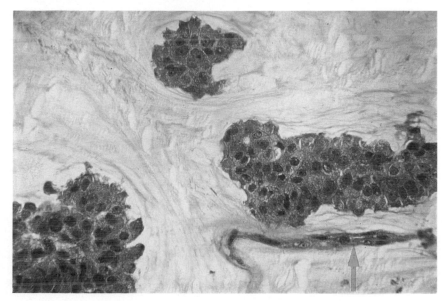

12.25 Lobular carcinoma: breast

Lobular carcinoma has its origin in lobules and terminal ducts. It is often multicentric and both breasts are not infrequently involved. It starts off as carcinoma-in-situ within the lobules. A woman of 45 developed a lump in her right breast and it was resected. The ducts are completely filled with polygonal small cells with round or ovoid nuclei which show little pleomorphism. The cells have a moderate amount of cytoplasm but the cell boundaries are indistinct. There is no necrosis or mitotic activity. A few lymphocytes are also present in the ducts. There is a well-defined basement membrane around each duct. All the ducts of the lobule were similarly affected. The stroma around the ducts is loose fibrous tissue. There was no evidence of invasion and the lesion was diagnosed as lobular carcinoma-in-situ.   HE ×360

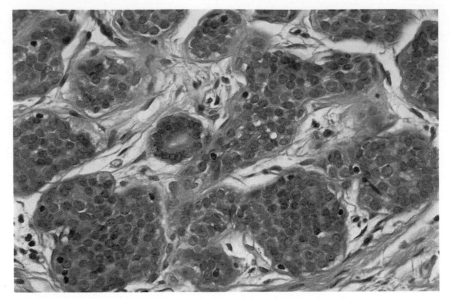

12.26 Lobular carcinoma: breast

In lobular carcinoma the terminal ducts fill with closely-packed cells. Eventually however, perhaps after many years, the malignant cells break out and infiltrating (invasive) carcinoma results. A woman of 56 had the left breast removed for suspected malignancy. A very firm mass (8.5 × 6 × 2cm) in the upper and outer quadrant proved to be an invasive (infiltrating) lobular carcinoma, with massive involvement of the axillary nodes. This part of the breast lesion consists of a small duct (arrow) surrounded by tumour cells arranged in concentric circles. The cells tend to line up one behind the other (Indian file) to form a 'dart-board' pattern in a loose connective tissue stroma. There is no gland formation. Tumour cells are also invading adjacent fatty tissue (right). The small duct is infiltrated with small lymphocytes.   HE ×150

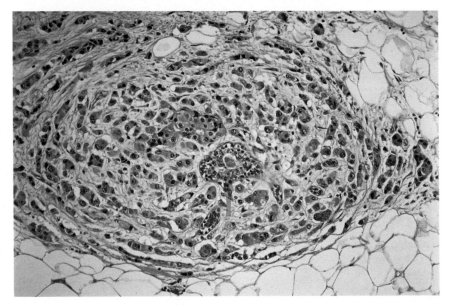

12.27 Lobular carcinoma: breast

This is the invasive lobular carcinoma shown in **12.26**. The tumour cells are lying in a densely collagenous stroma. They have a marked tendency to lie one behind the other in an Indian file. There is no attempt at gland formation. The tumour cells are small, with ovoid or elongated nuclei and fairly scanty cytoplasm. Nuclear pleomorphism is moderate and no mitoses are present in this field.   HE ×235

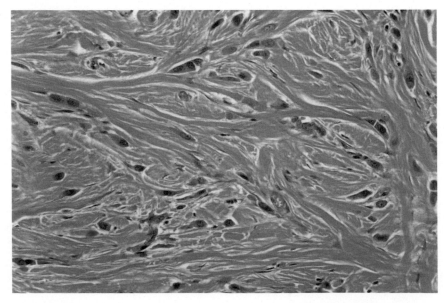

12.28 Lobular carcinoma: breast

This is an invasive infiltrating lobular carcinoma from a woman of 57. The tumour cells vary in size. They have pleomorphic nuclei and in some cells there is a large cytoplasmic vacuole which produces a signet-ring appearance (arrow). The stroma around the cells is colourless. Mucin stains showed that the vacuoles in the tumour cells contained mucin and the 'stroma' in this area also consisted of epithelial mucin, presumably secreted by the tumour cells. Intracellular mucin is detectable in most cases of intralobular carcinoma. A small blood vessel is also present.   HE ×360

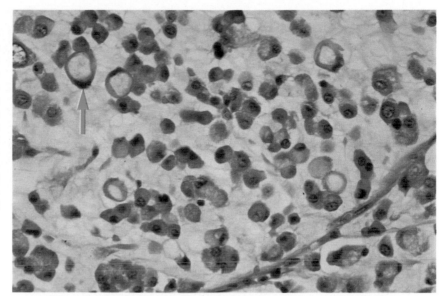

12.29 Paget's disease: breast

A woman of 64 had suffered from a red painful 'eczematous' left nipple for some months. An ellipse of skin 7 × 2.5cm bearing the nipple was resected along with a wedge of underlying breast 5cm deep. This is a section of the squamous epithelium of the nipple. Large atypical cells with slightly basophilic cytoplasm (Paget's cells) are present in large numbers in the basal layers of the squamous epithelium and in smaller numbers in the more superficial layers. The Paget's cells are not an integral part of the squamous epithelium and they are displacing and causing pressure atrophy of the adjacent epithelial cells. There is mild lymphocytic infiltration of the underlying dermis but no malignant cells are in the dermis. The cells of the squamous epithelium of the nipple show no signs of being tumorous, in contrast to Bowen's disease.   HE ×200

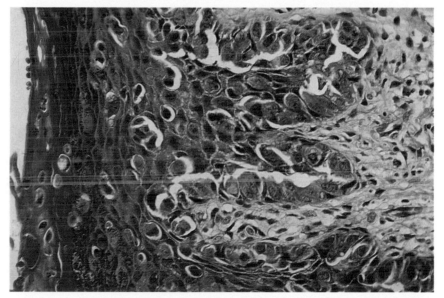

12.30 Paget's disease: breast

This is one of the several ducts within the resected portion of breast. The lumen is filled with closely packed tumour cells. The cells have large pleomorphic vesicular nuclei each containing one or more very prominent nucleoli and scanty cytoplasm. Some nuclei are pyknotic and there is considerable mitotic activity (arrows). The basement membrane of the duct is thickened and hyaline. The Paget's cells in the epithelium of the nipple are tumour cells which have migrated to the nipple via the ducts of the breast; and Paget's disease of the breast may be an early sign of a carcinoma that is still confined within the ducts (intraduct carcinoma). Unfortunately however at the time of clinical presentation the carcinoma is often already invasive. HE ×360

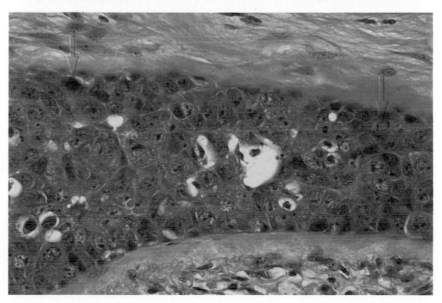

12.31 Lichenification (lichen simplex): vulva

The skin of the vulva may undergo lichenification as a reaction to chronic irritation such as scratching. The changes are fully reversible. Keratin formation is disorganized and the nuclei persists in the keratinized cells (parakeratosis) (left). There is epithelial hyperplasia of a selective nature, affecting only the rete ridges and not the intervening surface epithelium. The rete ridges are greatly elongated and the dermal papillae are correspondingly longer. There is an infiltrate of chronic inflammatory cells in the dermis (right).   HE ×135

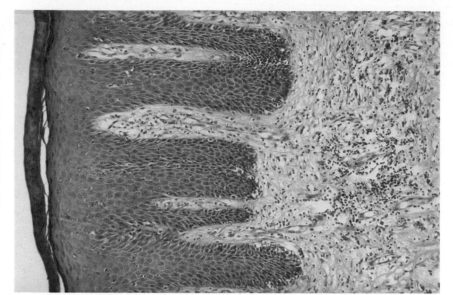

12.32 Keratosis and chronic inflammation ('leukoplakia'): vulva

The squamous epithelium of the vulva is thickened and hyperplastic. The keratin on the surface (left) is increased (hyperkeratosis) and the stratum granulosum prominent. The keratinocyte layer also is hyperplastic and thicker than normal (acanthosis) and the cells show some loss of polarity. The rete ridges are irregular in shape and generally elongated (at higher magnification there is increased mitotic activity in the keratinocyte and basal layers of the epithelium). The papillary processes of the dermis are also irregular in shape and the dermal connective tissues are heavily infiltrated with chronic inflammatory cells. Dysplasia of the epithelial cells may be sufficiently great as to suggest carcinoma-in-situ but malignant change seems to develop only rarely.   HE ×135

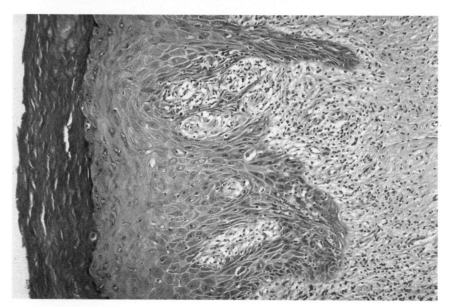

12.33 Lichen sclerosis et atrophicus ('kraurosis'): vulva

Lichen sclerosis et atrophicus may affect the skin of any part of the body. As the term suggests there is shrinkage and atrophy of the epithelial and dermal components of the skin. The epithelium (left) is flat and atrophic but intact. There is almost complete loss of the rete ridges. There is moderate increase in the surface keratin (left). The most noteworthy change is the presence of a band of hyalinized collagen, with loss of the dermal papillae, beneath the epidermis. This tissue is edematous and vacuolated, and a sharply-defined band of chronic inflammatory cells (centre) separates it from the deeper dermis (right). See also 11.13.
     HE ×135

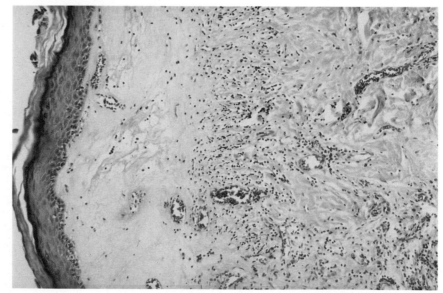

12.34 Condyloma acuminatum: vulva

Condyloma acuminatum occurs in moist regions and may be found on the vulva, penis or in the anal region where it forms soft polypoid (cauliflower-like) masses. The lesion is caused by a papovavirus of a different strain from that causing the common viral wart of skin. The virus is usually transmitted sexually. This shows the surface of one of the papillary structures. The squamous epithelium is well-differentiated. There is some parakeratosis but no hyperkeratosis. A striking feature is the enormous hyperplasia of the keratinocytes (acanthosis) with elongation of the rete ridges (centre) and a corresponding increase in the length of the dermal papillae. Many of the keratinocytes are swollen and vacuolated. The dermis (right) is heavily infiltrated with chronic inflammatory cells.　　　　　HE ×80

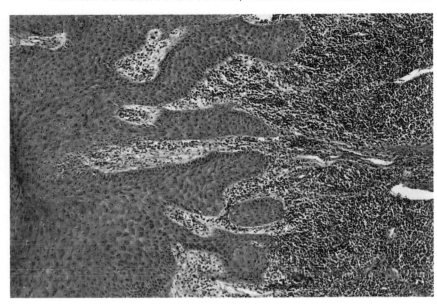

12.35 Bartholin's cyst: vulva

If the excretory duct of a Bartholin's gland is blocked, for example by inspissated secretion, a cyst may form. The cyst which may measure up to 5cm dia is filled with mucoid secretion. If the contents get infected, a Bartholin's abscess forms. This cyst was in a 24-year-old woman. It measured 2.5 × 1.5 × 1.0cm and was not acutely inflamed. It is lined by transitional epithelium (top) in which there are numerous pigment-laden macrophages. The fibrous tissue wall is infiltrated by large numbers of mature plasma cells and smaller numbers of small lymphocytes. The pigment in the macrophages was a mixture of hemosiderin and lipofuscin.　　　　　HE ×360

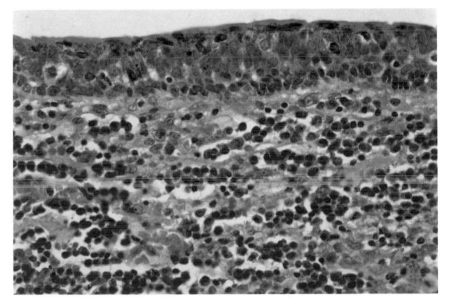

12.36 Hidradenoma: vulva

Hidradenoma of the vulva is a benign neoplasm of the epithelium of the apocrine sweat glands of the vulva. It takes the form of a small well-defined nodule in the subcutaneous tissues. The surface of the skin is on the right and although the squamous epithelium is stretched over the lesion it is intact. The tumour is a papilliferous mass lying within a cystic space which is demarcated from the adjoining tissues by a 'capsule' of compressed dermal connective tissue. The papillae of the tumour have a well-formed fibrous stroma and are covered with a layer of epithelial cells. The structure of the lesion is similar to that of an intraduct papilloma of breast.　　　　　HE ×55

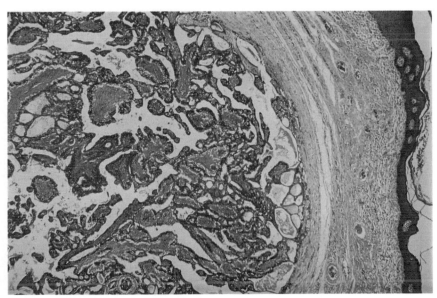

12.43 Contraceptive pill: endometrium

The contraceptive pill is a mixture of an estrogen and a progestogen. It is given cyclically for contraceptive purposes but it may be taken continuously (every day) for therapeutic reasons. This shows the effect of the contraceptive pill on the structure of the endometrium. There is a fairly well-developed pseudodecidual reaction in the stroma, whereas the few endometrial glands are small simple structures lined by flat or cuboidal hypoplastic non-secretory cells. After prolonged use of the pill, the stroma tends to lose the pseudodecidual reaction and become more compact and atrophic. When the woman stops taking the contraceptive, the endometrium soon resumes its normal cycle of activity.   HE ×235

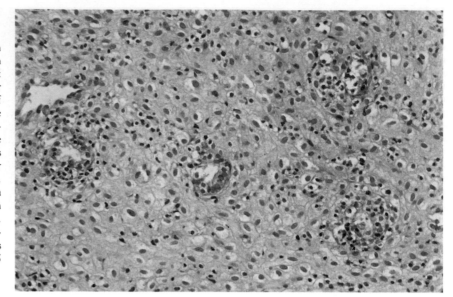

12.44 Endometrial hyperplasia: uterus

Hyperplasia of the endometrial glands varies greatly in its intensity. At its most extreme (atypical hyperplasia) it is difficult to distinguish from adenocarcinoma; and the risk of malignant change is significant in severe (atypical) hyperplasia. In this example the glands are moderately dilated and lined by a thick layer of hyperplastic epithelial cells. They show no evidence of secretory activity. Elsewhere the glands were more dilated and cystic. The stroma also is intensely cellular and numerous mitoses are present among both the epithelial and stromal cells.   HE ×135

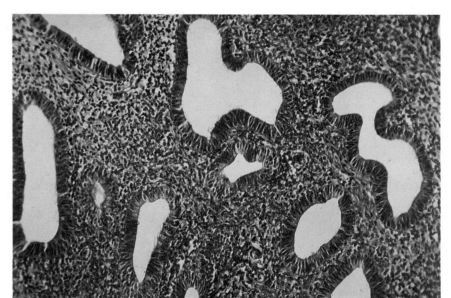

12.45 Adenomyosis: uterus

In adenomyosis, endometrial tissue is present in the wall of the uterus deep to the normal endometrium. The islands of endometrial tissue are however in direct continuity with the endometrial lining. The ectopic tissue may be in localized sites or distributed diffusely throughout the myometrium. Its presence causes thickening of the wall of the uterus and the wall in the region of this nodule of ectopic endometrial tissue was very thick. Part of the muscle coat is visible on the right. There is no fibrous capsule around the nodule which consists of glands and stroma clearly of endometrial origin. Some glands contain eosinophilic secretion but there is no evidence of previous hemorrhage in or around the lesion, perhaps because the glands are typically basal in type. The endometrial stroma is fairly fibrous.   HE ×80

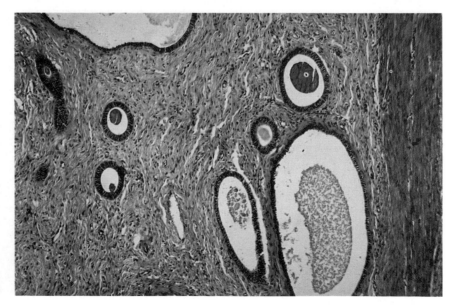

12.46 Adenocarcinoma of endometrium: uterus

Adenocarcinoma of the endometrium occurs mainly in elderly women, and nulliparous women have a greater tendency to develop the tumour than parous women. The tumour may be localized to one part of the uterine cavity but more often the whole of the cavity is lined by a thick rather nodular layer of tissue and the uterus may be enlarged. The tumour tends to bleed and there may be considerable necrosis. This is part of the bulky tissue that filled the uterine cavity in this case. It consists of very abnormal and closely-packed glands arranged 'back-to-back'. There is considerable mitotic activity in the tall columnar epithelial cells forming the glands and some glands are full of cell debris. The stroma between the glands is infiltrated with inflammatory cells.
HE ×120

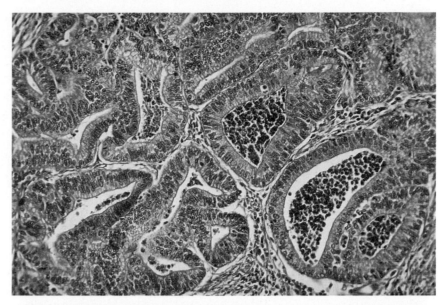

12.47 Adenoacanthoma: uterus

In about a quarter of the cases of adenocarcinoma of endometrium, the columnar epithelial cells undergo metaplasia to squamous epithelium. In this example a well-differentiated adenocarcinoma (bottom) is invading the cellular myometrium (right). Accompanying the adenocarcinoma are sheets of squamous epithelium (left) that does not look unduly malignant. The prognosis of this type of tumour is that of the adenocarcinomatous component and is not influenced by the presence of the metaplastic squamous epithelium.   HE ×135

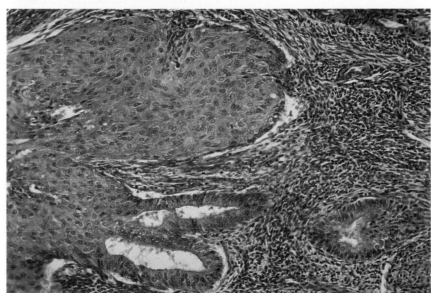

12.48 Clear-cell (mesonephroid) carcinoma of endometrium: uterus

Clear-cell carcinoma is another variant of adenocarcinoma of the endometrium. Despite a superficial resemblance histologically to the mesonephric carcinomas of ovary, cervix and vagina, the endometrial tumour is unlikely to arise in mesonephric remnants. A Mullerian origin is more probable. In this lesion the closely-packed ('back-to-back') glands of the tumour are lined by tall columnar cells. The cytoplasm is vacuolated and in many cells there is a large supranuclear vacuole, caused by glycogen in the cytoplasm. The nuclei of the tumour cells have a large central nucleolus and show fairly marked pleomorphism. Many are pyknotic. Pale-staining secretion is present in the lumen of the glands and there is a small quantity in the stroma (arrow).   HE ×150

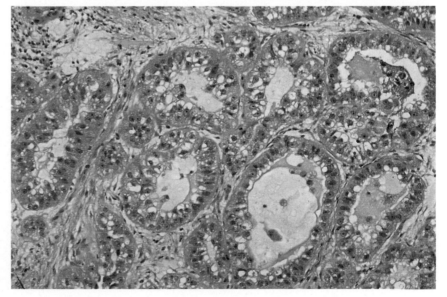

12.49 Mesodermal mixed tumour: uterus

Mesodermal mixed tumour is a rare tumour of post-menopausal women. It usually forms a large soft mass filling the uterine cavity and may protrude through the cervical os. It arises in remnants of the Mullerian ducts and contains both carcinomatous and sarcomatous elements. The carcinoma is generally an adenocarcinoma. The prognosis is poor, the tumour being radioresistant and tending to spread by the lymphatic or bloodstreams. This tumour consists of cords and sheets of epithelial cells with round basophilic nuclei (left). There are small gland-like spaces in the larger sheets of cells (lower left). Chondrosarcomatous tissue (right) is present, the chondrocytes forming clusters in a deeply basophilic matrix. The connective tissue stroma is tending to become hyalinized (arrow). HE ×150

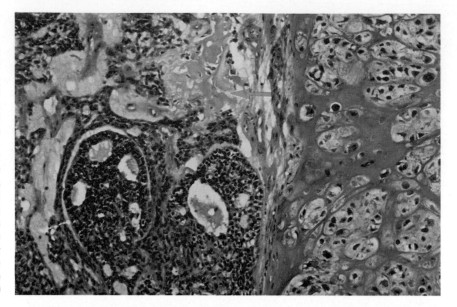

12.50 Mesodermal mixed tumour: uterus

This tumour was in a woman of 69. There was extensive necrosis. It consists of groups of cells with large round pleomorphic nuclei containing one or more prominent nucleoli. Their cytoplasm is vacuolated and the cell boundaries are indistinct. Between the cells is hyaline amorphous material not unlike cartilage matrix. Vacuoles are present also in this material and the tissue has a generally chondroid appearance. Many tumour cell nuclei are pyknotic. There are no mitotic figures in this field but many were present elsewhere in the tumour. The sarcomatous part of mesodemal mixed tumours is generally less conspicuous than the epithelial element, often resembling leiomyosarcoma or fibrosarcoma but rhabdomyosarcomatous cells and other malignant mesenchymal tissues may be present. HE ×360

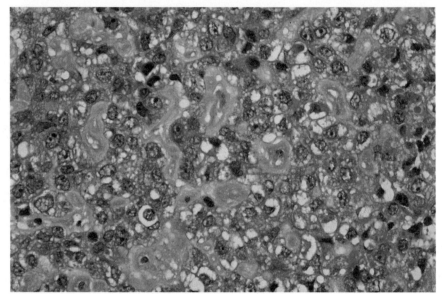

12.51 Leiomyoma: uterus

Leiomyoma of uterus is one of the commonest tumours, being found in a majority of women. A large tumour may have significant mechanical effects, whereas a small lesion projecting into the lumen of the uterus may interfere with conception or become ulcerated and cause heavy bleeding. Malignant change is very rare. The edge of the tumour is on the right and it is typically well-defined. There is however no fibrous capsule. The tumour consists of mature smooth-muscle cells which form bundles that run at various angles, so that some are seen in cross-section (thin arrow) and others in longitudinal section (thick arrow). There is some fibrous tissue between the muscle fibrils and this usually increases, so that older tumours become hard and fibrous (fibromyomas or 'fibroids'). Some tumours eventually calcify. HE ×200

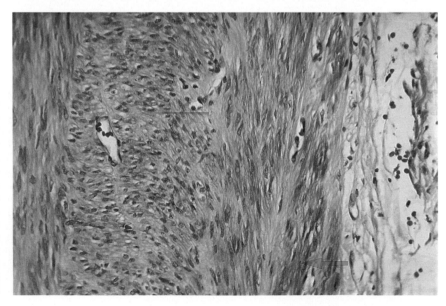

12.52 Ectopic (tubal) pregnancy: Fallopian tube

Fallopian tubes which are fibrosed as a result of chronic inflammation or which are congenitally abnormal may delay or prevent the passage of the ovum. The retained ovum may undergo fertilization and implant itself in the wall of the tube to constitute an ectopic (tubal) pregnancy. A woman of 22 was diagnosed as having an ectopic pregnancy. 5cm of her left Fallopian tube were resected surgically. The tube was expanded by blood clot. The muscular wall of the tube is on the right. The lumen of the tube contains blood clot and chorionic villi (thin arrows). The villi have a myxoid core covered with basophilic trophoblast. Sheets of trophoblast (thick arrow) lie free in the lumen. The epithelial lining of the tube has been replaced by well-developed decidua (double arrows).　　　　HE ×60

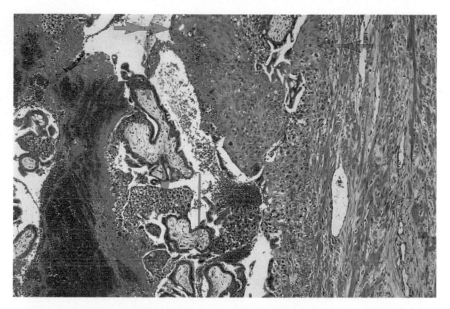

12.53 Ectopic (tubal) pregnancy: Fallopian tube

At higher magnification the muscle of the Fallopian tube (right) is vascular and edematous. The tube is lined with decidua (thin arrow) and decidual tissue lies free within the lumen (left). Adherent to the latter are a sheet of cytotrophoblast and a chorionic villus (thick arrow) with its pale-staining core of myxoid tissue covered with a thin (outer) layer of syncytiotrophoblast and (inner) cytotrophoblast. Other fragments of trophoblast are attached to the tube lining (double arrow) or lie free in the lumen. No fetal elements were found among the contents of the tube. A decidual reaction usually develops simultaneously in the uterus. Only a very small proportion of ectopic pregnancies go on to term.　　HE ×150

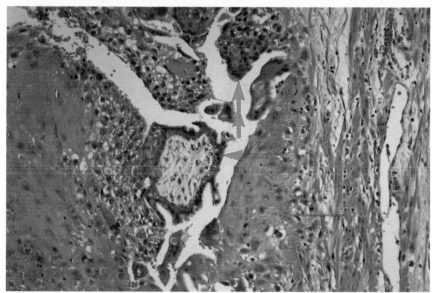

12.54 Hydatidiform mole: uterus

A hydatidiform mole consists of large numbers of hydropic chorionic villi and it arises in the placenta (the embryo is usually absent). The mole forms a large mass which causes the uterus to enlarge beyond the size appropriate for the stage of pregnancy. In most cases the cystic grape-like villi (up to 2cm dia) do not invade the muscular wall of the uterus, but in a small minority of 'invasive' moles they do invade and may even 'metastasize'. The 'metastases' regress however after removal of the mole. Histologically the core of the cystic chorionic villi is greatly expanded by edema fluid, with wide separation of the connective tissue cells; and there are no fetal blood vessels. The surface of each villus is covered with a thin layer of trophoblast from which small nodules of syncytiotrophoblast project. Trophoblastic proliferation is often considerable.　　HE ×55

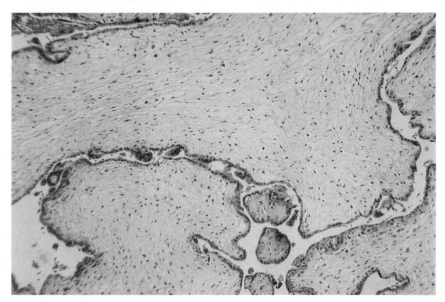

12.55 Choriocarcinoma: uterus

Choriocarcinoma (chorionepithelioma) is a rare highly malignant neoplasm of the trophoblast and therefore of fetal origin. It usually follows pregnancy or abortion. The presence of a hydatidiform mole predisposes to its development but chorionic villi are not present in choriocarcinoma. The tumour tends to invade blood vessels and it characteristically forms round hemorrhagic nodules and is prone to metastasize by the bloodstream. It consists of cells derived from the cytotrophoblast, with clear cytoplasm and well-defined cytoplasmic borders (thin arrow); and syncytiotrophoblast, consisting of sheets of multinucleated cytoplasm (thick arrow). The tumour produces chorionic gonadotrophin and the levels of this hormone in the serum may serve as a marker for the presence of viable tumour in the body.   HE ×135

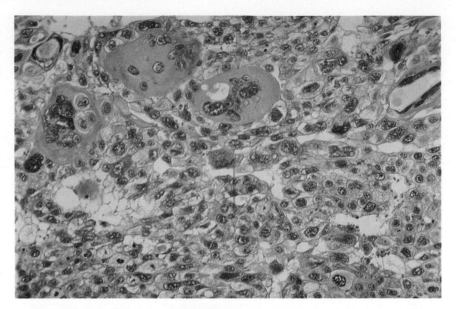

12.56 Stein-Leventhal syndrome: ovary

In the Stein-Leventhal syndrome there is a defect of steroid synthesis. Both ovaries are enlarged and polycystic. Each ovary has a 'capsule' consisting of thick fibrous ovarian cortex. The patient has amenorrhea or menstrual irregularity and is usually sterile. Numerous atretic follicles are also present but corpora lutea and corpora albicantia are absent. Wedge resection of the ovary usually restores the menstrual cycle. This ovary has been cut in longitudinal section. Multiple round cysts are present throughout the cortex of this ovary. They are lined by a single layer of flattened cells. They are simple follicular cysts and several contain eosinophilic fluid. The most superficial layer of the cortex (arrows) is fibrous, the so-called capsule.   HE ×5

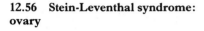

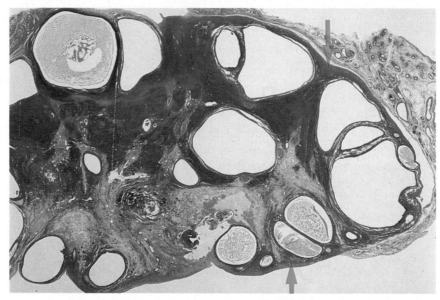

12.57 Stein-Leventhal syndrome: ovary

This shows the cortex of the ovary between two of the cysts. The surface of the ovary is on the left. The most superficial part of the cortex consists of interlacing bundles of collagenous fibrous tissue (left) which form the 'capsule'. Deep to this in the cellular basophilic ovarian stroma are a number of atretic follicles. Large numbers of similar follicles were present in both ovaries.   HE ×150

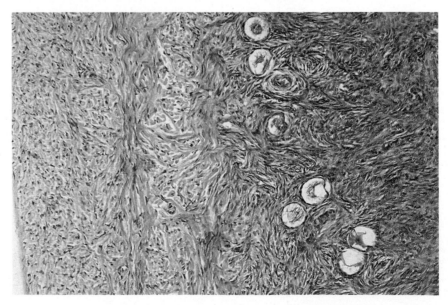

12.58   Corpus luteum cyst: ovary

The normal corpus luteum is itself slightly cystic but occasionally a much larger (up to 6cm dia) cystic structure develops at the end of the menstrual cycle (or in pregnancy). Its wall is luteinized granulosa (bottom) which consists of cells with granular (lipid-rich) cytoplasm. This gives the wall of the cyst a yellow colour macroscopically. The contents of the cyst (top) include a considerable amount of blood, and a corpus luteum cyst may bleed into the peritoneal cavity.

HE ×55

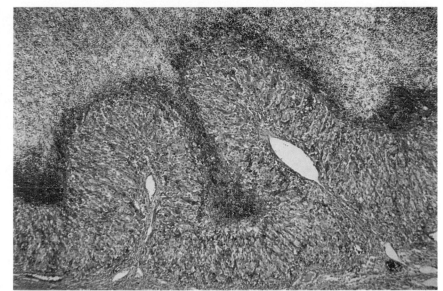

12.59   Endometriosis: ovary

Not infrequently endometrial tissue is found in sites other than the uterus. The sites tend to be in the pelvis and include the ovaries and peritoneal cavity. The endometrial tissue and the associated hemorrhage usually provoke a fibrous reaction, and the fibrous nodules that form may interfere with the functioning of adjacent organs. In the ovaries, however, a cyst filled with brown debris from previous hemorrhages tends to develop. The cysts are sometimes called 'chocolate' cysts. This lesion was in the ovary. Endometrial glands lined with columnar epithelium (arrows) and endometrial stroma (right) are both present. There is evidence of hemorrhage. The brown debris (left) in the cystic space has been produced by old hemorrhage and there is new hemorrhage into the stroma around the gland (centre).   HE ×200

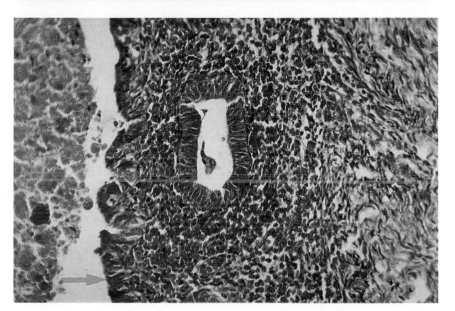

12.60   Endometriosis: umbilicus

Spontaneous endometriosis of the skin is found only at the umbilicus and in the inguinal regions although it may develop in surgical scars in many other sites, particularly after caesarean section. This lesion presented as a firm nodule at the umbilicus in a woman of 25. The glands are large and lined by a single layer of cuboidal or columnar epithelium and some contain a small amount of pale-staining secretion. Some are surrounded by a loose spindle-cell stroma but around others it is dense and fibrous (left). There is a dense infiltrate of chronic inflammatory cells in the vicinity of some of the glands (right). Elsewhere there were hemorrhages but only small amounts of hemosiderin.   HE ×60

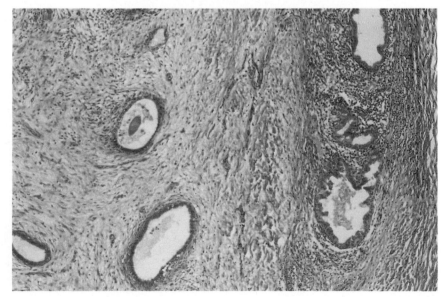

12.61 Benign cystic teratoma (dermoid cyst): ovary

A teratoma contains a wide range of tissues in a disorderly arrangement. The ovary and testis are the commonest sites, but teratomas occur in other sites (retroperitoneal, mediastinal, etc.). Most ovarian teratomas consist of mature tissues and are benign. Skin and its appendages are usually the dominant tissue and result in a cyst full of sebaceous debris. This teratoma was a large unilocular ('dermoid') cyst full of sebaceous material and hair. This part of the wall consists of mature sebaceous glands which open into numerous ducts discharging on to a 'surface' covered with stratified squamous epithelium (left). A hair follicle is also present, cut in cross-section (arrow). There were various other tissues, all fully differentiated, elsewhere in the wall of the cyst.

HE ×55

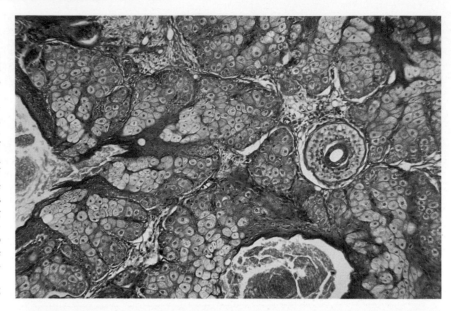

12.62 Struma ovarii: ovary

Rarely in a benign cystic teratoma one type tissue may outgrow the other tissues and become completely predominant. Thyroid tissue is present in about 10% of benign cystic teratomas of ovary, but in struma ovarii the 'teratoma' consists entirely of thyroid tissue, as happened in this case. The thyroid follicles are filled with eosinophilic colloid and lined by flat inactive-looking epithelial cells. The thyroid tissue in struma ovarii, however, can undergo all the pathological changes that affect the normal thyroid gland (hyperplasia, malignancy, etc.). HE ×135

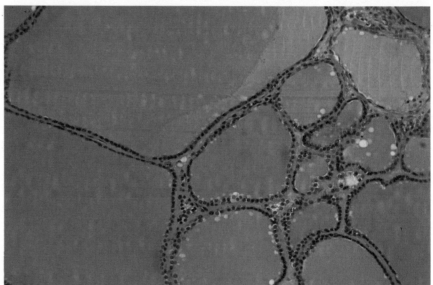

12.63 Benign cystic teratoma of ovary: malignant change

Occasionally one type of tissue in a benign teratoma becomes malignant, and when metastases develop they consist of the same type of tissue. A squamous cell carcinoma has developed in the epithelial elements of this benign cystic teratoma. The cells have large and very basophilic nuclei which show considerable pleomorphism but the tumour is well-differentiated with numerous cell nests containing clusters of keratinized cells at their centres (arrows). Between the sheets of malignant epithelial cells there is abundant fibrous stroma. HE ×120

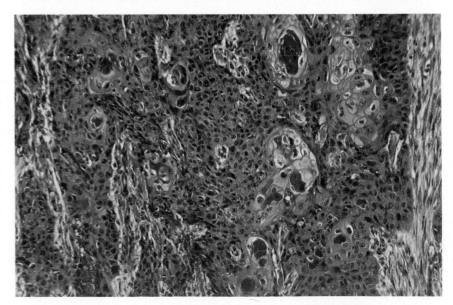

12.64 Immature (malignant) teratoma: ovary

Immature teratomas of the ovary contain a mixture of embryonic and adult tissues of all three germ layers. They generally occur in children and adolescents and the prognosis is poor. The tumour may have a solid appearance or microcysts may be visible. The more malignant tumours contain few or no mature tissues and vice versa. This tumour was in a girl of 13 and very few mature tissues were present. The tissue here however resembles small intestine fairly closely. Small tubular glands with crypts lined by columnar cells (arrows) open on to a surface covered with mucin-secreting goblet cells. The lining epithelium forms rudimentary villi. The stroma consists of loose cellular connective tissue.
HE ×200

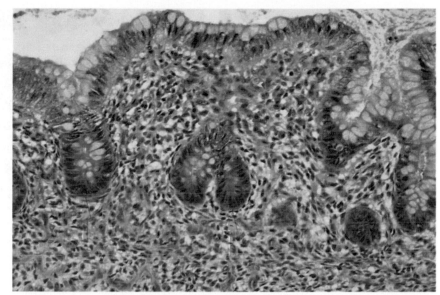

12.65 Immature (malignant) teratoma: ovary

This tissue consists of glandular structures (thin arrows) lined by columnar cells with large infranuclear vacuoles. One of these is in a group of vacuolated polyhedral cells bounded by a basement membrane (thick arrow). The epithelial structures lie in embryonic (immature) connective tissue which is loose and myxomatous. HE ×270

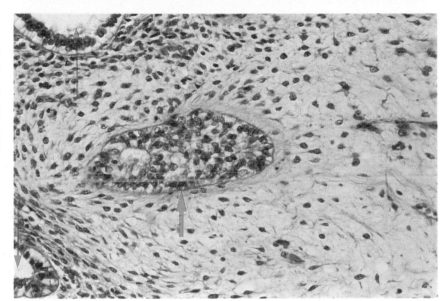

12.66 Immature (malignant) teratoma: ovary

This is the same teratoma as that shown in **12.65**. This part consists of cells with round deeply basophilic nuclei which exhibit considerable mitotic activity. Cells with similar but more elongated nuclei and associated with fine eosinophilic fibrils surround clefts in an arrangement suggestive of embryonic neural tissue. There is also a solid cord (arrow) of cells which resembles primitive cartilage. Loose connective tissue is also present. The main component of immature teratomas of ovary is generally neurogenic tissue.
HE ×215

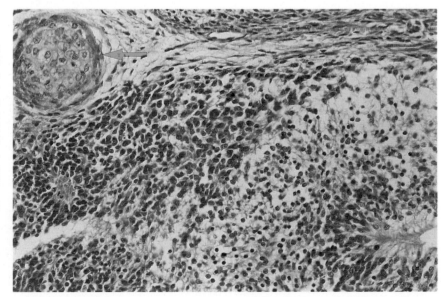

12.67 Serous cystadenocarcinoma: ovary

Serous cystadenoma and serous cystadeno-carcinoma are fairly common tumours of the surface epithelium of the ovary. The smaller lesions tend to be unilocular but as they enlarge they generally become multilocular. They are not infrequently bilateral. The epithelium lining the cysts varies from a single layer of flat inactive cells (cystadenomas) to a thick lining of highly malignant cells which form innumerable papillomatous out-growths into the cystic spaces (cystadeno-carcinomas). This was a large tumour with many cystic cavities filled with watery fluid. One cystic cavity is shown. It is lined by deeply basophilic epithelium which forms long papilliferous processes. The epi-thelium has also grown outwards, penetrat-ing the fibrous wall of the cyst to reach the serosa (arrow). HE ×55

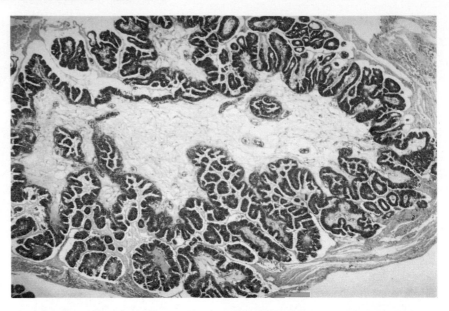

12.68 Serous cystadenocarcinoma: ovary

This is the tumour shown in **12.67**. The fibrous wall of the cystic cavity is on the right. The epithelial cells lining the cavity are tall columnar cells with large basophilic nuclei and eosinophilic cytoplasm. They form long papilliform processes, each with a delicate core of vascular connective tissue, and there is considerable mitotic activity among them. They do not contain large droplets of mucin and the contents of the cavity are weakly basophilic. Many serous cystic tumours are difficult to categorize as benign or malignant and are regarded as bor-derline lesions. HE ×270

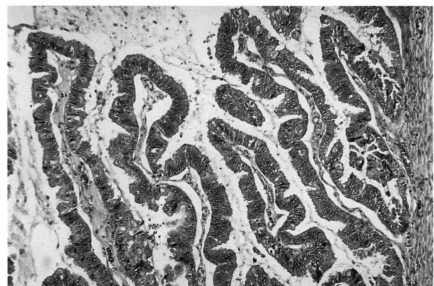

12.69 Mucinous cystadenoma/cystadenocarcinoma: ovary

Mucinous cystadenomas and cystadenocar-cinomas are less often bilateral than serous tumours. The epithelium lining the cysts secretes much mucus and is less papilliferous than that in serous tumours. It varies from a single layer of tall cells filled with mucin to a thick layer of anaplastic cells. A smooth-sur-faced ovarian cyst (15 × 12 × 10cm) weigh-ing 990g was removed from a woman of 28. It was multiloculated. The loculi have fibrous walls lined by columnar mucus-secreting epithelium with multiple papilliform pro-jections. The epithelium does not invade the wall and the serosa (bottom) is intact. Each loculus is filled with mucus. Although there was no firm evidence of malignancy, the lesion was diagnosed as a 'borderline' muci-nous cystadenoma/cystadenocarcinoma. HE ×60

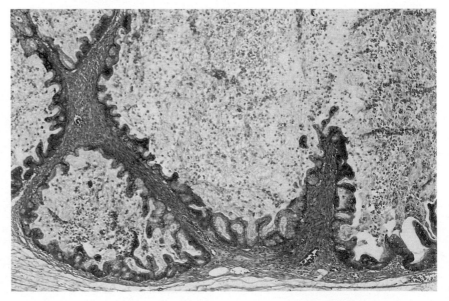

12.70 Mucinous cystadenocarcinoma: ovary

An ovarian cyst measuring 15 × 11 × 9cm was removed by surgical operation from a woman of 65. It was multilocular, the walls of the loculi varying in thickness from 0.1cm to 1cm. There was also a polypoid gelatinous mass 5cm dia projecting into one loculus. This shows parts of three loculi. One is lined by a single layer of short mucin-secreting columnar cells (thin arrow) and another by much taller columnar cells (thick arrow). Epithelium of this type is sometimes called 'picket-fence' epithelium. In the third loculus (below left) the epithelium forms many short papillae; and there is layering of the nuclei which show moderate basophilia and pleomorphism (double arrow). The loculus contains cell debris as well as mucin. The stroma between the loculi consists of cellular fibrous tissue.　　HE ×150

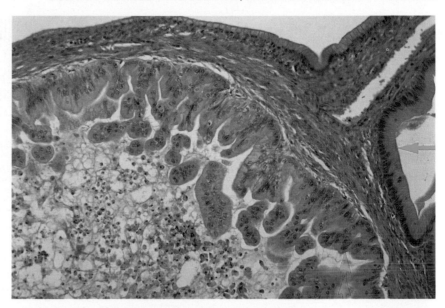

12.71 Mucinous cystadenocarcinoma: ovary

This is the tumour shown in **12.70** at higher magnification. In this part of tumour the epithelium is forming both large loculi and smaller gland-like spaces. These are lined by closely-packed columnar cells with ovoid hyperchromatic nuclei which show considerable mitotic activity (arrows). The nuclei of the epithelial cells form several layers. There are no droplets of mucin in the cytoplasm of the epithelial cells. The stroma (bottom right) consists of cellular fibrous tissue.　　HE ×235

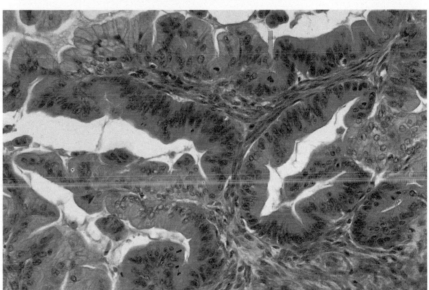

12.72 Clear-cell carcinoma: ovary

Clear-cell (mesonephroid) carcinoma of ovary owes its name to the large vacuoles in the cytoplasm of the malignant cells. The vacuoles are caused by the presence of glycogen and they give the tumour a distinctive histological appearance which is reminiscent of carcinoma of kidney. The tumour has no link with mesonephric structures, however, and arises from the surface epithelium of the ovary (see also **12.48**). In this example the cells with their abundant 'clear' cytoplasm are polyhedral and form sheets separated by scanty strands of fibrous stroma (arrow). The nuclei are hyperchromatic and extremely pleomorphic but there are no mitoses in this field. The tumour cells may also arrange themselves in tubules and when they do so they may have a 'peg' shape.　　HE ×335

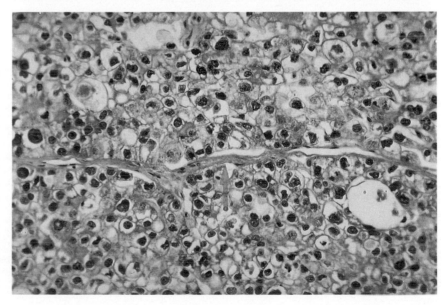

12.73 Brenner tumour: ovary

The Brenner tumour is an uncommon lesion which is almost always benign. It is probably derived from the surface epithelium of the ovary. It is associated with endometrial hyperplasia (hyperestrinism) and uterine bleeding in post-menopausal women. It is usually a single firm mass with a white cut surface. Histologically the appearance is unmistakable: large 'nests' of epithelial cells resembling transitional or squamous epithelium lie in a dense fibrous stroma. The cytoplasm of the epithelial cells is 'clear' from the presence of glycogen but occasionally they are accompanied by mucin-secreting cells.

HE ×120

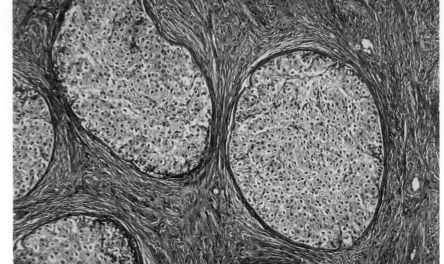

12.74 Granulosa cell tumour: ovary

Granulosa cell tumours originate in the stromal cells of the ovary and they secrete large amounts of estrogen. In children precocious puberty may be induced and in adults endometrial hyperplasia and excessive bleeding. The tumour is usually solid and smooth-surfaced. The composition of granulosa cell tumours varies from pure granulosa cell tumour with a few thecal elements through thecomas to luteinomas. This tumour consists of closely-packed polyhedral granulosa cells and a number of small abortive follicles (Call–Exner bodies). The presence of the latter makes the diagnosis of granulosa cell tumour easier; and another feature which helps is the presence in the tumour cells of nuclear folds or grooves (visible in this tumour at higher magnification). HE ×335

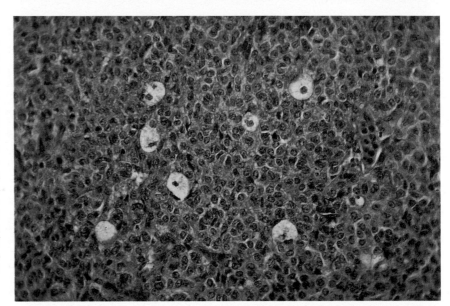

12.75 Granulosa cell tumour: ovary

Granulosa cell tumours are generally benign although a minority tend to recur. This granulosa cell tumour consists of a fairly uniform population of ovoid or elongated cells with basophilic nuclei and scanty cytoplasm. There is no nuclear pleomorphism and no mitotic activity. The cells are arranged in long cords separated by narrow clefts. HE ×335

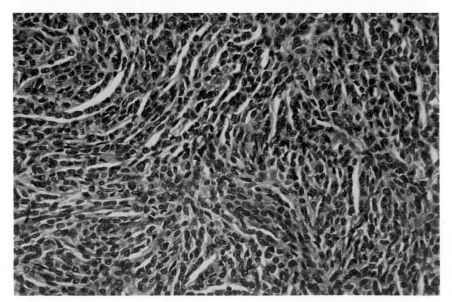

12.76 Thecoma: ovary

Thecoma of ovary also is derived from the stromal tissue of the ovary. It is usually a single lesion, and is an encapsulated benign neoplasm of older women (the large majority are post-menopausal). This one occurred in a woman of 70. The cut surface of the tumour had a distinct yellow colour. The tumour is composed of cells with elongated irregular nuclei and abundant pale finely granular cytoplasm, an appearance produced by the lipid (steroids) present. Strands of dense fibrous tissue (arrows) are also present and a reticulin stain usually reveals delicate connective tissue fibres around individual tumour cells. Deposits of hyaline material are also found occasionally in thecomas.   HE ×235

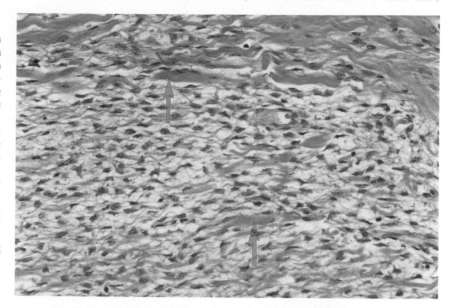

12.77 Luteinoma: ovary

In this form of granulosa cell tumour the neoplastic cells contain so much lipid (steroids) that they resemble the lutein cells of the mature corpus luteum. The lipid gives the cut surface of the neoplasm a yellow colour. Histologically the cells resemble epithelial cells, with very regular round nuclei and abundant granular (foamy) cytoplasm. A band of fibrous stroma is also present. There is no mitotic activity.   HE ×335

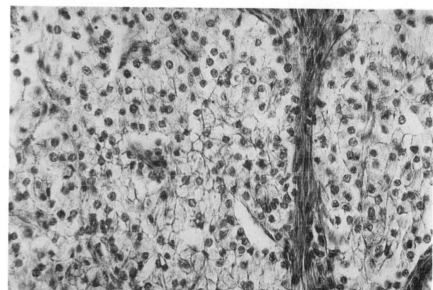

12.78 Fibroma: ovary

Fibroma of ovary is a common benign tumour also derived from the ovarian stromal tissue. Many are undoubtedly inactive thecomas and differentiation from thecoma may be difficult. Generally a fibroma is a solid white mass although myxomatous change is sometimes present. About 1 in 5 of patients with a fibroma of ovary have abdominal ascites and sometimes also a pleural effusion on the same side as the tumour (Meigs' syndrome). The fluid disappears after removal of the tumour which is usually large. This tumour (13 × 10 × 6cm) weighed 650g and had a smooth bossellated surface. The cut surface was yellowish-white and whorled. The tumour consists of interlacing bundles of cellular fibrous connective tissue. There was no evidence of malignancy and the serosal surface was intact.   HE ×235

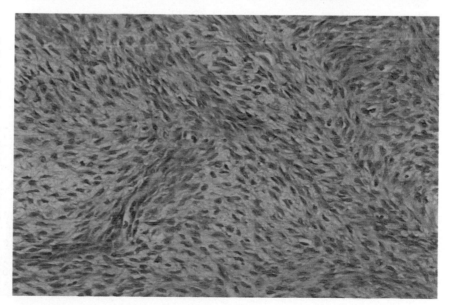

12.79   Sertoli cell tumour: ovary

A 'pure' Sertoli cell tumour is a rare neo-plasm. Generally Sertoli cell tumours do not secrete significant amounts of steroid hormones and the small minority that do, secrete estrogens rather than androgens. This tumour is composed of long well-formed 'tubules' of well-differentiated cells bounded by a thick basement membrane. The cells have ovoid nuclei and abundant eosinophilic cytoplasm and they are arranged with their long axis at right angles to the basement membrane.   HE ×150

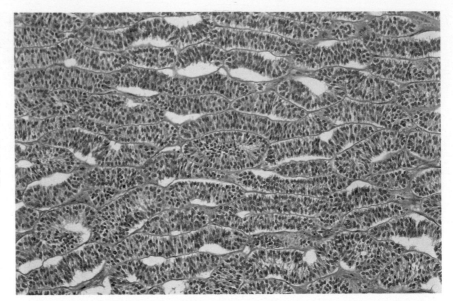

12.80   Sertoli cell tumour: ovary

This is the tumour shown in **12.79**. At higher magnification the tubules are bounded by a well-formed basement membrane. The tumour cells have uniform ovoid nuclei with one or more nucleoli. They are tall columnar cells and are closely packed, with some layering of their nuclei. Their apices are tapered (arrows) and drawn out into 'fibrils' which meet in the lumen of the tubules. There is no mitotic activity in this field but occasional mitoses were present throughout the tumour. Sertoli cell tumours are relatively benign however and rarely recur after surgical excision.

HE ×360

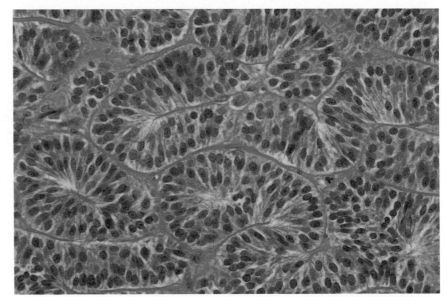

12.81   Sertoli-Leydig tumour (arrhenoblastoma): ovary

Sertoli-Leydig tumours usually cause de-feminization and virilization. The baso-philic cells of this tumour form long branching cords which resemble the sex cords of the embryonic testis before they acquire a lumen. Leydig (interstitial) cells, inconspicuous in this field, were present elsewhere in the tumour in the stroma between the tubules.   HE ×235

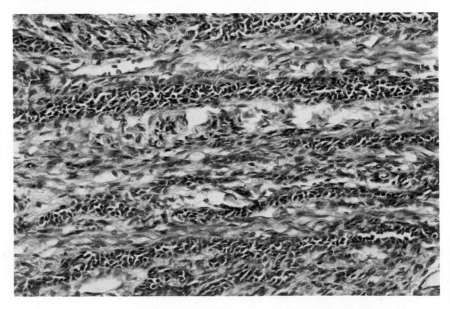

12.82 Dysgerminoma: ovary

Dysgerminoma is an undifferentiated germ
cell tumour which is histologically identical
with seminoma of testis. Like seminoma it
spreads readily to the regional lymph nodes
and is very radiosensitive. The tumour cells
are large with vacuolated cytoplasm and
resemble the primordial germ cells of the
sexually indifferent embryonic gonad.
They are arranged in groups and cords and
the round nuclei show some pleomorphism.
No mitoses are present in this field but there
were scattered mitotic figures elsewhere in
the tumour. The fibrous stroma is infiltrated
with lymphocytes.　　　　HE ×235

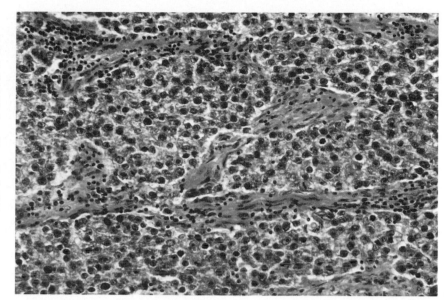

12.83 Krukenberg tumour: ovary

The ovaries are a common site for metastatic
deposits from tumours in the alimentary
tract or the pelvic organs. Krukenberg
tumour is a metastatic mucin-secreting carcinoma. Both ovaries become enlarged and
solid. A woman of 61 presented with
obstruction of the large intestine. At operation both ovaries were enlarged and were
resected. One (6 × 6 × 4cm) weighed 35g;
the other measured 7 × 4.5 × 4cm. Their
cut surfaces were white, homogeneous and
slightly whorled. Scattered throughout an
abundant spindle-cell stroma (thin arrows)
are tumour cells with hyperchromatic
pleomorphic nuclei and large vacuoles in
their cytoplasm (thick arrows). Cells with
only one vacuole look like signet rings. No
mitoses are present in this field and few
could be detected elsewhere in the ovaries.
　　　　HE ×540

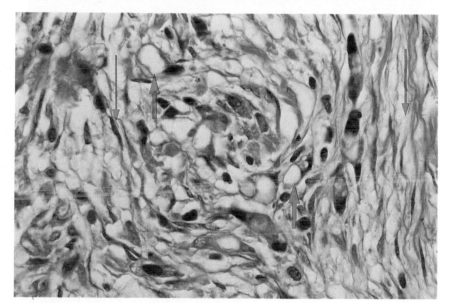

12.84 Krukenberg tumour: ovary

This section of one of the ovaries has been
stained by the periodic acid-Schiff method
which colours epithelial mucin purplish-
red. Large numbers of mucin-containing
tumour cells are present, some lying singly
and others in cords. The spindle-cell fibrous
stroma is also clearly shown (right). The
term Krukenberg was originally applied to
secondaries from a gastric carcinoma, but
has been extended to include other primary
sites; and the carcinoma of colon, resected at
the same time as the ovaries, was assumed to
be the source of the Krukenberg tumours in
this case. It was also believed that the
tumour cells crossed the peritoneal cavity to
reach the ovaries (transcelomic spread) but
this is not now generally accepted.
　　　　PAS ×360

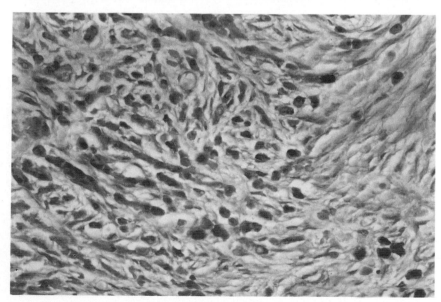

13.1 Osteopetrosis: vertebra

Osteopetrosis (marble bone disease) is a hereditary disorder. In its severe form the infant is affected in utero and often dies at an early age. In the less severe form the disease may not be manifest until near adolescence. All bones may be affected but membranous bones (skull, etc.) tend to be spared. Resorption of cartilage matrix and of cortical bone is defective. The bulky new bone may encroach on the marrow cavity and interfere with hemopoiesis and with the exit foramina of nerves with disturbance of function. Although the bones are thickened and dense they are brittle and fracture readily. This is a vertebral body. It consists of closely-packed thick curved trabeculae of eosinophilic bone (thin arrow) and pale cartilage (thick arrows). Islands of cellular hemopoietic tissue (double arrow) survive.
HE ×90

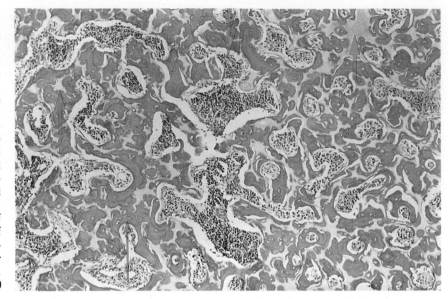

13.2 Osteopetrosis: vertebra

This shows the trabeculae at higher magnification. They are very broad: the surface layers consist of eosinophilic bone matrix (thin arrows) and the centres of pale-staining homogeneous material resembling cartilage matrix (thick arrow). Relatively few osteocytes are present in the bone matrix and none is detectable in the cartilage matrix. Between the trabeculae there are small cellular foci of hemopoietic tissue. As well as defective resorption of cartilage in osteopetrosis, osteoclasts are scarce. There seems also to be defective ossification, since the bone in the trabeculae is woven (coarse-fibred) bone and compact trabeculae of lamellar bone are not present.   HE ×215

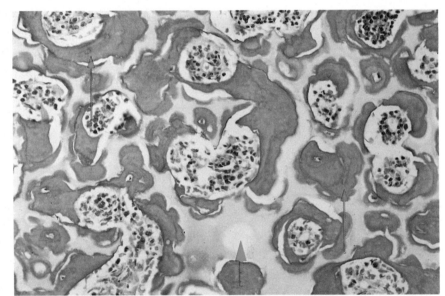

13.3 Fibrous dysplasia: bone

There are two distinct forms of fibrous dysplasia: the monostotic, in which only one bone is affected; and the polyostotic in which there are lesions in many bones. The lesions in the polyostotic form are on one side of the body and accompanied by endocrine abnormalities (including precocious puberty in the female). This is Albright's syndrome. There is often also patchy brown pigmentation of the skin on the same side as the bone lesions. The affected bones are expanded by a distinctive type of connective tissue, as shown in this material from the medullary cavity of an affected rib. It consists of loosely-textured highly cellular and vascular connective tissue in which there are spicules of woven (coarse-fibred) bone, characteristically curved or sickle-shaped. There are no osteoblasts on the surfaces of the trabeculae.   HE ×150

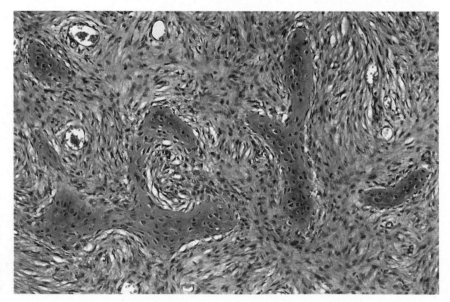

13.4 Osteomyelitis: bone

A wide range of bacteria can cause acute sup-purative lesions of bone but in most cases the organism is *Staphylococcus aureus*. The infection may reach the bone locally as in compound fractures or by the bloodstream. The latter is the route in young people (less than 20 years of age) in most of whom the lesion is at a metaphysis in a lower limb. If treatment is inadequate pus may spread in the medullary cavity and beneath the peri-osteum. The medullary cavity (left) is full of pus, mostly polymorph leukocytes and macrophages. The pus is eroding the tra-beculae of the cortical bone (right). Con-siderable necrosis of bone is liable to occur in osteomyelitis, often through damage to the blood supply, but in this lesion the osteocytes in the bone lacunae retain their nuclear staining.　　　　　HE ×150

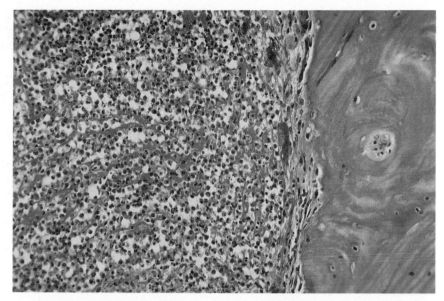

13.5 Osteomyelitis: bone

Inadequately treated osteomyelitic lesions tend to become chronic, with extensive nec-rosis of existing bone as well as formation of much new reactive bone by the periosteum and endosteum. This shows reactive bone formation in the vicinity of a chronic pyo-genic lesion of bone. The new bone consists of interlacing trabeculae of intensely baso-philic trabeculae of woven (coarse-fibred) bone (arrows). The osteocytes in woven bone are large and round. Woven bone is usually not a permanent tissue and is nor-mally removed by osteoclasts. Trabeculae of lamellar bone may replace it.　HE ×135

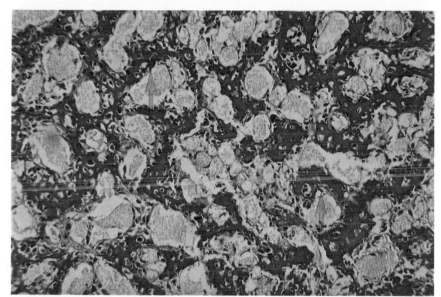

13.6 Tuberculous osteomyelitis: bone

The tubercle bacillus (*Mycobacterium tuberculosis*) infects bone via the blood-stream. The lesion tends to start at the meta-physis but synovial membrane is also sus-ceptible. It is a disease of children and young adults and unpasteurized milk was often the source of the infection. Destruction of bone and synovial membranes may be consider-able, and the necrotic tissue may liquefy to form a 'cold abscess'. In this example the bone trabecula (arrow) is necrotic and the osteocyte lacunae empty. The tissue on the right is tuberculous granulation tissue, rich in macrophages, lymphocytes and epi-thelioid cells but containing relatively few capillaries. Between this tissue and the bone trabecula there is deeply eosinophilic necro-tic tissue which would appear caseous on macroscopic examination.　　　HE ×55

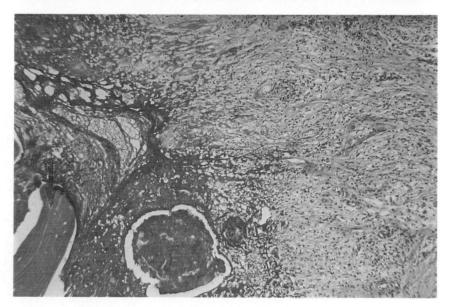

13.7 Necrosis (post-irradiation): cartilage

Necrosis of cartilage is a fairly common and often troublesome complication of X-ray therapy, particularly when high-pressure oxygen is used. In this case therapeutic X-irradiation had been given to the skin overlying the cartilage 3 months previously. Many of the cells of the perichondrium (right) are nucleated and probably viable. However although a few of the chondrocytes still have small pyknotic nuclei, the nuclei in the others do not stain. Moreover the matrix has lost its characteristic basophilia and stains red. The cartilage is therefore probably completely necrotic. Necrotic cartilage is removed by macrophages. This may take a long time however and result in considerable fibrous scarring. HE ×150

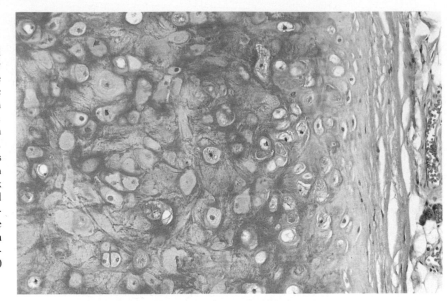

13.8 Chondrodermatitis nodularis helicis: ear

In this condition one or more small very tender nodules form on the apex of the helix of the ear. The nodules sometimes ulcerate. Necrosis of the underlying cartilage of the ear is usually present and was thought to be the prime lesion. Necrosis of dermal collagen may however be the first event. This lesion was in the left ear of a woman of 74. The squamous epithelium is on the left. It was not ulcerated. There is a perivascular infiltrate of chronic inflammatory cells in the dermis and deep to the auricular cartilage. The matrix of the cartilage is eosinophilic and many of the chondrocytes have either no nucleus or a small pyknotic nucleus. The cartilage is being invaded and eroded by cellular connective tissue (arrow). HE ×60

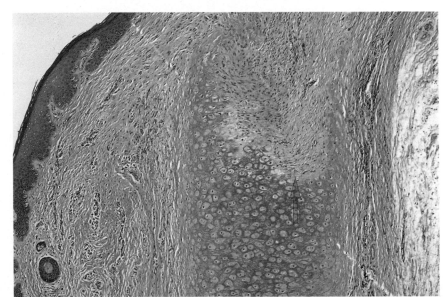

13.9 Chondrodermatitis nodularis helicis: ear

This shows the cartilage and the tissue invading it at higher magnification. The tissue (top) which is eroding the cartilage is delicate connective tissue containing many fibrocyte-type cells and occasional macrophages. The matrix of the cartilage is red and granular. Many of the chondrocytes lack nuclei and in others the nucleus is small and pyknotic. Living cartilage is very resistant to phagocytic attack and this auricular cartilage is probably completely necrotic, despite the presence of nuclei in some chondrocytes, the dead chondrocytes being protected by the dense cartilage matrix.
  HE ×150

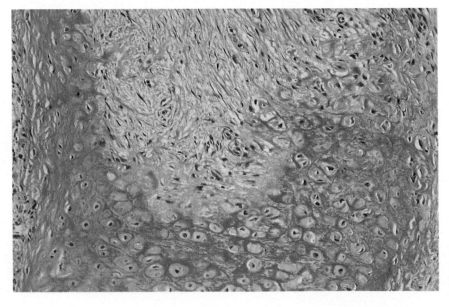

13.10 Hyperparathyroidism (generalized osteitis fibrosa cystica): bone

The effects of hyperparathyroidism on bone structure include both bone resorption by osteoclasts and new bone formation by osteoblasts. Destruction may be sufficient to lead to the formation of cysts and 'brown tumours' (brown colour is from the presence of hemosiderin). The number of giant-cells (osteoclasts) in these 'tumours' may lead to confusion with true giant-cell tumours of bone. In this example the normal bone has been replaced by vascular and cellular connective tissue within which there is fairly extensive hemorrhage. Basophilic trabeculae of new bone (arrows) project from one side of a long bone trabecula. The other side of the trabecula is under active attack by osteoclasts, and they have already removed most of it.   HE ×135

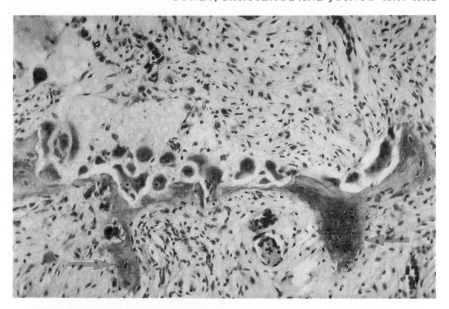

13.11 Rickets: costochondral junction

Rickets is caused by deficiency of vitamin D. Newly formed bone matrix fails to calcify and remains as osteoid. At the epiphysis, mineralization of the cartilage matrix fails to take place. The unmineralized cartilage resists resorption and replacement by bone can not take place. The epiphysis enlarges but growth of bone is retarded. The periosteum is at the top. The matrix of epiphyseal cartilage (left) is uncalcified and thick disorderly columns of chondrocytes (thin arrows) cause the epiphysis to bulge. At the metaphysis (right) there are small thin-walled blood vessels (thick arrows) and elongated trabeculae of widely varying size and shape which contain both basophilic cartilage and eosinophilic osteoid. Hemopoietic tissue in the marrow cavity is just visible (top right).   HE ×60

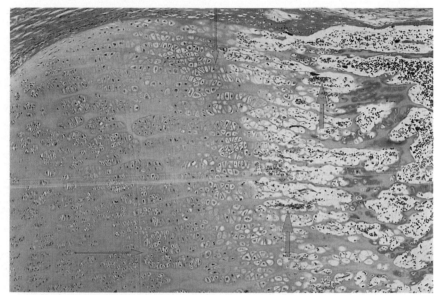

13.12 Rickets: costochondral junction

This shows the uncalcified matrix and the irregular columns of cartilage cells at the metaphysis. There is no evidence of calcification of the matrix of the epiphyseal cartilage (left) and the fully viable proliferating chondrocytes form broad columns which show increasing disorder as they approach the metaphysis (right). At the metaphysis the small blood vessels are very thin-walled and appear to penetrate between the columns of cartilage cells with difficulty. Few osteoblasts accompany them.   HE ×150

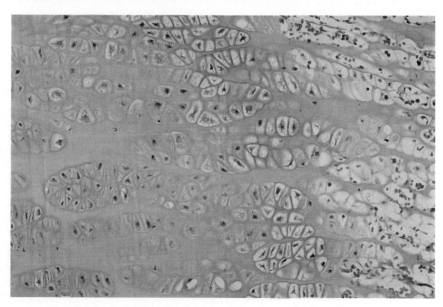

13.13 Paget's disease: skull

Paget's disease (osteitis deformans) is a fairly common condition affecting men and women equally. It is focal, and lesions may be found in the vertebrae, pelvis, skull and long bones (other than the ribs). In the early stages the bone is soft and vascular with active resorption of bone but later bone formation predominates and the bone becomes very thick and dense but brittle. Microscopically the trabeculae are thick and the irregular pattern (a 'mosaic') formed by the cement lines (thin arrows), which denote previous episodes of bone formation and resorption, shows that bone resorption and formation have been patchy and disordered. In the vascular connective tissue between the bone trabeculae there is bone formation by osteoblasts (thick arrows) and bone resorption by osteoclasts (double arrows) with 'notching' of the trabeculae.   HE ×150

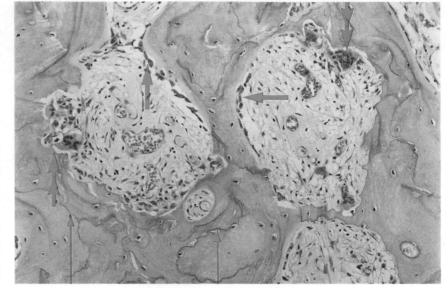

13.14 Osteomalacia: bone

Osteomalacia is the counterpart in adults of rickets in children. Where bone is being formed, it is inadequately calcified and non-calcified bone matrix (osteoid) is present in abnormally large amounts. This is a biopsy specimen of bone from the iliac crest of a man of 52 with adult-type Fanconi 'rickets'. The section has been cut without prior decalcification of the bone, and the bone salts (stained black) remain. Only parts of the several trabeculae of lamellar bone are calcified, the remaining tissue consisting of red-stained osteoid. Normal bone lamellae usually have a thin layer of osteoid on their surfaces but in this specimen the layer of osteoid is many times thicker than normal (up to 200μm thick), evidence of very severe osteomalacia.

Von Kossa–neutral red ×80

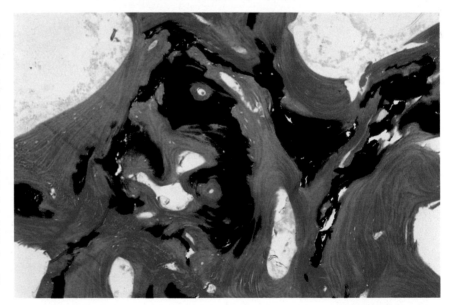

13.15 Osteoporosis: bone

In osteoporosis there is increased 'porosity' or rarefaction of the bone. The bone present is adequately calcified; that is, the quantity but not the quality of the bone is reduced. The condition frequently develops in women after the menopause. In this example, the trabeculae are thin and delicate and widely spaced. They were however fully calcified. There is no osteoclastic activity. The osteoblasts on the surface of the trabeculae are flat and inconspicuous. Nevertheless the cause of osteoporosis is more likely to be increased resorption of bone rather than decreased formation. Fatty marrow fills the spaces between the trabeculae.

HE ×55

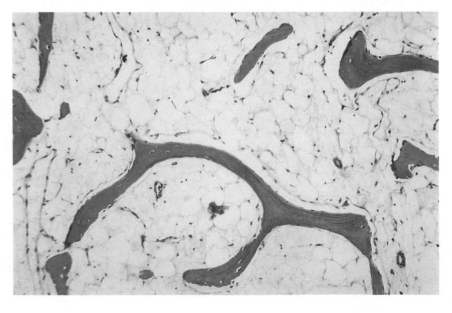

13.16 Aneurysmal bone cyst: vertebra

Aneurysmal bone cyst is a very uncommon benign lesion which affects young people, less than 20 years of age. It is found in a variety of sites (vertebra, long bones, etc.) and causes eccentric erosion and expansions of the bone, including the cortex. Macroscopically it consists of soft hemorrhagic tissue. This lesion was in a vertebral body. It consists of cellular connective tissue in which there are numerous large thin-walled blood vessels (top left). In the connective tissue there are many lymphocytes and multinucleated giant cells of osteoclast type (thin arrows), as well as trabeculae of eosinophilic osteoid (thick arrow). Extensive hemorrhage has occurred from the large blood vessels. HE ×60

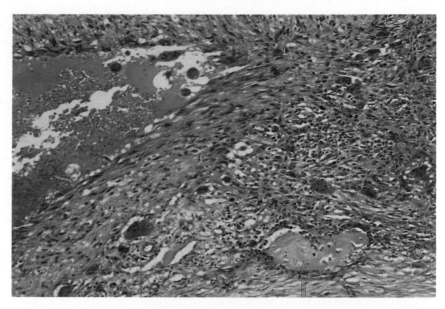

13.17 Aneurysmal bone cyst: vertebra

At higher magnification the large blood vessel (top left) is lined by an incomplete layer of flat cells, supported by a cellular connective tissue in which there are many multinucleated giant cells (osteoclasts). Many round cells are present throughout the connective tissue and there is a trabecula of eosinophilic osteoid (arrow). Aneurysmal bone cyst is curable by adequate surgical excision and it is therefore important to distinguish it from giant-cell tumour of bone or vascular osteosarcoma. HE ×150

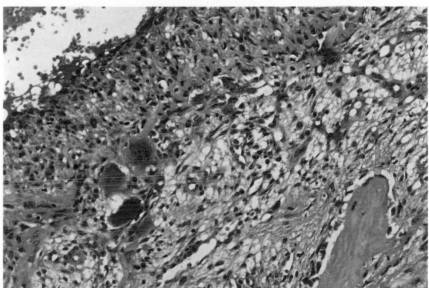

13.18 Hemangiopericytoma: femur

Hemangiopericytoma is believed to arise from the pericytes of Zimmerman. It may occur in any site including bone. Most are well-localized lesions but a minority recur locally and occasionally metastasize. This lesion in the femur is a well-differentiated tumour, reminiscent of a glomus tumour, and composed of cells with ovoid nuclei of regular size and shape. No mitotic activity is present and few mitoses could be found elsewhere. The cells have eosinophilic cytoplasm and form cords separated by a delicate stroma consisting largely of thin-walled blood vessels, some containing blood but others (arrows) collapsed and inconspicuous. The blood vessels are lined by endothelial cells, and a reticulin (silver) stain revealed a continuous basement membrane between the endothelial cells and the cells of the tumour. HE ×360

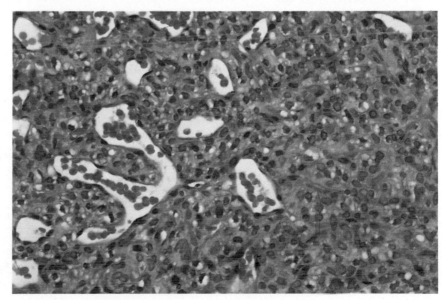

13.19 Eosinophilic granuloma: bone

Eosinophilic granuloma is one of the group of conditions included in the term histiocytosis X. The conditions are typified by the presence of Langerhans cells and a mixture of other cells including histiocytes, eosinophils, neutrophils and foamy cells (sometimes multinucleated). In eosinophilic granuloma a single lesion is present in bone, and this one was an osteolytic lesion (2cm dia) in the temporal bone of a man of 33. It is composed of histiocytes with foamy (finely vacuolated) cytoplasm (thin arrow), eosinophils (e.g. left), multinucleated giant cells with peripheral nuclei (Touton giant cells) (thick arrow) and lymphocytes. Strands of fibrous tissue are also present. There was also evidence of invasion of the temporal muscle. After adequate surgical excision an eosinophilic granuloma does not usually recur. HE ×270

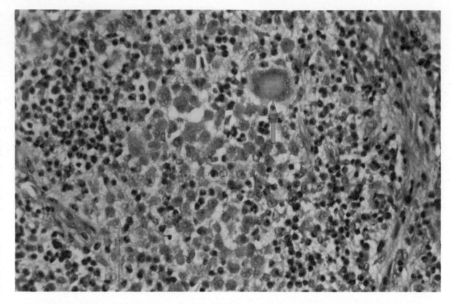

13.20 Osteochondroma: femur

Osteochondroma (cartilaginous exostosis) is a common lesion, generally occurring in subjects less than 20 years of age. It is benign but may grow large enough to press on local structures. The surface is often lobulated. This specimen was a small sessile very hard mass attached to the shaft of the femur near the knee-joint, a common site. It consists of trabeculae of mature bone (right) and an outer layer of cartilage (left). The surface of the cartilage is smooth and covered with connective tissue which was continuous with the periosteum of the femur. Thin-walled blood vessels are eroding the cartilage matrix (arrow). Growth of an osteochondroma is similar to that which occurs at the epiphysis, with the cap of cartilage acting like the epiphyseal plate. Rarely, malignant change (chondrosarcoma) supervenes. HE ×120

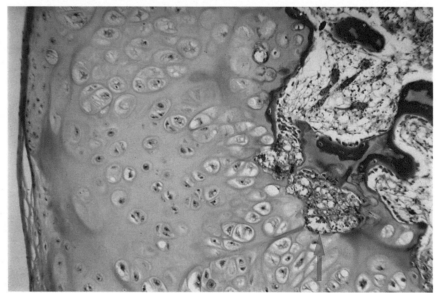

13.21 Giant-cell tumour : bone

Giant-cell tumour (osteoclastoma) of bone is a tumour of adults (more than 20 years of age) and tends to be located at the end of long bones, particularly around the knee-joint, where it forms an expansile locally destructive mass which may invade the joint. The tumour is usually red and hemorrhagic, with dark brown areas and fibrous-walled cystic spaces often filled with blood. This tumour was in the head of the fibula. It consists of sarcoma-like tissue scattered throughout which there are numerous multinucleated giant cells resembling osteoclasts. The connective tissue element does not form osteoid or bone. It is however actively growing and determines the degree of malignancy of the lesion. Giant-cell tumours are malignant and inadequately resected lesions recur locally. About 10% metastasize. HE ×215

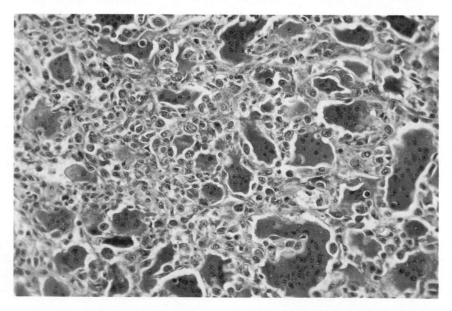

13.22 Osteoid osteoma: bone

Osteoid osteoma is a small (less than 1.5cm) benign tumour which occurs in the bones of the limbs, hands and feet or vertebral column. It causes intense pain and induces an osteosclerotic reaction in the surrounding bone. This example was in the head of the ulna. It consists of irregular masses of eosinophilic osteoid (thin arrow) and densely-staining bone (mineralized osteoid) (thick arrow). There are many osteoblasts (double arrow) in the vascular connective tissue between the sheets of osteoid and bone.   HE ×120

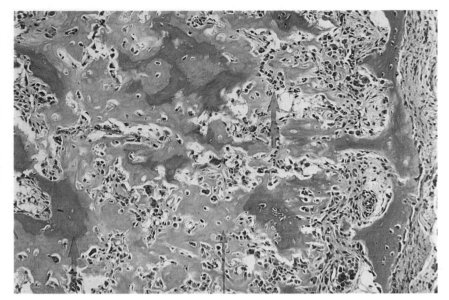

13.23 Osteoblastoma: bone

Osteoblastoma (giant osteoid osteoma) occurs in the same age group as osteoid osteoma and is located in the same sites but with different frequencies. Histologically it is also similar to osteoid osteoma. It is, however, usually a bigger lesion, does not cause pain, and does not induce a sclerotic reaction in the surrounding bone. This specimen is composed of small irregular trabeculae of eosinophilic osteoid, the surfaces of which are covered with closely-packed large osteoblasts with abundant basophilic cytoplasm. There are also many osteoblasts in the connective tissue between the trabeculae.   HE ×235

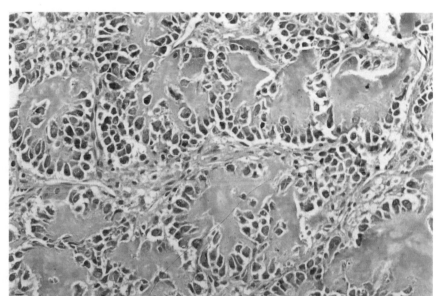

13.24 Osteosarcoma: bone

Apart from the cases developing in older people with Paget's disease of bone, osteosarcoma is a highly malignant tumour of children and young adults under 25 years of age. It tends to develop in the metaphysis of a long bone and many are located around the knee. The tumour invades the medullary cavity and expands the periosteum, sometimes breaking through it. It metastasizes readily by the bloodstream to the lungs and the prognosis is poor. The cells of the tumour have large pleomorphic nuclei, with numerous mitoses, often abnormal (thin arrow). They have formed small irregular trabeculae of osteoid. A few multinucleated giant cells are also present (thick arrow). The malignant cells also frequently produce bone and cartilage.   HE ×360

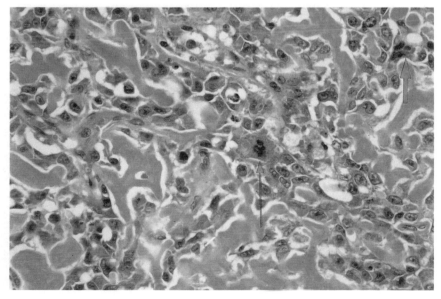

13.31 Rheumatoid arthritis

Rheumatoid arthritis is a systemic disease of unknown etiology. There is an inflammatory arthritis involving many joints and particularly those of the hands and feet. The synovial lining of an affected joint is inflamed and it proliferates to form villous processes which project into the joint space. The cartilaginous lining of the joint may be destroyed and fibrous ankylosis of the joint results. This is the capsule and synovial lining of an inflamed joint. The synovial cells (top) have proliferated and the joint is now lined by a thick multicelled layer of elongated cells with eosinophilic cytoplasm (thin arrows). The fibrous capsule of the joint is hyperemic with dilated blood vessels (thick arrow) and heavily infiltrated with chronic inflammatory cells, mainly lymphocytes which have formed a lymphoid follicle. HE ×150

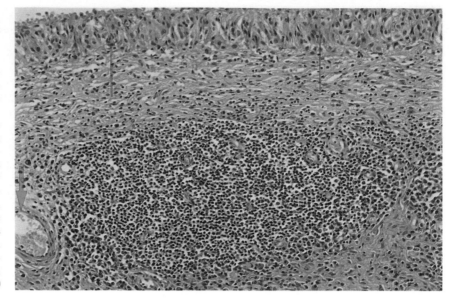

13.32 Rheumatoid arthritis

This is a synovial villus projecting into the joint shown in **13.31**. It consists of connective tissue covered with an irregular layer of flat cells with pleomorphic nuclei. It is vascular with numerous dilated small blood vessels (thin arrows) and its stroma is heavily infiltrated with chronic inflammatory cells, most of which are mature plasma cells. Two Russell bodies (thick arrows), effete plasma cells with red-stained cytoplasm, are present. HE ×310

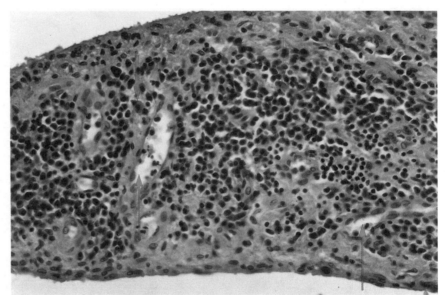

13.33 Rheumatoid arthritis: metacarpo-phalangeal joint

The joint was ankylosed and the head of the metacarpal is at the bottom. The articular cartilage which would normally have been present on its upper surface has been destroyed by fibrous connective tissue (pannus) which has grown across the surface of the metacarpal from the capsule of the joint and at the same time obliterated the joint space (top). A few small pale-staining fragments of cartilage matrix (arrow) remain in the fibrous tissue. HE ×200

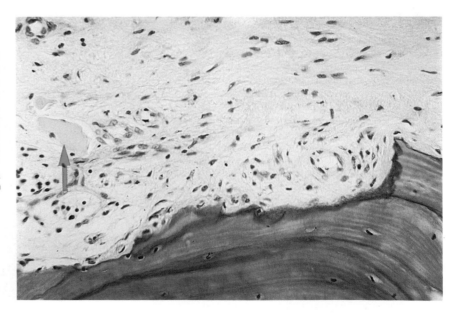

13.34   Synovial osteochondromatosis

Synovial osteochondromatosis is a comparatively rare condition which usually affects the knee- or hip-joints. The role of trauma is unknown. The synovial membrane lining the joint proliferates and within it cartilage and bone form. Fragments containing the new tissue may break off and become loose bodies within the joint. The condition is benign but local recurrence may take place. Most of the new tissue in this example consists of connective tissue resembling immature cartilage with large well-spaced ovoid cells and abundant basophilic intercellular matrix. In several areas the tissue is definitely cartilaginous, with mature chondrocyte-type cells, and in other parts the proliferating synovial tissue has undergone hyaline degeneration (arrows).

HE ×150

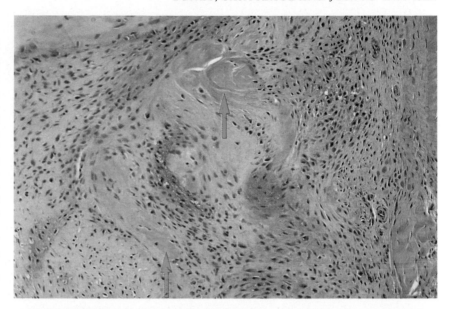

13.35   Pigmented villonodular synovitis: knee-joint

Villonodular synovitis usually involves a large joint (knee, ankle, hip, etc.) of a young adult. The synovial lining of the joint proliferates to fill the joint with papillary processes which are dark brown in colour from the presence of hemosiderin. It is a benign lesion but complete excision of all the papillary processes is difficult and recurrence is not infrequent. Malignant change does not take place. The fibrous capsule of the knee-joint is on the right. Large numbers of long villi project from it into the joint. The villi have a core of vascular connective tissue and are generally slender and delicate, lacking the bulky infiltrate of lymphocytes and plasma cells in the synovial villi in rheumatoid arthritis. The histological appearances resemble the so-called giant-cell tumour of tendon sheath. HE ×80

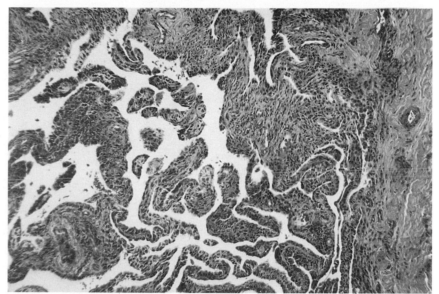

13.36   Giant-cell tumour (fibrous histiocytoma): tendon sheath

This lesion is usually located in the hands and feet. It is benign but may cause pressure atrophy of neighbouring structures. It often appears yellowish-brown from the presence of hemosiderin. Histologically it consists of cellular fibrous tissue in which there are numerous fibroblast-like cells, histiocytes, macrophages and multinucleated giant cells. Many of the macrophages are hemosiderin-laden (arrow). The surface of the tumour is covered with the flattened synovial cells which line the tendon sheath.

HE ×335

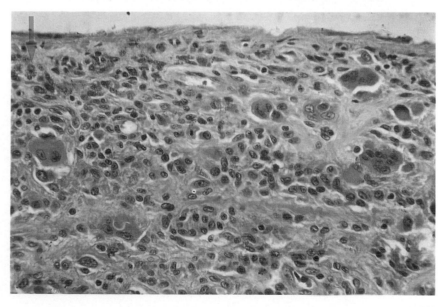

13.37 Olecranon bursitis

A bursa becomes inflamed as a result of trauma. Granulation tissue forms and the wall of the bursa becomes thick and vascular and eventually fibrosed. The bursa is distended with watery fluid at first but hemorrhage sometimes occurs, and altered blood and fibrin are often present. This bursa has a wall of vascular granulation tissue and young connective tissue infiltrated with chronic inflammatory cells (right) and it is lined with flattened synovial cells (centre). Projecting from the wall of the bursa into the lumen are long dense strands of fibrin (stained red). In chronic bursitis long strands of fibrous tissue may form in the lumen. HE ×135

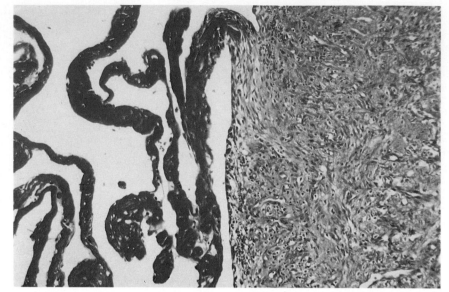

13.38 Osteoarthritis: hip joint

Osteoarthritis is not an inflammatory disease and the changes in an affected joint are apparently secondary to degenerative changes in the hyaline articular cartilage. In this joint the articular cartilage is thinner than normal and its surface (top) is uneven and 'rough'. Several clefts extend deeply into the cartilage, one reaching as far as the underlying bone trabeculae (thin arrow). This process of fissuring of the articular cartilage is termed fibrillation. Fissuring exposes the chondrocytes in the cartilage and many have disappeared, leaving acellular matrix (thick arrows). Some of the remaining chondrocytes are however large and hypertrophic (double arrows). There are few changes in the bone (bottom) but in severe osteoarthritis the articular cartilage may be lost and the trabeculae of the exposed bone then tend to thicken. HE ×55

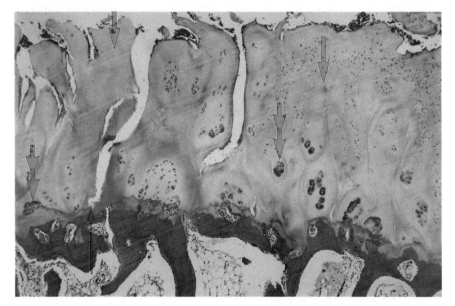

13.39 Fibromatosis

The term fibromatosis covers a number of related conditions in which fibroblastic proliferation occurs, with formation of tumour-like masses of collagenous tissue. This is an example of plantar fibromatosis which affects the plantar fascia of the feet. The patient was a man of 30. The dense collagenous tissue of the plantar fascia with its scanty content of slender fibrocytes (top) is being eroded and replaced by cellular fibrous tissue in which the fibroblasts are much larger and active-looking. When the new connective tissue matures to become collagenous and scar-like, it contracts and causes deformities of the toes. A similar lesion in the palms (palmar fibromatosis) has the same effect on the fingers (Dupuytren's contracture). HE ×360

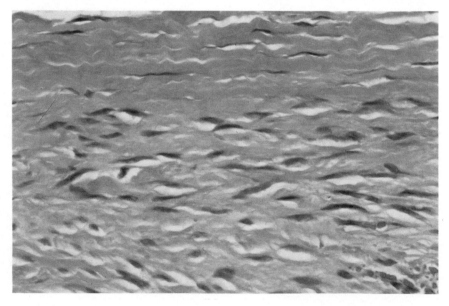

13.40 Fibromatosis

Many lesions arise in the fascia of a muscle (musculo-aponeurotic fibromatosis). The more aggressive types infiltrate locally and may be confused with fibrosarcoma. They may recur after excision but do not metastasize. This was an aggressive type of lesion in a 46-year-old man who developed a subcutaneous lump on the medial aspect of his left arm 8cm above the elbow. There is an aponeurotic band of fibrous tissue in the centre, from which the lesion appears to be arising. Cellular connective tissue rich in small blood vessels (arrows) and resembling granulation tissue is spreading from both surfaces of the fibrous tissue, more superficially into the overlying subcutaneous fat (left) and more deeply (right) towards the underlying muscle (not visible).   HE ×60

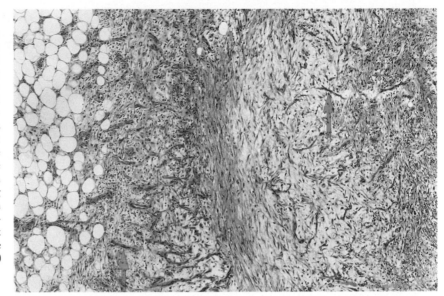

13.41 Fibromatosis

The collagenous fibrous tissue of the aponeurosis is on the right and the subcutaneous fat on the left. This is the edge of the lesion where it is invading fat. Loose highly cellular connective tissue containing large numbers of small blood vessels is growing out into the fat from the surface of the fibrous tissue. Present in the new tissue are macrophages (many foamy from ingested lipid), small lymphocytes and elongated fibroblasts. The fibroblasts among which there is considerable mitotic activity have already formed a considerable amount of collagenous tissue.   HE ×150

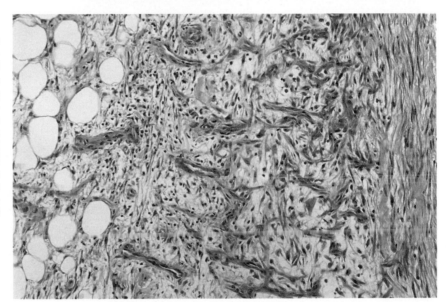

13.42 Fibromatosis

This is one of the more cellular 'fibroblastic' areas of the lesion shown in **13.40** and **13.41**. The nuclei of the large fibroblast-like cells show considerable pleomorphism. Occasional mitoses were present elsewhere among these cells. They have abundant eosinophilic cytoplasm and are forming collagenous tissue. The dense fibrous tissue (top left) is part of the aponeurosis. Despite these appearances this type of lesion is benign and should not be mistaken for fibrosarcoma.   HE ×360

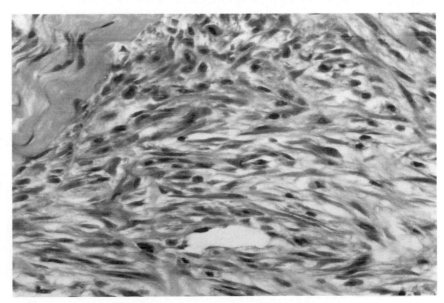

13.43 Synovial sarcoma

Synovial sarcoma is a neoplasm affecting mainly young adults. Most examples are located around the knee- or ankle-joint but the tumour occurs occasionally in many other sites. It may appear well-circumscribed but it is malignant and prone to metastasize to the regional lymph nodes. Two different types of malignant cell are present: one, large cells resembling epithelial cells with abundant eosinophilic cytoplasm and pleomorphic nuclei. These line elongated branching gland-like clefts some of which contain amorphous materials; and two, interlacing compact bundles of basophilic spindle cells with fibroblast-like nuclei. These form the stroma between the gland-like clefts. At higher magnification many mitotic figures could be seen among both the stromal cells and the cells forming the clefts.

HE ×150

13.44 Synovial sarcoma

At higher magnification the gland-like clefts are lined with several layers of cells with large round or ovoid vesicular nuclei containing one or more prominent nucleoli. Their cytoplasm is abundant and the boundary between the cells is indistinct. No basement membrane is present between these cells and the close-packed spindle cells of the stroma. Mitotic figures (arrows) are present among the cells lining the clefts and the stromal cells, and similar activity was pronounced throughout the tumour.

HE ×360

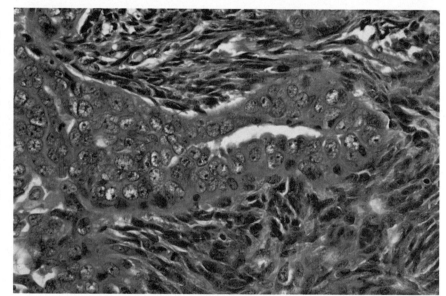

13.45 Fibrolipoma: rectus abdominis

A lobulated smooth-surfaced mass (9.5 × 7 × 5cm) weighing 186g was removed from the lower rectus abdominis muscle of a man of 54. It was firm and the cut surface was fibrous with a whorled pattern. This section has been stained by the van Gieson method which colours collagen fibres purplish-red. The tumour consists of interlacing bundles of densely collagenous fibrous tissue. It is also very cellular, the cells being elongated mature fibroblasts. At higher magnification scanty mitotic figures were present among these cells. Interspersed with the fibrous tissue there are strands of normal neutral fat. The lesion was diagnosed as a benign fibrolipoma. It is possible however that it is a so-called desmoid tumour (musculoaponeurotic fibromatosis) particularly in view of its situation. Such a lesion is liable to recur.

Van Gieson ×60

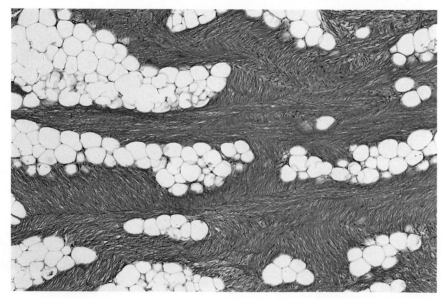

13.46 Elastofibroma: scapula

Elastofibroma is a distinctive lesion that is almost always confined to the subscapular region of elderly people who have worked as manual labourers. It is not a neoplasm but a degenerative process caused by movement of the scapula. A mass of fibrofatty tissue 8cm dia was removed from the scapular region of a man of 60. There was no capsule around the mass. It consists of collagen fibres (thin arrows) and deeply eosinophilic fibres resembling elastic tissue (thick arrows). The latter fibres appear to be disintegrating, and there are numerous small fragments and globules of varying size of the same material (e.g. top right). Neutral fat was also present in considerable amounts in the mass.   HE ×360

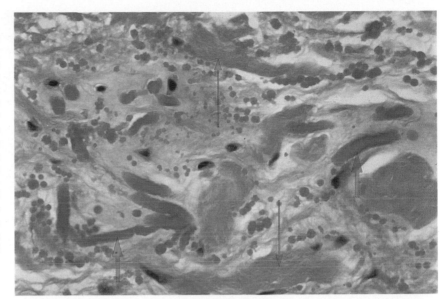

13.47 Elastofibroma: scapula

This section of the lesion shown in **13.46** has been stained by a sequence which colours collagen purplish-red and elastic tissue black. The mass consists of mature collagen fibres and globular bodies which stain like elastic tissue. The 'elastic' fibres are in fact degenerate collagen and the globular bodies are presumably fragments of the same material.   Elastic-van Gieson ×360

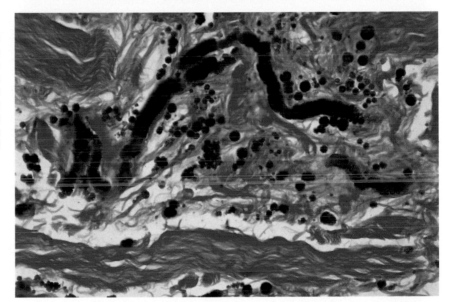

13.48 Intramuscular myxoma

A man of 35 had a 'unilocular cyst' 4cm dia containing copious mucoid material removed from his neck. It is a mass of myxomatous tissue bounded by striated muscle fibres (right). In the myxomatous tissue there are fairly numerous small blood vessels and a sparse but evenly-distributed population of fibrocyte-like cells. No mitotic activity was detectable in these at higher magnification. Most of the tumour is very pale-staining from the presence of abundant connective tissue mucin. There are however occasional collagen fibres and a reticulin stain revealed a network of reticulin fibres throughout the tumour. Tumours of this type arise as a rule in skeletal muscle and although they may infiltrate neighbouring structures, the prognosis is excellent after adequate resection.   HE ×60

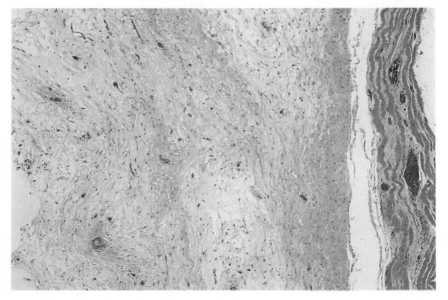

13.49 Lipomyxosarcoma: thigh

A mass (8 × 8 × 6cm) was removed from the thigh of a man of 70. It consisted of firm white tissue with areas of necrosis and calcification. Histologically it consists of connective tissue and fat cells of various types including mature fat cells. Most of the cells of the connective tissue are spindle-shaped fibrocytes and fibroblasts but a variety of other forms are present, including small multinucleated giant cells (thin arrows). Some of the latter have very finely granular ('frosted-glass') cytoplasm (thick arrow). Mononuclear cells with finely-dispersed fat in their cytoplasm elsewhere in the tumour were considered to be lipoblasts. The connective tissue is 'loose' and special stains demonstrated much connective tissue mucin. In other parts it was myxomatous. The tumour was diagnosed as a lipomyxosarcoma of low-grade malignancy.   HE ×235

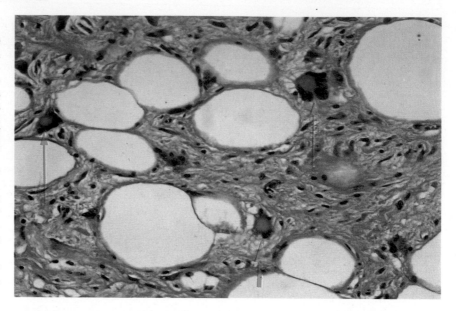

13.50 Fibroliposarcoma: thigh

A large tumour (3,165g) was removed surgically from the right thigh of a woman of 65. Its cut surface showed that it was composed of lobules of creamy-white, partly necrotic soft tissue. On its deeper aspect it was attached to striated muscle. Microscopically the lesion was well-circumscribed but there was no true capsule. It consists of a mixture of pleomorphic spindle cells (fibroblastic) and cells with one or more large vacuoles in their cytoplasm (lipoblasts). In this area the lipoblastic component predominates. Other parts of the tumour were more fibrosarcomatous.   HE ×360

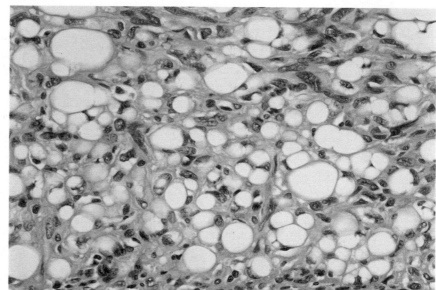

13.51 Fibroliposarcoma: thigh

This is another part of the tumour shown in **13.50**. It consists mainly of fibroblast-type cells with elongated nuclei and abundant eosinophilic cytoplasm. Most of the nuclei have rounded ends but many are pleomorphic and hyperchromatic. Numerous mitotic figures are present. Some of the cells have one or more clear vacuoles (which contained fat) in their cytoplasm. The tumour was diagnosed as a high-grade (highly malignant) fibroliposarcoma of mixed fibroblastic and lipoblastic type.   HE ×360

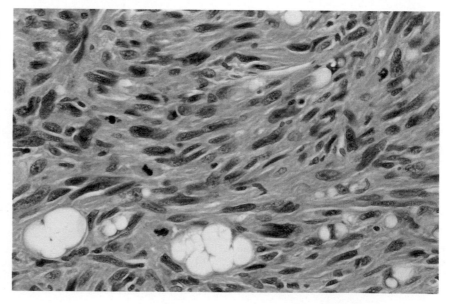

13.52 Desmoplastic fibroma: femur

Demoplastic fibroma is a rare tumour derived from myofibroblasts but fibroblasts are usually also present. Despite its name it may be locally destructive and unless resection is adequate tends to recur. It does not metastasize. This tumour caused a pathological fracture of the femur of a woman of 82. Macroscopically it was composed of firm white tissue and microscopically it consists of interlacing bundles of densely collagenous fibrous tissue. The tumour cells are elongated and fibrocyte-like, with slender basophilic nuclei which show only slight pleomorphism and no mitotic activity.

HE ×360

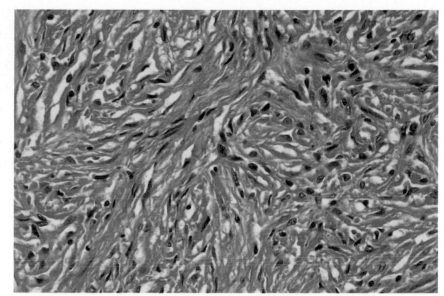

13.53 Fibrosarcoma

Fibrosarcoma may arise anywhere but skin and fascia are favoured sites. It varies in malignancy from the 'desmoplastic fibroma' to highly aggressive anaplastic tumours which rapidly metastasize by the bloodstream to the lungs. Collagen formation varies inversely with the degree of differentiation. This tumour was located at the origin of the pulmonary artery in a woman of 59. It is composed of interlacing bundles of fibroblast-type cells. Most of the nuclei are elongated and vesicular and contain one or more prominent nucleoli. There is mitotic activity (arrows). The cells have a moderate amount of cytoplasm but the cell boundaries are ill-defined. There is little evidence of collagen formation but a reticulin stain demonstrated many fibres. Small lymphocytes are scattered throughout the tumour.

HE ×235

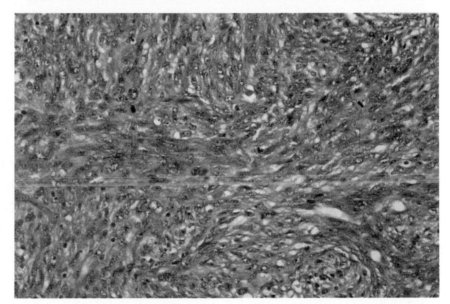

13.54 Fibrosarcoma: wrist

This tumour was on the dorsal aspect of the forearm of a woman of 63 just above the wrist. It appeared to originate in the periosteum of a bone. Fibrosarcoma of bone is a rare tumour which generally starts in the endosteum rather than the periosteum; and the wrist is an unusual site for a fibrosarcoma arising in bone, since most arise around the knee-joint, in the femur or tibia. Fibrosarcoma of bone is osteolytic and may spread widely locally, in and around the bone. The nuclei of the tumour cells in this case are uniformly large and ovoid with inconspicuous nucleoli and their cytoplasm in vacuolated. The vacuoles probably contain connective tissue mucin but formation of collagen is relatively slight. There is considerable mitotic activity (arrow). HE ×360

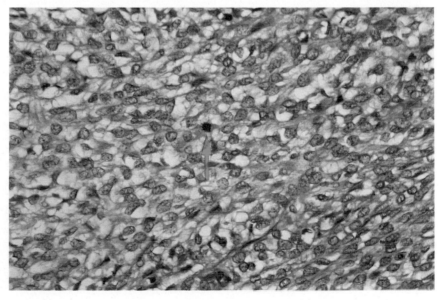

13.55 Malignant fibrous histiocytoma: scapula

Malignant fibrous histiocytoma generally occurs in skeletal muscle or deep fascia. This tumour was located in the scapular region. Other parts of the lesion had the structure of a malignant spindle-cell tumour. In this area however it is composed of large rounded histiocytes. These cells have extremely pleomorphic nuclei containing one or more prominent nucleoli and abundant eosinophilic cytoplasm. Some of the histiocytes have two or more nuclei and many nuclei are hyperchromatic (arrow). Lymphocytes and plasma cells are present in moderate numbers but there are only a few fibroblast-type cells. HE ×360

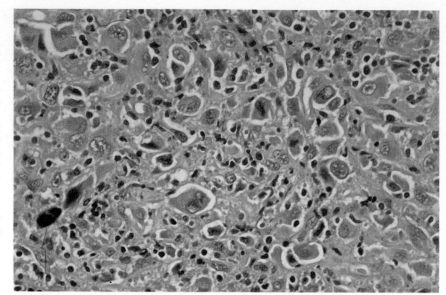

13.56 Malignant fibrous histiocytoma: scapula

In this part of the tumour shown in **13.55**, in addition to the closely-packed large pleomorphic histiocytes with abundant pale cytoplasm (thin arrows), there is a network of elongated fibroblast-like cells with eosinophilic cytoplasm (thick arrow). Multinucleated giant cells (centre) are also present. There is considerable mitotic activity among the histiocytes (double arrow). Elsewhere in the neoplasm the fibroblastic cells were the dominant component. The fibroblast component of malignant fibrous histiocytomas usually has a storiform (spoke-like) pattern. The lesion is liable to recur locally and may metastasize to the regional nodes and lungs. HE ×360

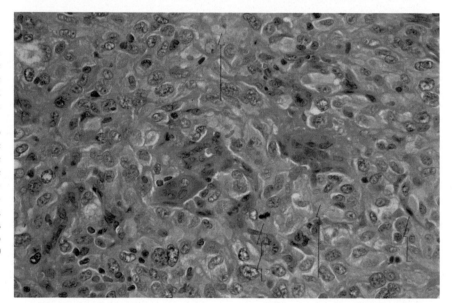

13.57 Alveolar soft part sarcoma: thigh

Alveolar soft part sarcoma is a malignant tumour of the deep soft tissues of the leg, usually the thigh, of young adults and especially women. It tends to recur locally and may metastasize to the lung. It presents as a large fairly firm mass apparently well-circumscribed. This specimen was a lobulated mass (10.5 × 5 × 5.5cm) in the right thigh of a woman of 27. It was removed surgically along with much adjacent skeletal muscle which it was invading. Histologically it consists of groups of cells bounded by a delicate fibrovascular stroma (arrow). The tumour cells are large with vesicular nuclei which show considerable pleomorphism, and pale granular cytoplasm. There are no mitotic figures in this field but occasional mitoses were present elsewhere. HE ×235

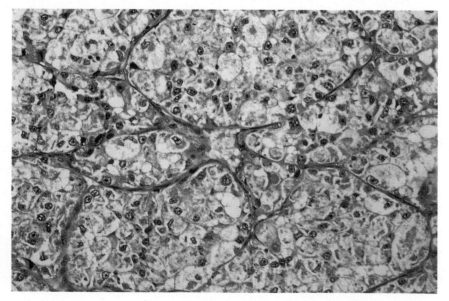

13.58 Onchocercosis: skin

Many different types of parasite may infest muscle, particularly in tropical countries where sanitation is unsatisfactory. The commonest parasites are worms. The parasite here is the filarial worm *Onchocerca volvulus*. The larvae are introduced into man by insect bites (by the gnat Simulium). The larvae mature in the dermis. They excite an inflammatory reaction which leads to the formation of nodules in the dermis (onchocercomas). This shows an adult female worm full of embryos in the subcutaneous tissues. It is enclosed in a capsule of dense but cellular fibrous tissue. It appears to be fully viable. When a fly bites someone with the parasite in the skin, the microfilarial larvae are ingested by the fly, in which they undergo the next stage of their development. *See also 1.10, 14.13*. HE ×120

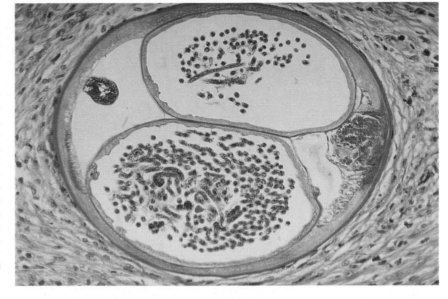

13.59 Cysticercosis: muscle

The parasite is the tapeworm *Taenia solium* which lives in the small intestine of man. The intermediate host is the pig but when man ingests food contaminated with feces containing the ova of *Taenia solium*, the larvae are liberated in the stomach. They penetrate the stomach wall and travel by the blood to various tissues including the brain. In the tissues each larva develops into a cysticercus, a small ovoid vesicle (5 × 10mm). When the cysticercus is alive, it evokes only a mild lymphocytic response but when it dies there is an acute inflammatory reaction. In this example the lumen of the cyst is at the top. The cysticercus has died and a pus-like exudate (bottom) has formed which is in contact with the smoothly convoluted surface of the cyst (arrows). Cysticerci can form in any organ and produce serious lesions. HE ×200

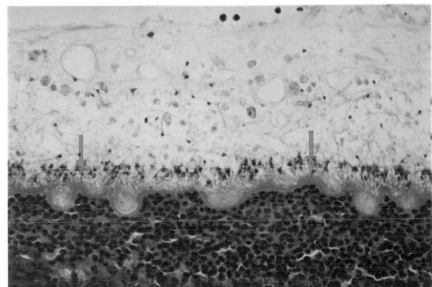

13.60 Trichinosis: muscle

The parasite is *Trichinella spiralis*, a common parasite of pigs in many countries, and man is infected by eating raw or undercooked pork containing encysted larvae. The larvae are released by the digestive enzymes, attach themselves to the mucosa of the small intestine and rapidly become sexually mature worms. The fertilized female penetrates the wall of the intestine and releases larvae which travel to all tissues. In muscle (but not in other tissues) the larva grows, coils in corkscrew fashion and forms a small cyst (about 1mm dia). This shows an encysted larva (centre) in striated muscle. It is enclosed in a pale-staining capsule formed by muscle cells (whose nuclei are visible). The muscle fibres are compressed and the interstitial tissue is infiltrated by chronic inflammatory cells. Larvae may remain alive for years in this state. HE ×110

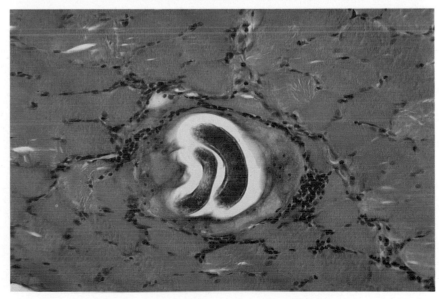

13.61 Infarct: muscle

Ischemic necrosis of muscle can be caused in a variety of ways, sometimes on a massive scale, e.g. in the 'crush syndrome' where compression of a limb may cut off the blood supply for a considerable time; and the muscles of the forearm may lose their blood supply following injury to, and swelling of, the tissues round the elbow (Volkmann's ischemic contracture). In this case the affected muscle is the gastrocnemius. The fibres are swollen and have lost their striations and their nuclei. Macrophages are digesting the necrotic fibres and fibrous tissue is forming (arrow). The fibrous tissue that replaces the dead muscle becomes densely collagenous and considerable distortion of structure may result, with pressure on neighbouring structures such as nerves. HE ×270

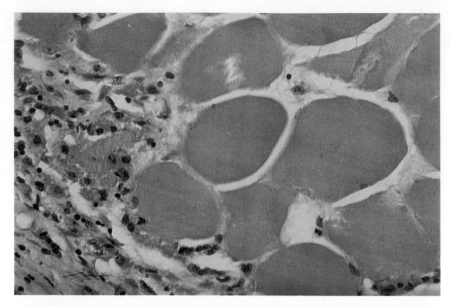

13.62 Anterior tibial compartment syndrome: muscle

Violent exercise, particularly in an untrained person, can rupture muscle sheaths and cause hemorrhage and edema. When this happens in a rigidly confined space such as the anterior tibial compartment, the increase in volume can raise the pressure to a level which produces ischemia and degeneration of the muscle. In this example the more acute phase has passed and although the muscle fibres have degenerated, there is evidence of regeneration, with large numbers of swollen sarcolemmal nuclei, some in mitosis. Nevertheless there is a marked tendency in this condition for the muscle to be replaced by fibrous tissue. HE ×270

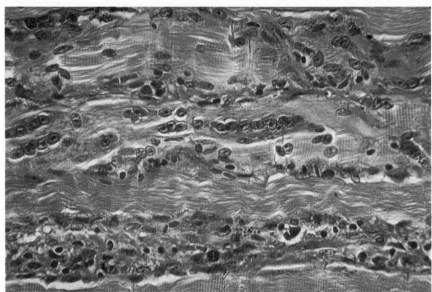

13.63 Progressive myositis ossificans: muscle

Richly cellular myxoid connective tissue is spreading between the muscle fibres (top). This tissue is osteogenic and bone is forming in it. Groups of cells have enlarged to become osteocytes and surround themselves with a deeply basophilic calcified matrix (arrows). These are trabeculae of woven (coarse-fibred) bone. Two such trabeculae are present. A similar process of bone formation sometimes occurs in the paralyzed muscles of paraplegic patients. HE ×200

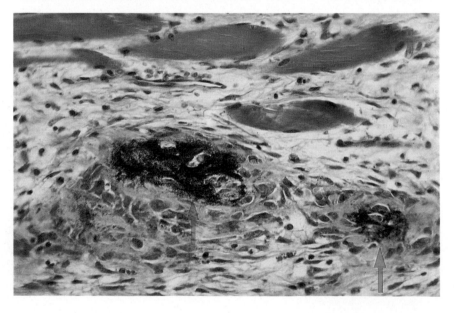

13.64 Rheumatoid arthritis: muscle

In rheumatoid arthritis there is atrophy and wasting of skeletal muscles, particularly those muscles which are immobilized by inflammation and fixation of the joints. Active polymyositis on the other hand is uncommon but there is often an inflammatory reaction in the endomysium and between the fascicles of muscle fibres. In this case lymphocytes and plasma cells have infiltrated the interstitial tissues around several muscle fibres and one of the fibres (centre) has undergone floccular degeneration. Macrophages are ingesting the necrotic fibre. HE ×335

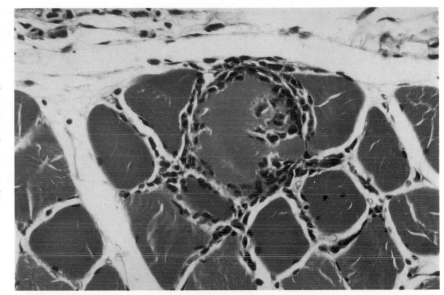

13.65 Dermatomyositis: muscle

In dermatomyositis both skin and muscles are inflamed. The polymyositis causes weakness and often also tenderness of the muscles. The disease may advance rapidly and if heart and respiratory muscles are involved, it may prove rapidly fatal. In most cases however remissions occur, followed by exacerbations. The fibre in the centre has undergone floccular degeneration. The fragment of the fibre that remains (arrow) is intensely eosinophilic and is being ingested by macrophages. Several of the other fibres are atrophic and in several the sarcolemmal nuclei are large and increased in number. This may be an attempt at regeneration. Phagocytosis of degenerating muscle is more characteristic of myositis than of muscular dystrophy. HE ×335

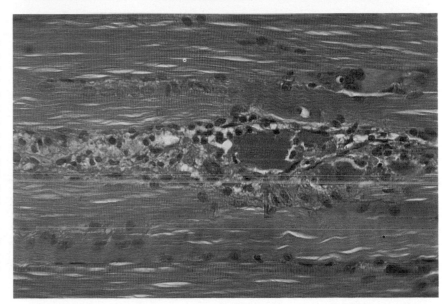

13.66 Dermatomyositis: muscle

In this case the muscle fibres are shrunken and atrophied to varying degrees and the myofibrils have retracted from the sarcolemmal sheaths. Several show central vacuolation, a change that is virtually pathognomonic of polymyositis. HE ×335

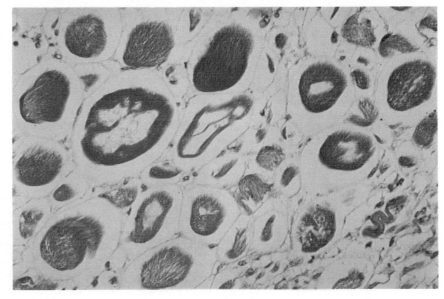

13.67 Muscular dystrophy: muscle

The various types of muscular dystrophy are inherited disorders. They begin in childhood and follow a progressive course. There is degeneration of skeletal muscle fibres and increasing muscle weakness. Each muscle fibre is affected individually and fibres showing a variety of changes are found in close association. The histological appearances are similar in the various syndromes and differentiation between them is based on the clinical presentation. The muscle fibres are rounded and vary considerably in size, several being very atrophic. The fibres are cut in cross-section. In practically every muscle fibre the sarcolemmal nuclei have migrated from the periphery of the fibre into its substance. Inward migration of nuclei particularly when widespread is suggestive of muscular dystrophy and especially myotonic dystrophy.   HE ×335

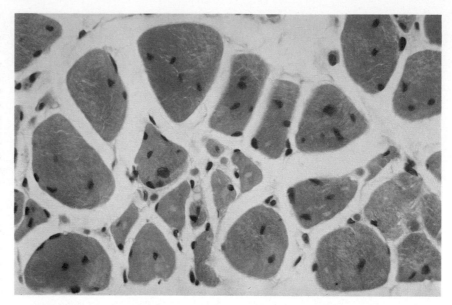

13.68 Muscular dystrophy: muscle

There is considerable variation in the size of the muscle fibres, and several of them are degenerate, with ragged margins and vacuolated weakly eosinophilic cytoplasm. One fibre shows increased eosinophilia. Many sarcolemmal nuclei are enlarged and in places they form chains (arrow). Clumps of pyknotic nuclei are also present and there are nuclei within some fibres. Many myofibrils have disappeared completely and been replaced by vascular fat (top).

HE ×150

13.69 Muscular dystrophy: muscle

One muscle fibre (arrow) has lost its eosinophilic myofibrils and the sarcolemmal nuclei are swollen and vesicular. This is a basophilic fibril which is attempting to regenerate. The sarcolemmal nuclei of adjacent fibres are enlarged but the remaining myofibrillary material appears normal, the 'waviness' being processing artefact.

HE ×335

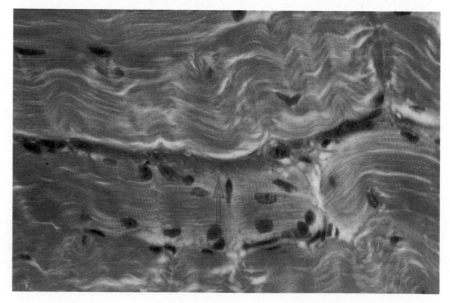

13.70 Neurogenic atrophy: muscle

Muscle fibres which lose their nerve supply atrophy. The fibres shrink and after 6 months may be less than 10μm in diameter. Their staining qualities do not change however and the cross-striations remain. With loss of sarcoplasm the sarcolemmal nuclei appear to increase in number. The fibres with an intact nerve supply do not hypertrophy. Fat may increase between the fascicles and later fibrous tissue may form. Intact fibres (right) and denervated fibres (left) lie side by side. There is some loose connective tissue between them and there is an infiltrate of fat (left) associated with the denervated fibres. The denervated fibres are very atrophic and narrow. The number of sarcolemmal nuclei in these fibres appears to be greatly increased, but in fact there is probably no absolute increase in their number.

HE ×335

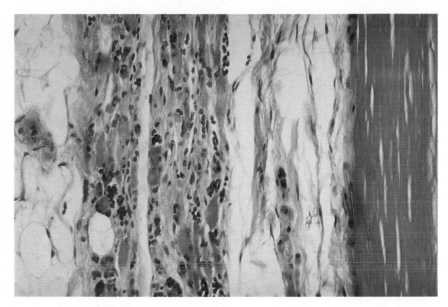

13.71 Neurogenic atrophy: muscle

Three denervated fibres are undergoing atrophy. The sarcoplasm at the periphery of each fibre is spongy and loose-textured, with loss of myofibrils, but the myofibrils that remain in the centre of the fibril still retain their cross-striations. The sarcolemmal nuclei are large, round and more prominent than normal. They appear to have increased in number because of the shrinkage of the fibres.

HE ×335

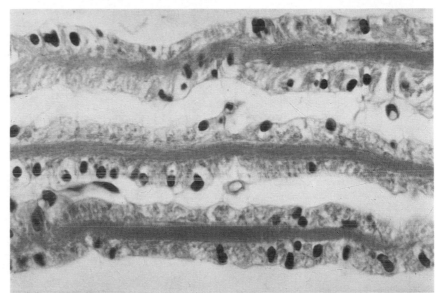

13.72 Idiopathic myoglobinuria: muscle

Myoglobinuria, the appearance of the low-molecular-weight protein myoglobin in the urine, is preceded by necrosis of muscle on a large scale and may occur in a wide range of disorders of muscle. After strenuous exercise a person with idiopathic myoglobinuria feels weak and the muscles are swollen and painful. The urine passed is dark-coloured from the presence of myoglobin. The patient in this case was a boy of 5. One fibre (above centre) has undergone floccular degeneration and necrosis, and macrophages are ingesting the degenerate myofibrillary material. Although death may occur from renal failure after a severe attack, idiopathic myoglobinuria is a relatively benign condition and since the basic structure of the muscles is maintained, recovery of muscle structure is usually complete.

HE ×360

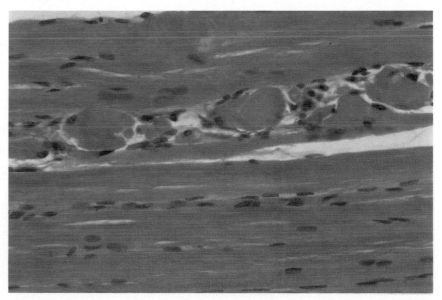

14.1 Epidermoid (epidermal) cyst: skin

Epidermoid cysts are situated in the dermis or subcutaneous tissue. They grow slowly to reach at most 5cm dia, elevating the overlying skin. They appear spontaneously as a rule but may arise from epidermis transplanted into the deeper layers of the skin by trauma. The cyst (left) is filled with deeply-eosinophilic laminated keratinous horny material. Its wall consists of fibrous tissue and it is lined by true epidermis which although stretched and thinned has recognizable squamous, granular and keratinocyte layers. The sweat gland (arrow) is a skin appendage and not related to the cyst.

HE ×335

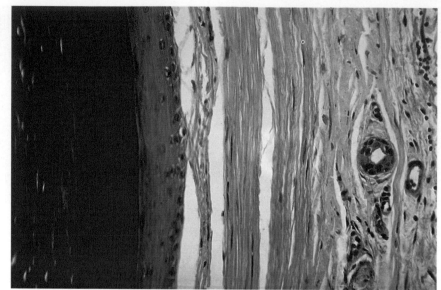

14.2 Actinic keratosis: skin

Actinic keratosis is a pre-cancerous lesion which is produced by the ultraviolet rays of the sun. It affects mainly older people and is commoner in those who are fair-skinned. In tropical countries it may be found in younger age groups. There is pronounced hyperkeratosis and parakeratosis, with alternating columns of hyperkeratotic (thin arrow) and parakeratotic cells (thick arrows). The granular layer of the epidermis is present in the hyperkeratotic area, but in the parakeratotic areas it is absent, and the cells of the keratinocytes and basal layers show some dysplasia. There is an infiltrate of chronic inflammatory cells in the papillary dermis.

HE ×190

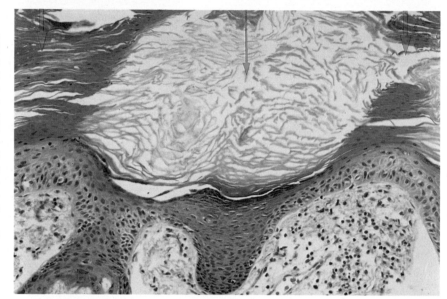

14.3 Radiodermatitis: skin

The total dose of ionizing radiation determines the severity of damage to the skin. A therapeutic dose produces an almost immediate acute reaction and very different effects later; and even fairly small doses may produce effects after a long period. This patient had received therapeutic X-irradiation to the arm 6 months previously. The epidermis (left) is hyperplastic and hyperkeratotic, and the rete ridges are elongated and irregular. The normal dermis and its appendages have been destroyed, and replaced by fibrous connective tissue in which there are many markedly dilated blood and lymphatic vessels (telangiectasis). There is a population of chronic inflammatory cells and fibroblasts in the upper dermis (centre) but the deeper collagenous tissue (right) is practically acellular.

HE ×120

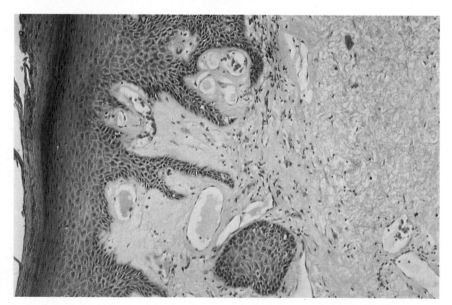

14.4 Pilonidal sinus

Fragments of epidermis can be driven by hairs through the epidermis into the dermis and subcutaneous tissues where they may form epidermoid (inclusion) cysts. This happens in hairdressers and those who clip dogs. It is probably also the mechanism which leads to the formation of pilonidal sinus in the sacral region. This is a section of the wall of such a sinus. The sinus is lined by epidermis and the lumen (left) contains many hairs and cell debris. The fibrous wall (right) is infiltrated by chronic inflammatory cells. Elsewhere the epithelial lining had broken down and the hairs, desquamated cells and bacteria had excited an acute inflammatory reaction in the surrounding tissues.   HE ×150

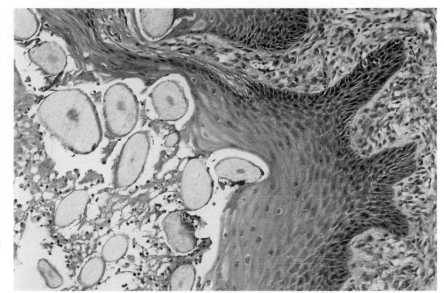

14.5 Keloid: skin

Occasionally, and particularly in black persons, connective tissue repair becomes overactive, and excessive amounts of collagen are produced. Some individuals are particularly prone to keloid formation and attempts to remove a keloid may induce recurrence. This is a healed wound of skin. The keloid consists of very large eosinophilic collagen fibres, lying between which are unusually large fibroblasts. The collagen fibres are much broader than normal collagen fibres and the margins of the fibres are irregular and ill-defined, appearing to merge with the cellular tissue between the fibres. Similarly the fibroblasts are greatly hypertrophied compared with normal fibroblasts. Most of them are elongated but some tend to be irregular in shape or stellate.   HE ×150

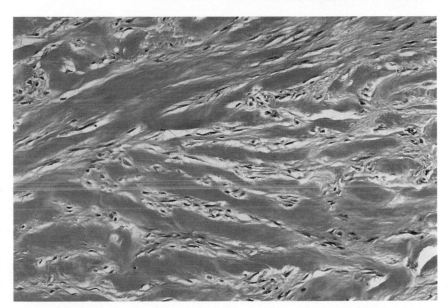

14.6 Keloid: skin

In this part of the keloid, the collagen fibres form broad bands between which the large fibroblasts lie. The nuclei of the fibroblasts are vesicular with prominent nucleoli and the cells are several times the length of a more normal fibroblast. Whereas the cellularity of ordinary scars grows less in time, a keloid usually retains its population of fibroblasts and they often become triangular in shape. *See also 1.33.*   HE ×360

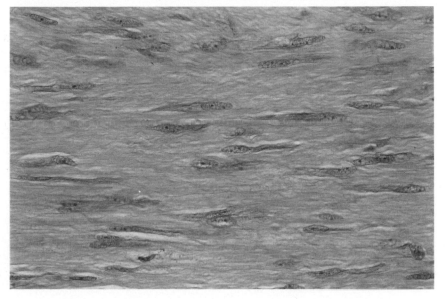

14.7 Secondary syphilis: skin

The skin is often involved in secondary and occasionally in tertiary syphilis. In secondary syphilis the histological changes are very variable and although plasma cells tend to be numerous and obliterative endarteritis is present, the appearances are not specific for syphilis. This is a biopsy specimen of skin from the right upper arm of a man of 50 suspected clinically of having secondary syphilis. There is hyperkeratosis but otherwise the structure of the epidermis (left) and the upper dermis shows little change. The deeper dermis and subcutaneous tissues are heavily infiltrated with chronic inflammatory cells. The cellular infiltrate is concentrated around the blood vessels and the skin appendages. The walls of the small vessels are thickened (arrows). HE ×80

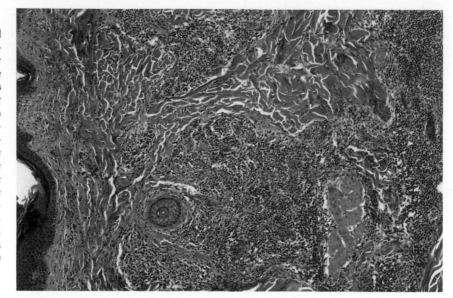

14.8 Secondary syphilis: skin

This is the same lesion as that shown in **14.7**, at higher magnification. It shows the cellular infiltrate around two small blood vessels in the dermis. Many of the cells are plasma cells but considerable numbers of lymphocytes and some histiocytes are also present. HE ×360

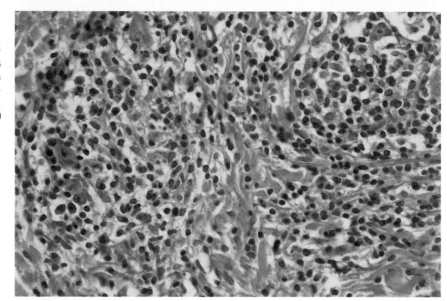

14.9 Tertiary syphilis: skin

This is a late-stage tertiary lesion in the skin. The infiltrate of chronic inflammatory cells seen in secondary-stage lesions is lacking and the most striking feature is the very marked fibrosis of the dermis and adjoining subcutaneous tissue with loss of the skin appendages. The fibrous tissue is vascular and fairly cellular and sparse lymphocytes are present throughout it. HE ×120

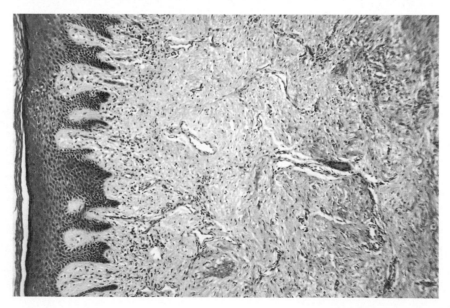

14.10 Sarcoidosis (Kveim test): skin

In the Kveim test for sarcoidosis, a sterilized suspension of sarcoid tissue is injected intradermally; and after 6–12 weeks the site of inoculation is excised and examined histologically. This is a positive result in a woman of 37, with two follicular granulomas in the dermis. The granulomas are composed of epithelioid cells, Langhans-type multinucleated giant cells (arrows) and small lymphocytes. The overlying epidermis is stretched but intact. It is essential that the material for the test be injected into the dermis and not into the fat of the subcutaneous tissues. Otherwise a granulomatous response to necrotic fat may give a false positive result.   HE ×360

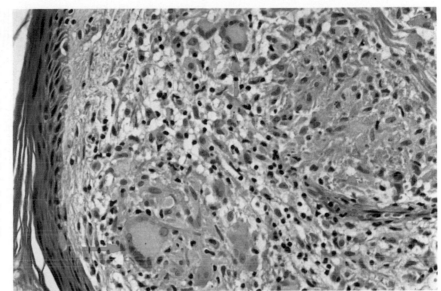

14.11 Leprosy: skin

Leprosy is an infection of the Schwann cells of peripheral nerves by *Mycobacterium leprae*. When the host's resistance is low, nodular lesions containing very large numbers of bacteria form in many tissues (lepromatous leprosy). When resistance is high, a follicular granulatomous reaction takes place, confined to the skin and peripheral nerves (tuberculoid leprosy). Some individuals have manifestations of both types of reaction. This is an example of lepromatous leprosy in the skin. The dermis is packed with large macrophages with fairly basophilic granular and vacuolated cytoplasm. Small dark-staining round bodies are present in some of the vacuoles (arrows). These round bodies are aggregates of lepra bacilli. Small lymphocytes are also present and several polymorphs. There is no necrosis or fibrosis.   HE ×360

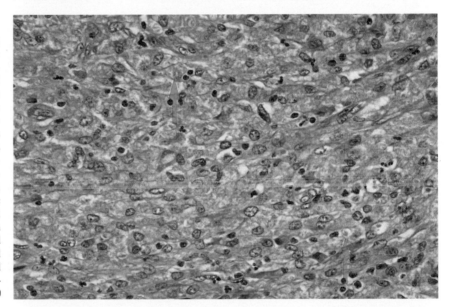

14.12 Lepromatous leprosy: skin

This shows a small nerve and part of a larger nerve in the subcutaneous tissues, separated by a band of fibrous tissue (centre). Both nerves are very swollen from the presence of large numbers of macrophages with vacuolated cytoplasm. These cells are laden with lepra bacilli. Clumps of bacilli (arrow) are visible in some vacuoles. Superficial nerves are often extensively involved in this way but anesthesia and paralysis are often late features of lepromatous leprosy, unlike tuberculoid leprosy where they are usually early manifestations.   HE ×235

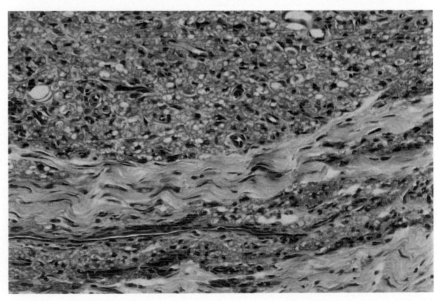

14.13 Onchocercosis: skin

In onchocercosis **(13.58)** microfilariae migrate into the adjacent dermis from the subcutaneous nodules (onchocercomas) containing the adult worms *(Onchocerca volvulus)*. This is the skin adjacent to a subcutaneous nodule. Microfilariae (arrows) are present in the dermis, probably in lymphatic channels. Migration of the microfilariae irritates the skin but there is no cellular response on the part of the host. There is however evidence of edema, with separation of the collagen fibres of the dermis. The cells of the basal layer of the epidermis contain melanin and there are melanin-laden macrophages in the dermis. *See also 1.10, 13.58.*

HE ×580

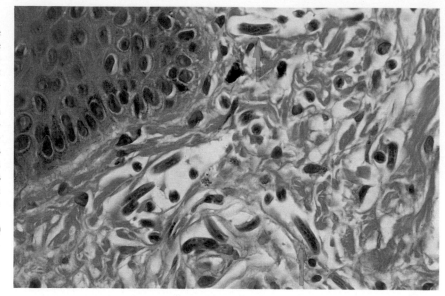

14.14 Ringworm: scalp

In ringworm of the scalp, the fungus *Microsporon audouini* invades the shafts of the hairs down to the zone of keratin formation, above the hair bulb. It does not invade the living cells of the hair bulb. The diseased hair shafts are fragile and when they break off, infected keratin remains in the follicle. Plucking the hairs does not therefore effect a cure. This is a section of a hair follicle stained by the periodic acid-Schiff method. The keratin of the hair shaft (centre) has been destroyed and replaced by a compact mass of purplish-red filamentous hyphae of the fungus. The basement membrane of the hair follicle also stains positively.

PAS ×335

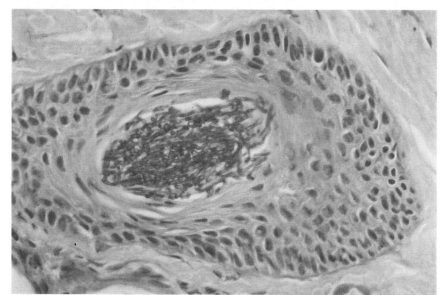

14.15 Molluscum contagiosum: skin

In molluscum contagiosum umbilicated nodules form in the skin as a result of infection by a DNA pox virus. The lesions heal spontaneously in several months. This was a nodule on the forehead of a woman of 52. It is a flask-shaped lesion and this is part of the wall of the flask with the lumen on the left. In the cytoplasm of the more superficial keratinocytes (left) there is a very large eosinophilic inclusion (thin arrows). The cell nuclei are flat and displaced to one side of the cell. The inclusions are aggregates of elementary bodies of the virus. Similar but much smaller round eosinophilic bodies in the nuclei of the keratinocytes deeper in the thickened epidermis (thick arrows) and in the basal cells are nucleoli and not viral inclusions.

HE ×360

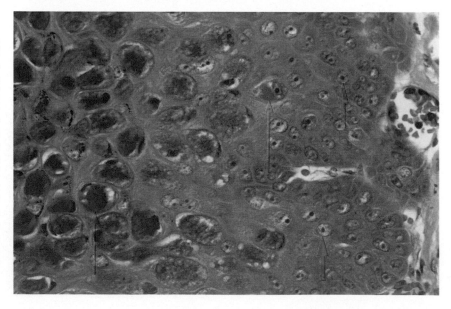

14.16 Varicella (chickenpox): skin

In the skin the varicella virus causes the nuclei of the keratinocytes to become hyperchromatic and the cytoplasm to swell (balloon degeneration). Cells affected in this way eventually die but the cell walls tend to remain, to become a lattice-work within the vesicle that forms in the epidermis (reticular degeneration). This is the margin of an intra-epidermal vesicle, part of which is visible on the right (thin arrow). The keratinocytes adjacent to the vesicle are swollen. Their cytoplasm is very edematous and pale-staining (thick arrow) (balloon degeneration) and in several cells large vacuoles have formed. The nuclei of the affected keratinocytes are also enlarged and some are hyperchromatic.　　　HE ×480

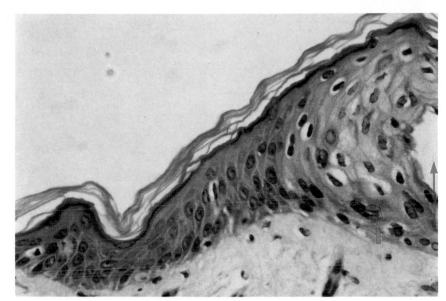

14.17 Varicella (chickenpox): skin

This shows the top of a fully-formed vesicle within the epidermis. The keratinized surface is still intact (top) and part of the lumen is visible at the bottom. Practically all the keratinocytes are rounded-up and swollen (balloon cells), those lying within the lumen of the vesicle (thin arrow) showing the most severe changes. The nuclei in most of the cells are pleomorphic and hyperchromatic (thick arrows). Changes similar to those shown here also occur in herpes simplex and herpes zoster.　　　HE ×830

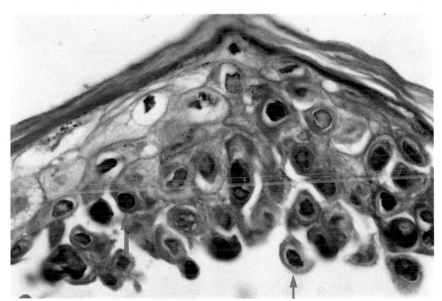

14.18 Herpes zoster: skin

Like varicella, herpes zoster is caused by a pox virus. The strain of virus is different but the tissue reactions are very similar in both diseases. This shows the base of an intra-epidermal vesicle in herpes zoster. All the remaining cells of the epidermis are degenerate and most of the cells have no nucleus or shrunken pyknotic nuclei and eosinophilic cytoplasm. Many rounded balloon cells are present, some in the lumen (thin arrows). Degenerate cells and cell debris form a network in the lumen (top right), the cell walls of necrotic keratinocytes tending to adhere to each other, to form strands traversing the vesicle (reticular degeneration). The small blood vessels in the underlying dermis are dilated (thick arrow) and there is infiltration of the dermis by polymorph leukocytes (bottom left). *See also 1.15.*　　　HE ×320

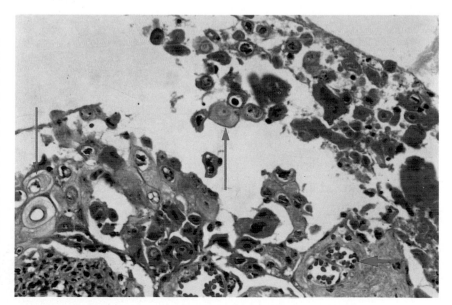

14.19 Eczema: skin

There are many causes of eczema (dermatitis) and the histological features tend to be the same irrespective of the cause. Initially the skin becomes red (erythema) and then vesicles tend to form. The vesicles may rupture and the serous exudate forms crusts over the lesions. If the etiological factor persists, the lesions will tend to become chronic. Histologically the main features are edema within the epidermis (spongiosis), with separation and disintegration of clumps of epithelial cells to form vesicles. In this lesion, the edema has led to separation of the keratinocytes and a vesicle (arrow) has formed high in the epidermis. The dermal papillae are swollen and edematous and their blood vessels dilated. The rete ridges are elongated. There is an infiltrate of small lymphocytes in both the dermis and epidermis. HE ×235

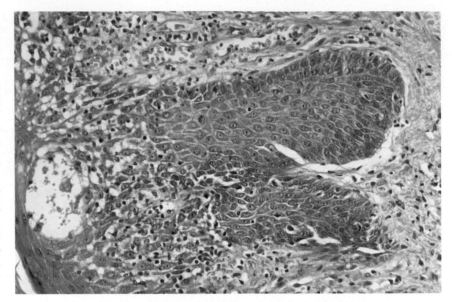

14.20 Lichen planus: skin

In lichen planus, crops of itchy lilac-coloured papules form on the flexor surfaces. In this lesion there is hyperkeratosis (top) the normal basket-weave keratin being replaced by dense laminated keratin; the keratinocyte layer is prominent (it often shows more marked focal thickening); and the rete ridges are narrowed and pointed, caused by their being stretched over broadened dermal papillae. This gives the lower border of the epidermis a saw-toothed appearance. The basal layer of the epidermis is intact. Sometimes it is destroyed. There is an infiltrate of lymphocytes and histiocytes in the upper dermis which goes right up to the epidermis but is sharply limited to the papillary and subpapillary layer of the dermis. HE ×190

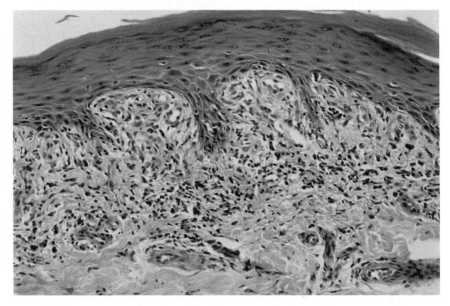

14.21 Urticaria pigmentosa: skin

In urticaria pigmentosa there is an excessive number of mast cells in the skin. In infants the upper dermis may be packed with them but when the condition starts in adult life, as in this case, the number is much smaller and the cells tend to be located around the blood vessels of the dermis. This is a section of the mid-dermis. The muscle fibres (thin arrows) belong to the arrectores pilorum muscle; and the large round cells with eosinophilic cytoplasm are mast cells (thick arrows) congregated around the small vessels. The granules in the cytoplasm of the mast cells are not visible in HE sections but are well demonstrated by methylene blue or toluidine blue, particularly when the tissues have been fixed in non-aqueous solutions. HE ×200

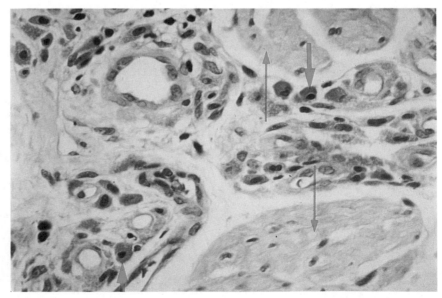

14.22 Pemphigus vulgaris: skin

In pemphigus the basic defect is acantholysis, i.e. loss of cohesion between the keratinocytes of the epidermis. Spaces form between the cells and fill with serous fluid. An intra-epidermal bulla is starting to form in this skin. Bulla formation is suprabasal and the keratinocytes are separating from the basal cells. The latter will form the floor of the bulla eventually. Round acantholytic cells have formed and are becoming detached (arrows), and will lie free in the lumen of the bulla. Small numbers of lymphocytes are present in the dermis.

HE ×360

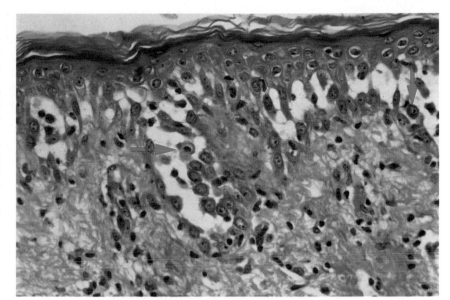

14.23 Dermatitis herpetiformis: skin

This condition is characterized by the formation of small clusters of intensely itchy vesicles or bullae which tend to form in certain sites, such as the scapular and sacral areas or in the region of the elbows. The papillary dermis (bottom) is edematous and vacuolated and there are collections of eosinophil leukocytes in the papillae ('eosinophil abscesses') (thin arrow). The eosinophils in these are degenerate and disintegrating and only a few have the characteristic eosinophilic cytoplasm. The cellular infiltrate is accompanied by eosinophilic material which is composed of serous exudate and debris from the eosinophils. Occasional eosinophils are present in the epidermis and also a small 'eosinophil abscess' (thick arrow) which is probably an extension of an abscess in a papilla.   HE ×235

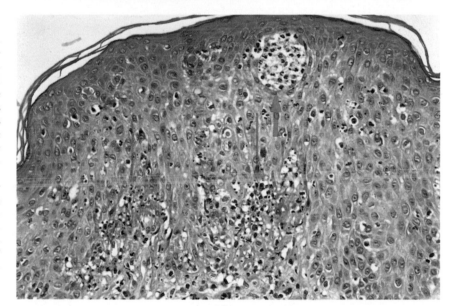

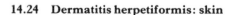

14.24 Dermatitis herpetiformis: skin

This shows the papillary dermis. In several dermal papillae are small 'abscesses' containing eosinophil leukocytes, degenerate polymorphs and eosinophilic debris (thin arrows). Although the cytoplasm of the degenerate polymorphs stains weakly, examination at higher magnification confirms that many are eosinophil leukocytes which have shed most of their granules. The eosinophilic debris contains fibrin and probably also collagen and debris from the degenerate eosinophils. Similar eosinophilic fibrillary material is present deeper in the dermis (thick arrows). The dermis (right) is edematous and infiltrated with polymorphs and macrophages. With further increase in size of these papillary microabscesses, the interpapillary ridges of the epidermis become detached and a subepidermal blister forms.   HE ×360

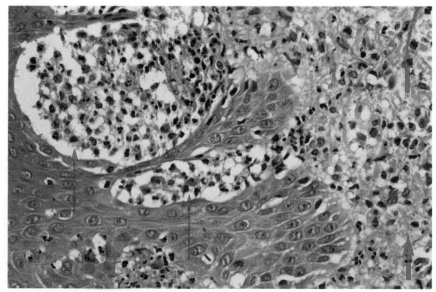

14.25 Pretibial myxedema: skin

Sometimes nodules form in the skin over the tibia in individuals who are myxedematous or have untreated thyrotoxicosis. The swelling of the skin is caused by a great increase in the amount of hydrophilic connective tissue mucin. Mucins of this type remain unstained in HE sections and are also sensitive to the type of tissue fixative used. In this section the spaces occupied by the mucins appear as clear spaces in the dermis between the widely separated components of the dermis (blood vessels, cells and connective tissue fibres). The epidermis (left) is hyperkeratotic. HE ×70

14.26 Pretibial myxedema: skin

This section of skin has been stained by the colloidal iron method which stains connective tissue mucins blue (Prussian blue). The epidermis (top) is coloured brownish-yellow, the collagen fibres are brownish-black, and the dermis and subcutaneous tissues are a deep blue colour from the presence of greatly increased amounts of connective tissue mucin. Prior treatment of the section with hyaluronidase removed almost all the material giving the reaction, confirming that most of it was hyaluronic acid, the most hydrophilic type of connective tissue mucin. Colloidal iron ×80

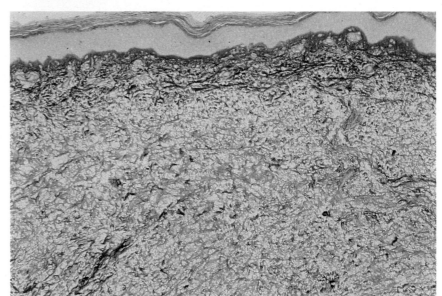

14.27 Lupus erythematosus: skin

In acute systemic lupus erythematosus (SLE) many tissues and organs are involved. In chronic (discoid) lupus erythematosus (CDLE) the lesions are confined to the skin and sometimes the mucous membranes. This is the skin below the eye of a man of 33 with chronic discoid lupus. The epidermis is thin, with a very atrophic stratum spinosum. There is hyperkeratosis which extends down into the pilosebaceous follicles (follicular plugging, thin arrows). The basal layer of the epidermis shows liquefaction degeneration (thick arrows). The dermis is edematous and hyaline, and there are deposits of eosinophilic fibrin in it. Beneath the damaged dermis there is a very intense infiltrate of chronic inflammatory cells which are mainly lymphocytes. The infiltrate is concentrated around blood vessels and pilosebaceous follicles. HE ×60

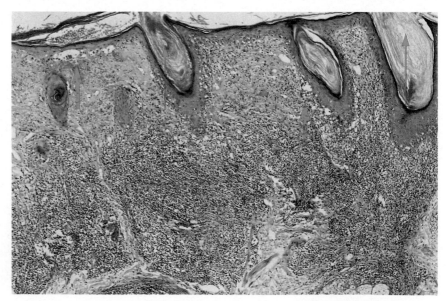

14.28　Lupus erythematosus: skin

This is the same lesion as that shown in **14.27**. The surface is hyperkeratotic and the opening of the pilosebaceous follicle (thin arrow) is plugged with keratin. The epidermis is atrophic, the change affecting particularly the keratinocyte layer (stratum spinosum). Liquefaction degeneration affects the whole basal layer but it is more severe in some parts where the basal layer is completely destroyed (thick arrows). The dermis is hyaline and hemorrhage has occurred in places (left) from the thin-walled blood vessels. Fibrin is also present. There is an infiltrate of lymphocytes and plasma cells around the pilosebaceous follicles.　　　　　　　HE ×150

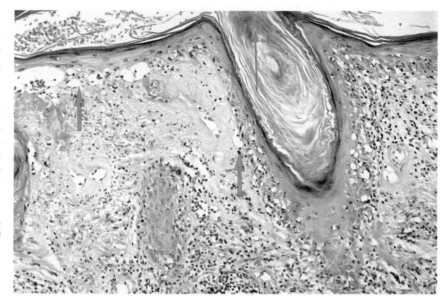

14.29　Rheumatoid arthritis: skin

In rheumatoid arthritis nodules form in the skin, generally in sites over bone and in the vicinity of joints. They are located in the subcutaneous tissues. This is part of a large nodule. The centre of the nodule is to the left. It consists of necrotic collagenous tissue which is strongly eosinophilic (fibrinoid necrosis) surrounded by elongated histiocytes and fibroblasts lying parallel to one another (palisading). The necrotic material in the centre of the nodule is resistant to phagocytosis and the nodules are very persistent. Rheumatoid nodules are commoner in the more severe forms of disease. They also occur in tissues other than the skin. *See also 1.17, 6.36.*　　　　　HE ×235

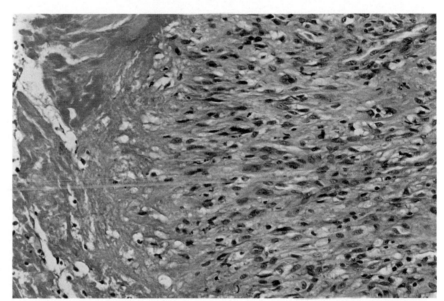

14.30　Scleroderma (systemic sclerosis): skin

Scleroderma is a connective tissue disorder with lesions in many organs and tissues. The skin shrinks and becomes stiff, and the taut skin restricts movement, particularly of joints. The main clinical symptoms of this patient, a man of 60 with scleroderma, were caused by lesions in the esophagus and skin. The epidermis (left) is hyperkeratotic. The sweat ducts and glands (right) are normal. The connective tissue of the dermis is homogeneous, swollen and poorly cellular (hyalinized). However this appearance is not caused by fibroblastic proliferation and fibrosis, but by marked atrophy of the dermis, the distance from the sweat glands in the subcutaneous tissues to the epidermis being only half the normal distance.

HE ×80

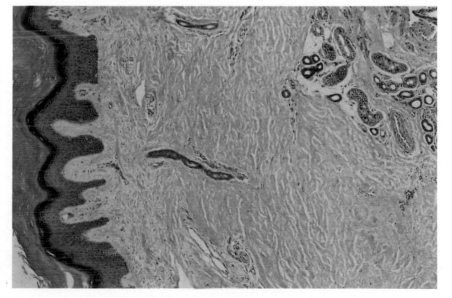

14.31 Verruca vulgaris: skin

Verruca vulgaris, the common wart, is caused by a papova virus. Papova viruses also cause verruca plantaris and condyloma acuminatum. Viral inclusions can be demonstrated in the nuclei of the cells of the verruca by special techniques but they are not visible in HE preparations. However in this active virus wart, as the epithelial cells moved outwards from the stratum spinosum (bottom) to form the granular layer, damage by the virus caused many of them to become vacuolated, and instead of forming normal keratohyalin granules they developed deeply eosinophilic bodies (arrows) within their cytoplasm. These 'inclusions' are probably tangled masses of tonofilaments. The keratin produced by the epithelial cells in this damaged condition is often nucleated (parakeratosis) (top).
HE ×270

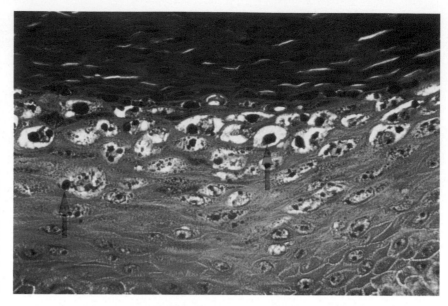

14.32 Verruca vulgaris: skin

As shown in **14.31** the papova virus causes focal epidermal hyperplasia, with acanthosis, hyperkeratosis (and parakeratosis) and papillomatosis. This however is an older lesion. The damage caused by the virus to the individual cells of the epidermis is no longer evident but the skin has a markedly abnormal papillary architecture and the surface of this 'warty' lesion is very hyperkeratotic.
HE ×55

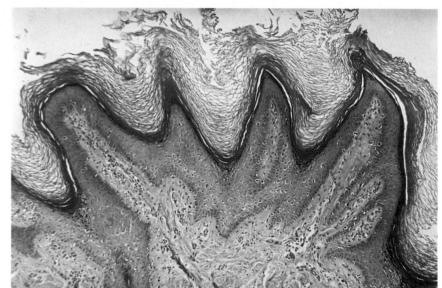

14.33 Keratoacanthoma: skin

Keratoacanthoma (molluscum sebaceum) is a benign tumour-like lesion of unknown etiology. It grows comparatively rapidly, projecting above the surface of the skin as a cup-shaped mass, undergoes keratinization in the centre, and then involutes. The whole cycle, to final scar formation, takes only 3–4 months. This shows a fully-formed lesion in cross-section. The crater in the centre is filled with keratin (top) and strands of proliferating epithelium. The 'wall' of the crater is composed of a thick layer of hyperplastic epithelium which is invading the dermis (thin arrows). The epidermis and the skin appendages on both sides of the lesion have been drawn up. There is an infiltrate of chronic inflammatory cells in the dermis around the margins of the lesions. It is patchy but intense in some areas (thick arrow).
HE ×9

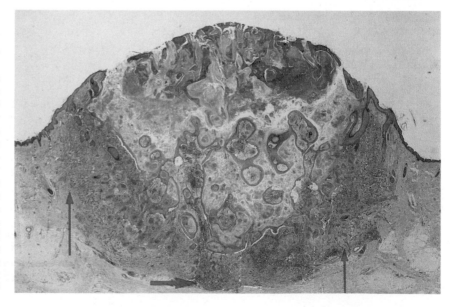

14.34 Keratoacanthoma: skin

This shows the epithelium of the kerato-acanthoma where it is invading the dermis. The invading cells, which are arranged in slender cords and clusters ('cell nests') with central keratin formation (arrow), have hyperchromatic pleomorphic nuclei which are mitotically active. There is a heavy infiltrate of chronic inflammatory cells around the tumour margin. The histological appearances are indistinguishable from squamous carcinoma **(14.48)**. Later in the evolution of the keratoacanthoma however inflammatory cells invade the epithelial strands, involution follows, and a keratin plug fills the crater.   HE ×150

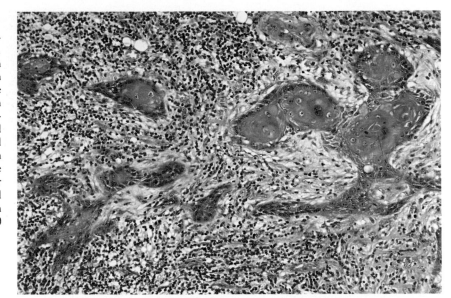

14.35 Basal cell papilloma: skin

Basal cell papilloma (seborrheic keratosis, seborrheic wart) is a common benign tumour which arises from the epidermis. It is raised above the surface of the skin and often being heavily pigmented it is liable to be mistaken for malignant melanoma. This is the solid type of basal-cell papilloma consisting of a mass of dark-staining cells within which there are small round cysts full of eosinophilic laminated keratin (horn cysts). The cysts are present throughout the tumour and the keratin in them is characteristically in concentric layers. They probably arise in pilosebaceous follicles and can often be seen to open on to the surface. The surface is usually hyperkeratotic but the keratin has been lost during processing of the tissues. There is an infiltrate of chronic inflammatory cells in the connective tissue stroma.   HE ×60

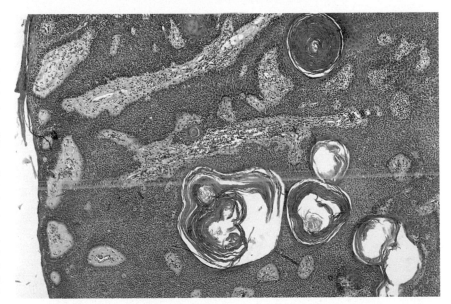

14.36 Basal cell papilloma: skin

This shows part of a keratin-filled horn cyst (left) and the cells of the papilloma at higher magnification. The cells are small and closely-packed 'basal' cells and they all contain melanin. The stratified epithelial lining of the horn cyst is attenuated and the transition from the 'basal' cells of the tumour to the laminated keratin in the horn cyst is abrupt, with only a barely detectable very thin stratum granulosum at the boundary. A small stromal blood vessel is shown (right).
HE ×360

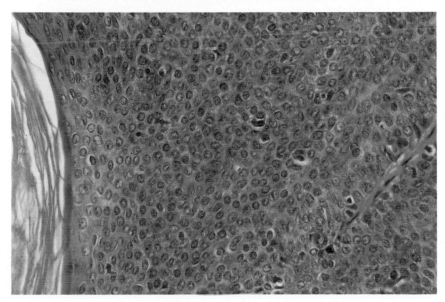

14.37 Eccrine poroma: skin

Eccrine poroma arises within the epidermis from the intra-epidermal sweat duct and grows downwards into the dermis. It is a benign tumour occurring most often on the palms and soles of the feet. This lesion was in the upper lip of a woman of 58. It was a raised white, hard nodule 2mm dia. It is cystic and flask-shaped. The 'neck' of the flask is at the top (arrow) and it is blocked by a plug of keratin (left). The lumen (top right) is full of keratinous cell debris. The tumour is composed of a uniform population of small cells with central ovoid nuclei. They are smaller than squamous cells but at higher magnification they are connected by intercellular bridges. The cells do not keratinize except at the surface where they have formed the keratin 'plug'. There is no mitotic activity.

HE ×60

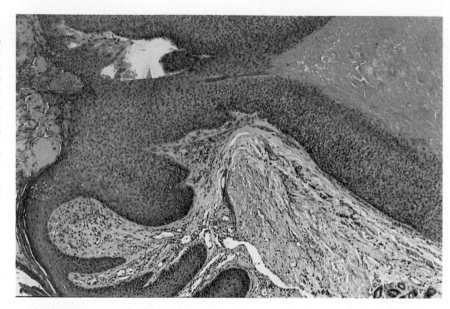

14.38 Sweat gland adenoma: skin

Sweat gland adenoma (syringoma) is a benign tumour which may arise from the dermal portion of the sweat duct, though an origin from apocrine glands is possible. It usually takes the form of multiple nodules on the neck and face of young adults. It consists of collections of small ducts lined by a double layer of epithelial cells and small cystic structures (arrows) filled with laminated material which gives a strongly positive reaction with the periodic acid-Schiff method. Some have characteristic comma-shaped extensions.

HE ×80

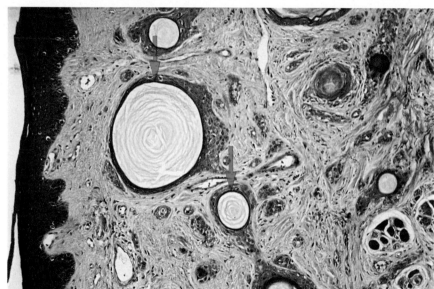

14.39 Naevus syringocyst-adenomatosus papilliferus: skin

This common lesion of the face or scalp is hamartomatous and probably arises from the ductal portion of an apocrine gland. The lesion is cystic and superficially situated, opening on to the surface. The wall (right) is partly lined by stratified squamous epithelium (thin arrow) which has spread down from the surface epidermis and a thick irregular epithelial layer (thick arrow). The cystic cavity is filled with eosinophilic material, into which papillary structures (top left) project. Each of the papillary processes has a fibrous core in which there are numerous chronic inflammatory cells and is covered with a double layer of epithelial cells.

HE ×60

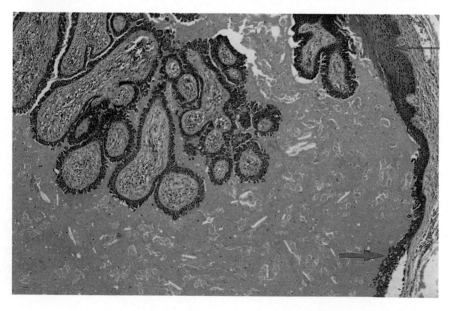

14.40 Naevus syringocyst-adenomatosus papilliferus: skin

This shows the tips of two of the papilliferous processes which project into the cystic space. The core of each consists of fibrous tissue in which there are plasma cells and lymphocytes. The surface is covered with a double layer of epithelial cells. The inner layer, in contact with the fibrous core, consists of small flat cells with dark-staining nuclei and the cells of the outer layer are tall and columnar, with ovoid nuclei and abundant eosinophilic cytoplasm. A few are vacuolated. The material in the lumen (lower half) is eosinophilic and amorphous with unstained fragments of debris floating in it.
HE ×235

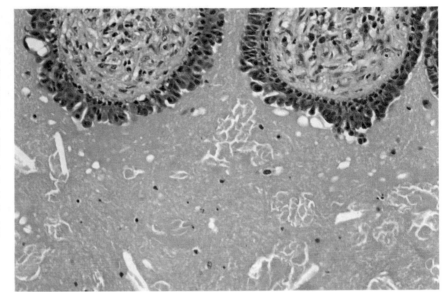

14.41 Eccrine cylindroma: scalp

Eccrine cylindroma (turban tumour) is a benign tumours of apocrine gland origin and most of them are located on the head and neck. They may be multiple and grow to a large size: hence the origin of the term turban tumour. The lesion consists of cords of cells enclosed by thin strands of eosinophilic stroma (thin arrow). Small blood vessels are present in the stroma which tends to be homogeneous and resembles basement membrane. The tumour cells in the centres of the cords are closely-packed and polyhedral with scanty cytoplasm and large vesicular nuclei containing a central nucleolus (thick arrow). At the periphery of the cords the cells have elongated basophilic nuclei which are palisaded where they rest on the eosinophilic stroma.
HE ×310

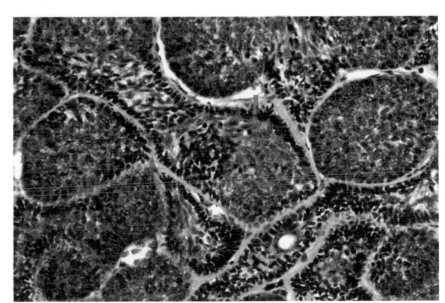

14.42 Eccrine cylindroma: scalp

Two lobulated tumours arose in the scalp of a woman of 66. The cut surface of each was brown and homogeneous. Histologically the two tumours were identical. Each consists of cords and sheets of epithelial cells. The cells in the centre of the cords have ovoid or round vesicular nuclei (thin arrows) and the cells at the periphery of the cords have smaller dark-staining nuclei (thick arrows). The latter are in contact with and tend to form a palisade on the thick strands of hyaline eosinophilic material which resembles basement membrane. The structure of this tumour is essentially the same as that in **14.41** apart from the much greater development of the basement membrane material which surrounds and permeates the columns and sheets of epithelial cells.
HE ×360

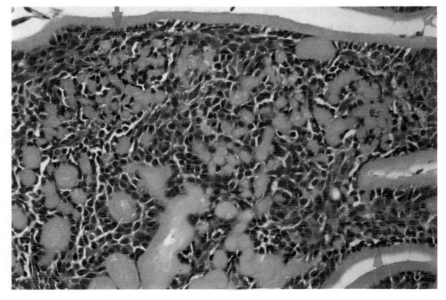

14.49 Benign pigmented naevus: skin

Melanocytes are dendritic but appear as round cells with 'clear' cytoplasm generally in the basal layer of the epidermis. Pigmented naevi are formed by proliferation and maturation of the melanocytes of the epidermis. This was a raised nodule 0.5cm dia on the face of a woman of 46. Melanocytes (thin arrow) in the basal layer of the epidermis (left) show no proliferative (junctional) activity. The dermis (right) is occupied by packets of mature naevus cells. There is some nuclear pleomorphism but most of the naevus cells have an ovoid basophilic nucleus and eosinophilic cytoplasm with ill-defined cell boundaries. Some are multinucleated (thick arrow). There is no mitotic activity and no melanin pigment is visible. Strands of fibrous tissue (double arrow) intersect the naevus cells. The lesion is a mature intradermal naevus.

HE ×235

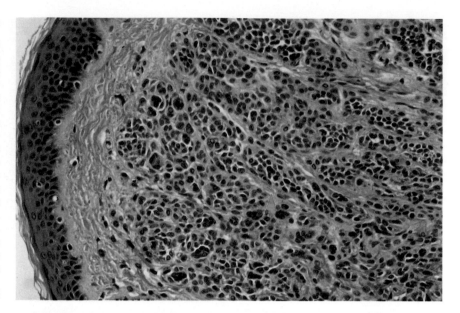

14.50 Juvenile melanoma: skin

Despite its name, juvenile melanoma (Spitz naevus) is a benign lesion. It is now recognized that it is not confined to children and a considerable proportion occur in older people. It is vascular and often looks pink. It is a compound naevus as a rule and liable to be mistaken for a malignant melanoma. It usually consists of large cells which have the form of both spindle cells and epithelioid cells. In this naevus there is active junctional activity, with nests of large melanocytes (thin arrows) in the epidermis and also in immediately adjacent dermis (thick arrows). The melanocytes in these nests are mostly round and epithelioid. The cells in the epidermis (left) adjacent to the naevus cells contain melanin and the heavily-pigmented cells in the dermis are macrophages laden with melanin.

HE ×360

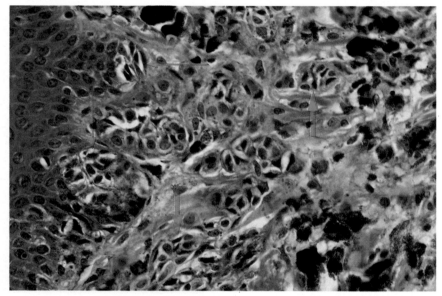

14.51 Juvenile melanoma: skin

Pseudoepitheliomatous hyperplasia is present in the epidermis in a minority of juvenile melanomas and the rete ridge (thin arrow) is elongated. The naevus cells in the dermis are a mixture of epithelioid and spindle cells. The cells are large and have abundant eosinophilic cytoplasm and pleomorphic vesicular nuclei in which one or more prominent nucleoli are present. Some are multinucleated. The naevus cells are characteristically much larger than the keratinocytes of the epidermis. No mitoses are present. There is however often considerable mitotic activity in a juvenile melanoma. A diagnostically helpful feature of most juvenile melanomas is the presence of many small blood vessels (thick arrows). A lymphocytic infiltrate is also present.　　　HE ×235

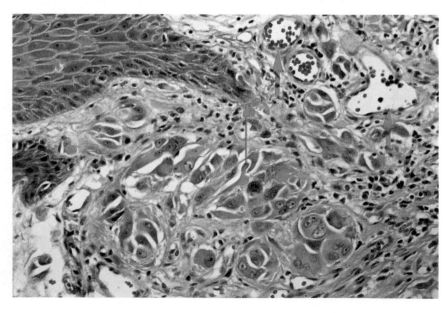

14.52 Intradermal naevus: skin

Many intradermal naevi 'mature' and eventually involute, with progressive fibrosis until the lesion is replaced by a fibrous nodule containing only a few naevus cells or structures resembling Meissner corpuscles. The epidermis (left) is attenuated. Beneath it in the dermis is the most superficial part of an intradermal naevus which has undergone a considerable degree of involution. The naevus cells have round vesicular nuclei in some of which there is a central nucleolus. The cell cytoplasm is homogeneous and hyalinized ('glassy'). Several naevus cells contain melanin and melanin-laden macrophages are also present. There is also an increase in fibrous tissue in the naevus and a reticulin stain would show a well-developed network of fine connective tissue fibres running between and around the individual naevus cells.　　　　HE ×400

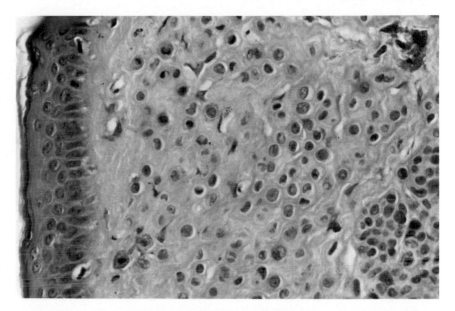

14.53 Blue naevus: skin

During development of the fetus, melanoblasts emigrating from the neuroectoderm to the epidermis may be retained in the dermis where they mature to form blue naevi. In the sacral region these cells form the Mongolian 'blue spots' present in some individuals at birth but they may also form bluish nodules elsewhere. A blue naevus is almost always benign. The epidermis (left) is normal (junctional activity is present in some lesions) and in the dermis there are many melanin-containing cells. The round cells are macrophages full of ingested melanin and the cells which produce the pigment are the elongated spindle-shaped cells, the melanocytic naevus cells. The naevus cells tend to be closely associated with the dermal collagen bundles.

HE ×150

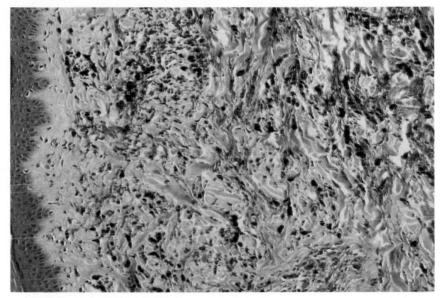

14.54 Blue naevus: skin

At higher magnification the branching dendritic process of the spindle-shaped melanocytes (naevus cells) are clearly visible. They contain melanin granules in their cytoplasm, and accompanying them are many large round or ovoid macrophages (melanophages) heavily laden with ingested melanin produced by the naevus cells.

HE ×335

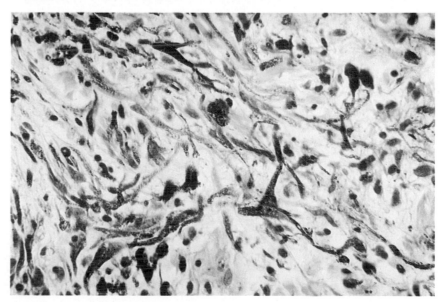

14.55 Lentigo maligna: skin

Lentigo maligna is a premalignant lesion which tends to develop on skin that has been heavily exposed to sunlight. It is therefore found on the faces of older people. It starts as a small macule which extends, sometimes to cover a large area, and eventually one or more nodules of malignant melanoma may develop within it. The basal layer of the epidermis and the pilosebaceous follicle have been replaced by a row, several cells thick, of large and aberrant melanocytes with hyperchromatic pleomorphic nuclei. They have not invaded the dermis however and the epidermal basement membrane is intact. Pigment-laden melanophages are present in the adjacent dermis (thin arrows). The collagen fibres of the dermis (thick arrows) are basophilic and fragmented ('elastotic degeneration'), another effect of ultraviolet light.   HE ×235

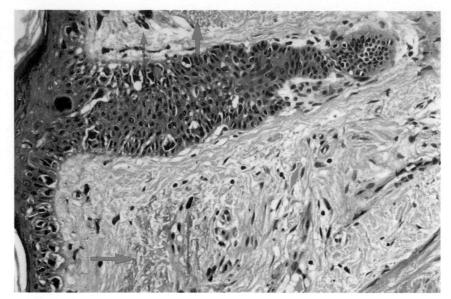

14.56 Lentigo maligna: skin

This is another part of the lesion shown in **14.55** showing an area of particularly active junctional activity. These are 'nests' of large melanocytes with abundant clear cytoplasm extending downwards into the dermis (right) and also upwards towards the surface of the epidermis. There are melanin granules within many of the melanocytes but melanin is present in much higher concentration in the elongated macrophages. Most of the macrophages are in the dermis but several are in the epidermis. The packets of melanocytes are bulging the basement membrane of the epidermis downwards, but it is almost certainly still intact and there is no invasion of the dermis. Not all cases of lentigo maligna go on to malignant melanoma.   HE ×360

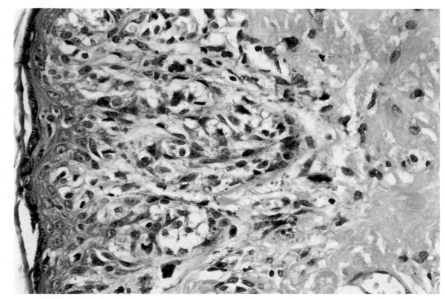

14.57 Melanocarcinoma: skin

Melanocarcinoma (malignant melanoma) is a highly malignant tumour of skin. This lesion arose in the nail-bed of the index finger of a woman of 63. Occasional pigment-laden cells are present in the stratum corneum. Junctional activity affects the whole of the epidermis (left) and particularly the rete ridges. The melanocytes in the 'nests' in the epidermis are abnormal, with hyperchromatic pleomorphic nuclei. Melanocytes have migrated from the epidermis into the dermis where they form a closely-packed mass of tumour cells, mostly spindle-shaped. Many macrophages laden with melanin (melanophages) are present, particularly in the deeper parts of the tumour (right). This was diagnosed as a superficial (spreading) type of melanocarcinoma with early dermal invasion (to a depth of 0.7mm).   HE ×150

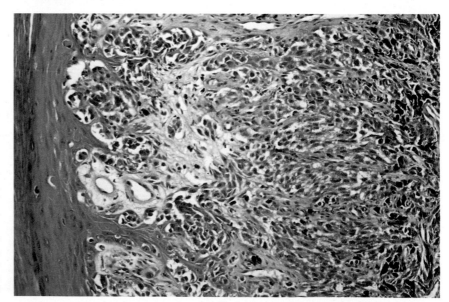

14.58 Melanocarcinoma: skin

Two main types of tumour cell are found in melanocarcinomas: spindle cells and epithelioid cells. The latter are large and round or polygonal with abundant cytoplasm. A tumour may consist more or less exclusively of one or other type of cell but both types are often present in the same lesion. In this example the epithelioid cells form clusters (thin arrows). The cytoplasm of the cells is pinkish-brown, from the presence of very fine melanin granules. The spindle cells are relatively few (thick arrow) and lie between the clusters of epithelioid cells. They have much less cytoplasm than the epithelioid cells and contain less melanin. The stroma (double arrow) is inconspicuous but within it there are many pigment-laden macrophages (melanophages). No mitotic figures are present in this field but there were many elsewhere. HE ×335

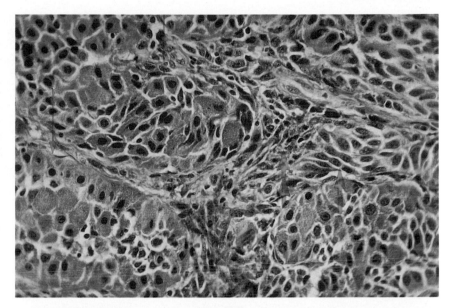

14.59 Melanocarcinoma: skin

Melanocarcinoma may arise in an existing benign pigmented naevus (junctional or compound but not intradermal) or arise de novo. Tumours on light-exposed skin tend to be flat (superficial melanocarcinoma) and probably carry a better prognosis than tumours in other areas which tend to be nodular (nodular melanocarcinoma). A useful sign of malignant change in a cellular naevus is the large size of the nuclei of the melanocytes which often exceeds that of the keratinocytes. Another helpful indicator is the upward penetration of the epidermis by abnormal melanocytes as far as the stratum granulosum. In this example the very active proliferation of melanocytes within the epidermis (bottom) has led to their being shed either singly or as clumps through the upper layers of the epidermis and into the keratin (top). HE ×335

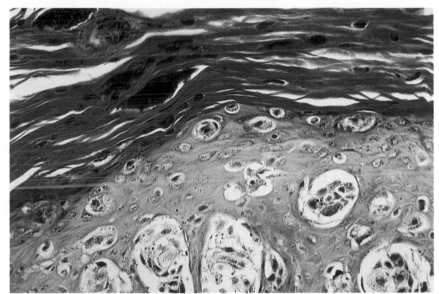

14.60 Melanocarcinoma (malignant melanoma): conjunctiva

A man of 29 developed a small lesion at the limbus of the eye which was resected. It consists of elongated spindle cells with deeply basophilic hyperchromatic nuclei. The tumour has an 'alveolar' pattern with groups of cells separated by delicate strands of stromal connective tissue (thin arrows). There is considerable mitotic activity (thick arrow). There is no melanin and only very small amounts could be detected elsewhere in the tumour. Dilated thin-walled blood vessels are present. The conjunctiva (top) is intact but was ulcerated over other parts of the tumour. There was a history of previous injury to the eye. HE ×350

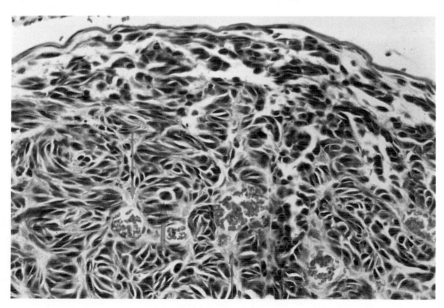

14.61 Gouty tophus: skin

In gout a disturbance of purine metabolism causes the uric acid level in the blood to be raised. Sodium urate crystals are deposited in many sites, including the skin around joints. The mass of crystals is large enough to produce a nodule (tophus) in the subcutaneous tissues. This tophus was situated over the olecranon of a man of 68. It consists of clusters of pale-staining urate crystals (arrows) surrounded by histiocytes and macrophages, many of which are multinucleated. The crystals are very small and the deposits appear amorphous. They were birefringent, however, in polarized light. There is no evidence of an acute inflammatory reaction but collagenous fibrous tissue surrounds the deposits.　　　HE ×150

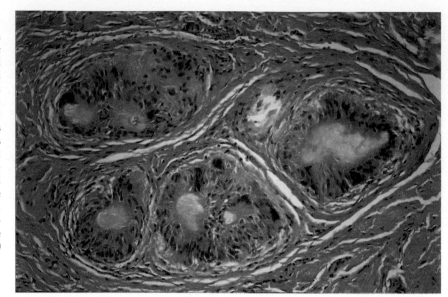

14.62 Lipoma ('hibernoma'): skin

Lipomas are common benign tumours which arise in any part of the body and are not infrequently multiple. Most are composed of large 'clear' adult fat cells. This tumour, however, is a rare variant in that although ordinary fat cells are present, most of the fat is 'brown fat'. The cells of brown fat are smaller and the fat is distributed throughout the cytoplasm in multiple small vacuoles (mulberry cells). The nucleus is generally located centrally. Brown fat is found in the fetus but only in small amounts in later life. It is also abundant in hibernating animals; hence the name of the lesion.　　　HE ×235

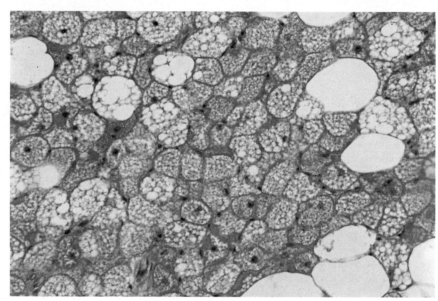

14.63 Xanthoma (xanthelasma): skin

When the levels of lipids and lipoproteins in the serum are raised, lipid may be deposited in the tissues where it is ingested by macrophages. This is the basis for the formation of xanthoma. The lipid-laden macrophages take a variety of forms and in time there is usually progressive fibrosis of the deposit. The term xanthelasma is applied to the yellow plaques that form in the eyelids of some individuals. The fat is neutral and in only a minority of cases is there an association with hyperlipoproteinemia. This lesion was situated at the inner canthus and it consists of closely-packed large discrete lipid-laden macrophages, some multinucleated. Xanthelasmas are noteworthy for their superficial location in the upper dermis and for the absence of fibrosis.　　　HE ×335

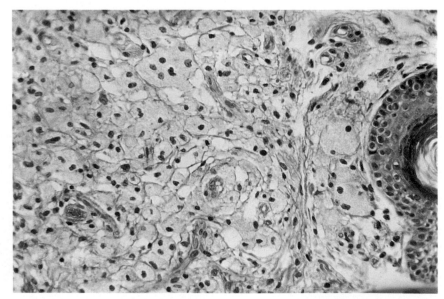

14.64 Xanthoma tuberosum: skin

In xanthoma tuberosum large nodules form in the skin on extensor surfaces (elbows, knees, fingers and buttocks). There is a close association with raised levels of beta-lipoproteins in the plasma. Early lesions consist of aggregates of xanthoma (foam) cells mixed with smaller numbers of histiocytes, lymphocytes and neutrophil polymorphs. As the lesions mature, fibroblasts appear in increasing numbers. This is a mature lesion. The foam cells are mostly large multinucleated cells. These are Touton giant cells, in which the nuclei are grouped around eosinophilic cytoplasm and there is a zone of foamy cytoplasm external to the nuclei (arrows). There is already considerable fibrosis, with elongated fibroblasts and much collagenous fibrous tissue.

HE ×360

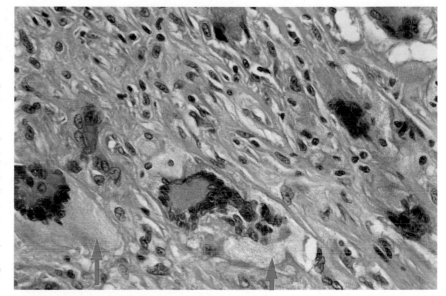

14.65 Xanthoma tuberosum: skin

Tuberous xanthomas contain higher concentrations of cholesterol esters than other types of xanthoma. In this lesion foam cells have been largely replaced by collagenous fibrous tissue and between the bundles of collagen fibres there are many clefts which contained crystals of cholesterol (cholesterol crystals dissolve in the reagents used for processing tissues). Elongated foreign-body giant cells lie alongside many of crystals. There is also a large collection of cholesterol-rich debris (top). Fairly numerous lymphocytes are also present in the lesion.

HE ×235

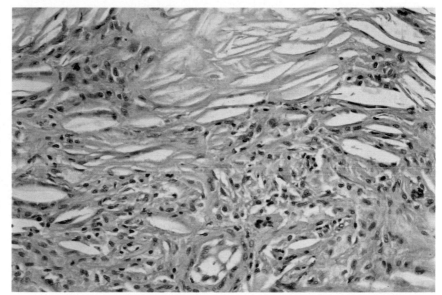

14.66 Reticulohistiocytic granuloma: skin

This lesion, previously called reticulohistiocytoma, is granulomatous and not neoplastic. It takes the form of one or several smooth nodules in the skin, usually of the head or neck. Apart from a small group of lymphocytes (bottom), the lesion consists of histiocytes, both mononuclear and multinucleated. The multinucleated cells are giant cells and each cell has finely granular (lipid-rich) cytoplasm and several vesicular nuclei. As the lesion matures, the number of mononuclear histiocytes decreases and the multinucleated forms increase. Eventually the lesion tends to fibrose. HE ×150

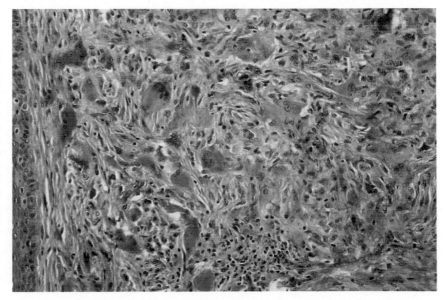

14.67 Nodular subepidermal fibrosis: skin

Nodular subepidermal fibrosis (sclerosing hemangioma, histiocytoma cutis, dermatofibroma) is probably a reaction to some form of trauma such as an insect bite. The fact that it goes through a series of stages accounts for the various names attached to it. The earliest lesion is a collection in the mid-dermis of mature histiocytes, many of which appear foamy. This was a large mushroom-shaped 'papilloma' on the knee of a woman of 52. In this frozen section in which lipid is retained, the histiocytes contain numerous droplets of fat (stained orange-red). They lie between collagenous fibres which are unstained. The cells with elongated nuclei and no lipid in their cytoplasm are fibroblasts and endothelial cells. Hemosiderin was also present in the lesion.

Sudan IV ×150

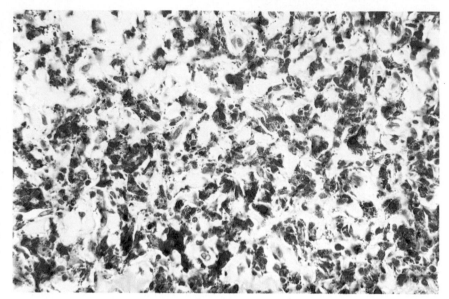

14.68 Nodular subepidermal fibrosis: skin

In nodular subepidermal fibrosis highly vascular connective tissue often forms, from which hemorrhage occurs. This type of lesion is sometimes given the name of sclerosing hemangioma. This example of sclerosing hemangioma was a raised smooth nodule 1.2cm dia on the back of the thigh of a woman of 23. The cut surface of the nodule looked characteristically brownish-yellow. The lesion is a network of strands of cellular connective tissue enclosing vascular spaces which are large and lack a well-formed endothelial lining. Some of the blood is extravascular and within the connective tissue there are many macrophages heavily laden with hemosiderin. A frozen section would reveal considerable numbers of lipid-laden cells, the lipid presumably having the same origin as the hemosiderin. HE ×215

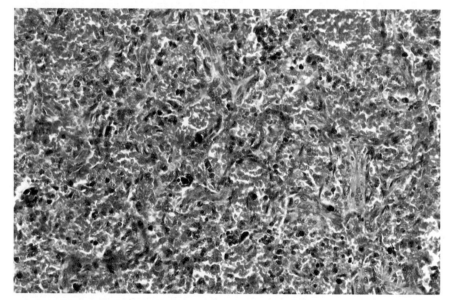

14.69 Nodular subepidermal fibrosis: skin

This was a firm nodule 1.5cm dia on the leg of a 43-year-old man and it represents the end stage in the evolution of nodular epidermal fibrosis. It is a fibrous nodule, a 'dermatofibroma' or 'sclerosing histiocytoma cutis'. The curved bundles of cells and tissue produce a storiform (basket-weave) pattern (see 14.71) consisting of interlacing strands of collagenous fibrous tissue closely associated with which are characteristically curved or stellate fibroblasts. Many vacuolated cells are present and special stains would reveal the continued presence of histiocytes containing lipid and hemosiderin. Although the nodule appeared well-demarcated to the naked eye, microscopically it merged gradually into the surrounding dermis and subcutaneous fat. It is nevertheless benign. HE ×235

14.70 Dermatofibrosarcoma protuberans: skin

Dermatofibrosarcoma is a slow-growing neoplasm which is not encapsulated and forms a nodule, usually on the trunk, which protrudes above the skin surface. It also grows downwards so that early lesions have an hour-glass shape. Although its malignancy is low-grade it is notoriously liable to recur. This was a recurrent tumour in the skin of the back of a woman of 46. It was a firm white nodule 2 × 2 × 1.5cm. Histologically it is a well-differentiated fibrosarcoma, consisting of intertwining bands of collagenous fibrous tissue. The fibrous tissue is populated by fibrocyte/fibroblast-like cells with spindle-shaped nuclei. Nuclear pleomorphism is slight but occasional mitoses are present. The tumour was invading the adjacent subcutaneous fat.

HE ×150

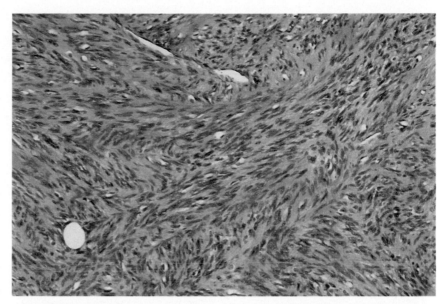

14.71 Dermatofibrosarcoma protuberans: skin

This was a raised nodule 1cm wide on the right shoulder of a man of 26. It consists of well-differentiated spindle cells which lie in tight whorls and interlacing bundles and so produce a storiform (basket-weave) pattern (see 14.69). There is moderate mitotic activity. Special stains revealed abundant reticulin but only small amounts of collagen. The tumour was invading the subdermal tissues. It has been suggested that dermatofibrosarcoma protuberans has a histiocytic origin, but this is not certain.

HE ×235

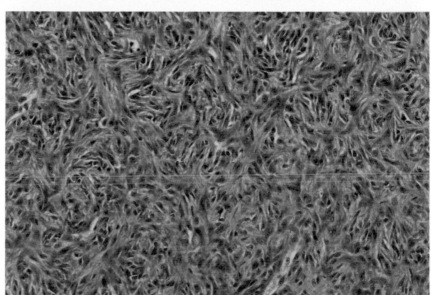

14.72 Kaposi's sarcoma: skin

Kaposi's sarcoma is common in Africa but was rare until recently in the USA and Europe. The red or purplish lesions may be confined to the skin or widely distributed throughout the body, occurring particularly in lymph nodes and gastrointestinal tract. The lesion is of uncertain histogenesis. Histologically its structure varies considerably, containing angiosarcomatous and fibrosarcomatous elements in varying proportions. This tumour is mainly angiosarcomatous. The cells are spindle-shaped and resemble poorly-differentiated fibroblasts or smooth muscle cells and form narrow blood-filled channels. Mitotic figures are numerous among them. There is considerable hemorrhage throughout the tumour, and a fibrinous exudate containing red cells is present beneath the epidermis (left).

HE ×200

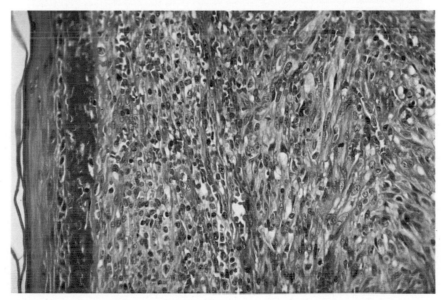

14.73 Neurofibroma: skin

Neurofibroma arises from the connective tissue of the peripheral nerve sheaths, probably from Schwann cells. It may be a single lesion in the skin, usually in adult life; or it may be one of many in neurofibromatosis (von Recklinghausen's disease) which is usually manifest in childhood. The lesion is generally well-circumscribed but it is not encapsulated. This is a plexiform neuroma which consists of thickened tortuous nerves and is typically large, pendulous and flabby. This one, however, was a small lesion in the upper lip of a woman of 22. The epidermis (left) is stretched over a plexus of enormously thickened small nerves. The sheaths (thin arrows) of the nerves are intact but each nerve is greatly expanded by the presence of pale-staining myxoid connective tissue produced by the Schwann cells. A sebaceous gland (thick arrow) is present. HE ×70

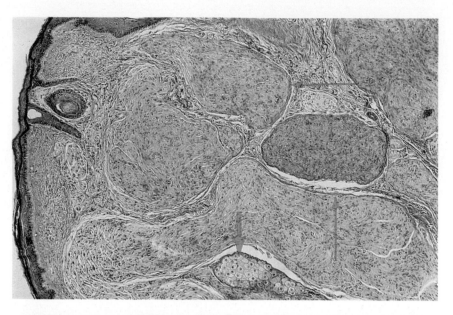

14.74 Neurofibroma: skin

This lesion has the structure of the more common type of neurofibroma. The epidermis is hyperkeratotic and the papillary dermis is spared. In the deeper dermis the lesion consists of cells with ovoid or spindle-shaped nuclei of fairly uniform size and shape. There is no mitotic activity. The cells lie randomly in a loosely-textured connective tissue. Special stains showed the presence of considerable amounts of delicate connective tissue (reticulin) but little collagen. There is no fibrous capsule. Nerve fibres are present in most neurofibromas but require special stains for their demonstration. HE ×150

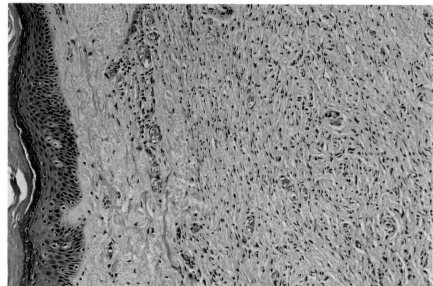

14.75 Neurofibroma: skin

This was one of two subcutaneous nodules each 1cm dia in the subcutaneous tissues of the chest wall of a man of 34. Histologically they were identical, consisting of irregular spindle-shaped cells in an abundant loose extracellular matrix. There is no mitotic activity. Neither lesion was encapsulated and some dermal appendages were caught up in both. HE ×235

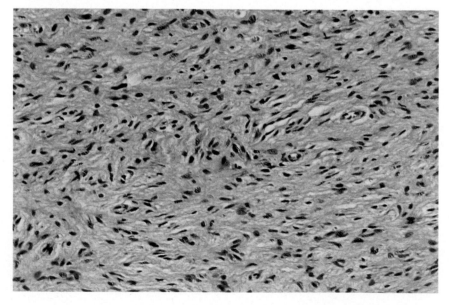

14.76 Neurofibrosarcoma (malignant neurofibroma): skin

Very uncommonly malignant change takes place in a neurofibroma in a person with neurofibromatosis. This is an example of this development which occurred on a neurofibroma in the axilla of a woman of 38. The tumour is highly cellular. The cells are elongated, with vacuolated weakly eosinophilic cytoplasm and indistinct cell boundaries. The nuclei of many of the tumour cells are pleomorphic and hyperchromatic; and they show considerable mitotic activity, much of it atypical (arrows). HE ×360

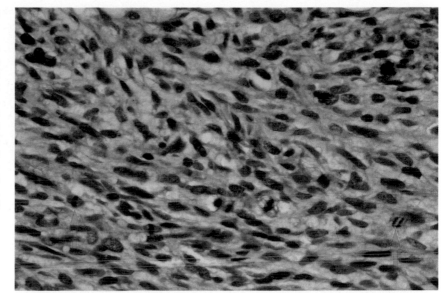

14.77 Traumatic neuroma: skin

When a peripheral nerve trunk is breached by trauma, fibroblasts and Schwann cells proliferate and the axons grow out from the cut ends of the nerve fibres. If the cut ends of the nerve are not accurately apposed, these elements continue to grow, to form a disorderly mass of fibrous tissue and nerve fibres. A firm white mass 2cm in its long axis was removed from the side of the neck from a woman of 55. It consists of irregular bundles of nerve fibres mingled with bands of dense fibrous tissue. The whole mass is enclosed in a fibrous capsule (right). HE ×235

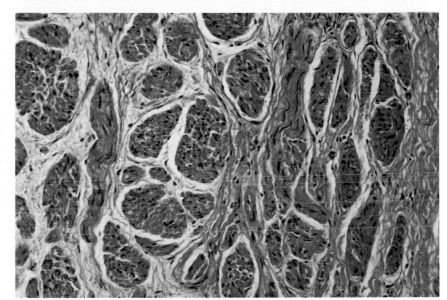

14.78 Hemangioma: skin

Hemangioma is a vascular hamartoma and it takes a variety of forms. This one was on the right cheek of a girl of 6. It is a capillary hemangioma, consisting of distended capillary-type vessels lined by prominent endothelial cells, lying in fat. Large feeding vessels were present elsewhere in the lesion. The lesion extended laterally for a considerable distance in the subcutaneous fat of the cheek. *See also 1.49, 5.5.* HE ×150

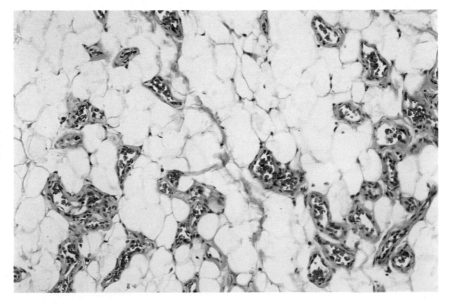

14.79 Glomus tumour: skin

Glomangiomas arise from the glomus bodies, the small arteriovenous anastomoses which control blood flow and temperature, especially in the fingers and toes. The tumour is usually a small purplish nodule. It is an encapsulated benign lesion but it is sensitive to the touch and may cause paroxysms of severe pain. This lesion was a nodule 2mm dia in the subcutaneous tissue of the knee of a 33-year-old man. It consists of groups of round or cuboidal glomus cells ('myoid') and angiomatous vascular spaces lined by flat endothelium. The glomus cells have pale-staining round or ovoid nuclei which show no pleomorphism or mitotic activity. They are closely associated with the vascular channels but some lie in the eosinophilic stroma. Special stains would reveal large numbers of medullated and non-medullated nerve fibres.   HE ×235

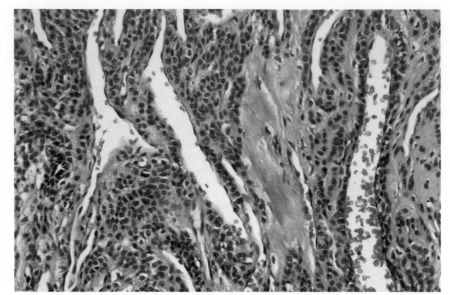

14.80 Metastatic renal carcinoma: skin

Metastases in the skin are usually multiple and a sign of widely disseminated malignancy. Occasionally, however, a solitary metastasis occurs, and one tumour with a reputation for producing a solitary metastasis is renal carcinoma. The distinctive histological structure of renal carcinoma may also indicate the source of the primary, as happened in this instance in which a solitary deposit appeared in the skin. The tumour cells are large with abundant unstained vacuolated (clear) cytoplasm and a highly vascular (sinusoidal) stroma. The cytoplasm appears clear because of the presence of abundant glycogen, although lipid is also present. Further clinical investigations confirmed the presence of a primary carcinoma of kidney.   HE ×235

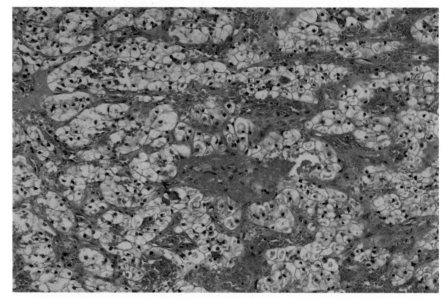

14.81 Extramammary Paget's disease: skin

In Paget's disease adenocarcinomatous cells are present in the epidermis. In addition to adenocarcinoma of the breast, which gives rise to Paget's disease of the nipple, carcinomas in other sites and particularly neoplasms arising in apocrine or modified apocrine glands in the scrotum, perineum and labia majora may cause extramammary Paget's disease. In this case the lesion was in the anal canal. The surface of the epidermis is out of the field on the left. In the rete ridges there are large pale vacuolated cells with well-defined boundaries. The cells, singly or in small clusters, tend to lie close to the basal layers of the epidermis but are extending into the stratum spinosum. Mucin was demonstrated in their cytoplasm. In some cases the underlying adenocarcinoma cannot be detected.   HE ×375

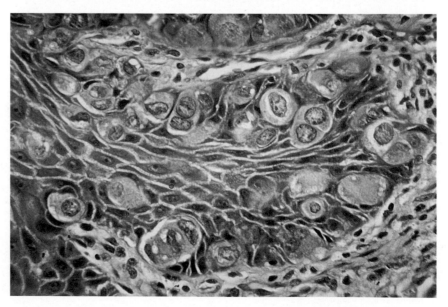

14.82 Chronic lymphocytic leukemia: vulva

Deposits of leukemic cells may form in any tissue but infiltration of the skin and mucous membranes is fairly common. It takes the form usually of macules or papules. In this case a woman with long-standing chronic lymphocytic leukemia developed lesions on the vulva. The epidermis (left) appears normal apart from the presence of small numbers of lymphocytes in the basal layers but in the deeper dermis (right) there is a very dense infiltrate of small dark-staining lymphocytes. However in the papillary dermis the majority of cells are plasma cells (arrow).

HE ×150

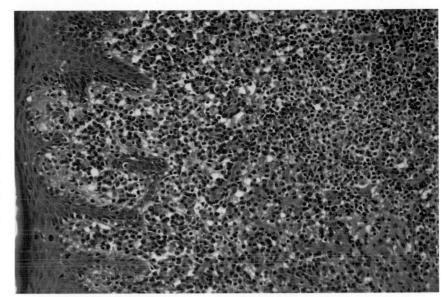

14.83 Mycosis fungoides: skin

Mycosis fungoides is a malignant T cell lymphoma of a very chronic nature and the course of the disease may last more than 20 years. The lesions are mainly in the skin but deposits also occur in lymph nodes and viscera. The condition may declare itself through non-specific eruptions in the skin but eventually plaques form which slowly grow and become elevated, even to the extent of becoming mushroom-shaped, hence the name. Adjacent plaques may coalesce. The patient in this instance was a man of 57. This is part of a plaque on the left upper arm. There is a dense infiltrate of a mixed population of cells in the upper dermis and around the dermal appendages. It is well-demarcated from the subcutaneous tissue (right). There are also several collections of similar cells (Pautrier abscesses) within the epidermis.

HE ×150

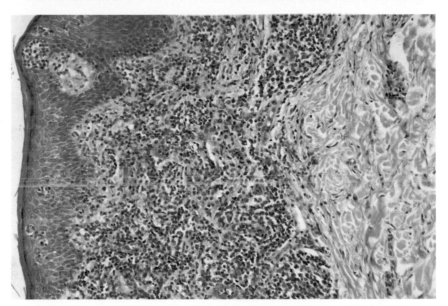

14.84 Mycosis fungoides: skin

This is the same case as that shown in **14.83**. The majority of the cells in the dermal infiltrate are lymphocytes with compact dark nuclei and very little cytoplasm. They are accompanied by histiocytes with larger vesicular nuclei. Small numbers of eosinophil leukocytes and plasma cells are also present. There are groups of cells within the epidermis (Pautrier abscesses) (arrows). A pleomorphic infiltrate consisting of a mixture of cell types is characteristic of mycosis fungoides.

HE ×360

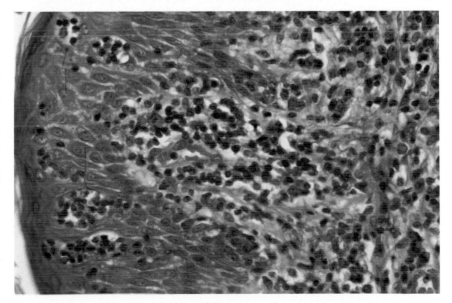

Index